Ceramic and Metallic Biomaterials. Application in Medical Sciences

Ceramic and Metallic Biomaterials. Application in Medical Sciences

Guest Editor

Oana-Georgiana Dragos-Pinzaru

MDPI

Basel • Beijing • Wuhan • Barcelona • Belgrade • Novi Sad • Cluj • Manchester

Guest Editor
Oana-Georgiana Dragos-Pinzaru
National Institute of Research
and Development for
Technical Physics
Iasi
Romania

Editorial Office
MDPI AG
Grosspeteranlage 5
4052 Basel, Switzerland

This is a reprint of the Special Issue, published open access by the journal *Coatings* (ISSN 2079-6412), freely accessible at: https://www.mdpi.com/journal/coatings/special_issues/ceramic_metallic.

For citation purposes, cite each article independently as indicated on the article page online and as indicated below:

Lastname, A.A.; Lastname, B.B. Article Title. *Journal Name* **Year**, *Volume Number*, Page Range.

ISBN 978-3-7258-5531-5 (Hbk)
ISBN 978-3-7258-5532-2 (PDF)
https://doi.org/10.3390/books978-3-7258-5532-2

Contents

About the Editor

Oana-Georgiana Dragos-Pinzaru

Oana-Georgiana Dragos-Pinzaru is a researcher at the National Institute of Research and Development for Technical Physics in Iasi, Romania, where her work focuses on materials synthesis and characterization, with a strong expertise in nanoscale research. She earned her MSc degree in 2004 from Université Paris-Sud XI and INSTN-CEA Saclay, followed by a PhD in 2007 at Université Paris-Sud XI, France. Her current research is centered on electrochemistry and the development of nanowires for applications in nanomedicine and catalysis. Over the past 18 years, she has contributed to numerous national and international projects, working on the synthesis of nanowires, nanopowders, single crystals, and other advanced materials using methods such as electrochemical deposition, sol-gel, co-precipitation, and thermal techniques. She has consolidated the electrochemistry working group at her institute and achieved notable results, documented in scientific papers, book chapters, and conference presentations. Her expertise spans a wide range of characterization techniques, including optical and electron microscopy, atomic force and magnetic force microscopy, thermal analysis, X-ray diffraction, magnetic and electrical property measurements, and impedance spectroscopy.

Preface

The Reprint "Ceramic and Metallic Biomaterials Application in Medical Sciences" presents recent advances in the development, characterization, and application of innovative materials designed to address pressing challenges in modern healthcare. The subject of this collection spans from fundamental investigations of coatings, alloys, and composites to practical strategies for improving implant integration, combating microbial colonization, and enabling new diagnostic and therapeutic approaches.

The scope of the Reprint is deliberately broad, encompassing both ceramic and metallic systems, as well as hybrid and nanostructured materials. Research topics include antibacterial surface treatments, bioactive and biocompatible coatings, porous titanium alloys for bone regeneration, magnetic nanomaterials for theranostic applications, and advanced manufacturing techniques such as additive manufacturing and sol-gel deposition. In addition, the Reprint features comprehensive review articles that consolidate current knowledge on dental implants, orthopedic devices, and structural optimization of biomaterials.

The aim of this Reprint is to highlight how the design and engineering of biomaterials can bridge laboratory innovation with clinical application. By gathering diverse yet complementary contributions, it seeks to provide a platform for advancing our understanding of how material properties—mechanical, chemical, and biological—can be optimized to improve patient outcomes.

The motivation for compiling this work lies in the growing demand for safer, more effective, and more personalized medical solutions. As populations age and chronic conditions become increasingly prevalent, the need for biomaterials that support tissue regeneration, resist infection, and integrate seamlessly into the body is more urgent than ever.

This Reprint is addressed to researchers, clinicians, engineers, and industry professionals who are engaged in the fields of biomaterials science, biomedical engineering, and medical device development. It will also serve as a useful reference for graduate students and academics seeking to understand the state of the art in this rapidly evolving domain.

Oana-Georgiana Dragos-Pinzaru
Guest Editor

coatings

MDPI

Editorial

Special Issue: Ceramic and Metallic Biomaterials Nanoparticles for Applications in Medical Sciences

Gabriela Buema, Dumitru-Daniel Herea and Oana-Georgiana Dragos-Pinzaru *

National Institute of R&D for Technical Physics, 700050 Iasi, Romania; gbuema@phys-iasi.ro (G.B.); dherea@phys-iasi.ro (D.-D.I.I.)

* Correspondence: odragos@phys-iasi.ro

Nowadays, the development of new materials that can be used to treat, repair, diagnose, replace, or restore a function of the human body represents one of the key research topics for the worldwide scientists and medical industries. For example, metallic nanoparticles are studied for possible applications in biomedicine such as the manipulation of living cells (transportation, displacement, positioning, or cell separation) or cancer detection and treatment. Another important class of compounds that has attracted the attention of the scientific community is represented by ceramic biomaterials, which can be used for the replacement of various types of tissues (as implants or for the repair and reconstruction of various diseased parts of the body). The outperformance of a biomaterial is determined by its crystalline structure, microstructure, (such as grain size or porosity), biocompatibility, corrosion resistance, or mechanical properties. In this context, the development of new biomaterials with improved physical and mechanical properties and a low production cost, high availability, and good esthetics is imperative. This short review aimed to explore the innovative progresses in the field of biomaterials used for biodetections, treatment, or replacement of injured tissues in order to develop practical solutions for clinical practice.

Ceramic biomaterials can be used in a wide range of biomedical procedures, such as tooth replacement or restoration [1,2], knees, ligaments, or tendons [3], bone implants [4], maxillofacial reconstruction [5,6], or particle transport inside the body to the vicinity of the target by applying an external magnetic field [7], etc.

Together with magnetic nanoparticles, ceramic materials can be used to treat cancer diseases by destroying tissues by hyperthermia or magneto-mechanical effects. The advantages to use material in "nano" form for applications in medicine are related to the fact that these materials retain both "bulk" properties, but also gain a number of special properties due to their small size, and large surface area or "aspect ratio" (shape) [8,9]. Interesting new properties of nanoparticles can be obtained by preparing bi-metallic nanomaterials in the form of alloys or core-shell structures. The performance of these types of materials often exceeds those of simple metals. The form and structure of the nanomaterial, as well as its physical and chemical characteristics, vary as the metal components are changed.

The large number of studies presented in the literature related to the synthesis and properties of magnetic nanoparticles demonstrate the importance of nanomaterials whose properties make them useful both in diagnosis and in treatment of many diseases. One of the main characteristic of magnetic nanomaterials is represented by the fact that such materials can be guided (moved) or held in place by a magnetic field. Another important property that makes this type of material suitable to be used in bio-medicine is that magnetic nanomaterials can be heated by an applied magnetic field. Thus, the data presented in the literature show that magnetic nanoparticles can be used to transport drugs in the body, for the treatment of cancer (by hyperthermia or magneto-mechanical effect), and as contrast agents for cancer cell visualization [10–12], etc.

One of the disadvantages of the use of magnetic nanoparticles for biomedical applications is related to the cytotoxicity of these materials and to the fact that these materials can be degraded in the human body. To avoid degradation and toxicity, the surface of the nanomaterials is coated with a layer of organic compound (for example polyethylene glycol (PEG), dextran, or oxide layer [13,14]). The encapsulation of magnetic nanoparticles in an organic compound layer increases the circulation time of nanoparticles in the body, making the nanomaterials biocompatible [15]. However, due to the high reactivity of the organic compounds in the blood, the organic shell can be degraded by the body, the magnetic materials re-becoming toxic to the human body. In order to avoid the biodegradation and reduce their toxicity, a new method has been developed, involving the coat of the nanoparticles with a noble metal shell, which is non-toxic to the human body and manifests a low reactivity in the blood.

Given the limitations described above, biological applications need the employment of magnetic core nanoparticles (with high magnetic moment) encapsulated in a biocompatible shell. The use of magnetic materials with a high magnetic moment leads to a decrease in the applied external magnetic field or an increase in the distance from the external magnet. In this context, nanomaterials containing noble metals and transition simple metals and alloys can be used for biological testing [16–19]. The unique physical properties of nanomaterials to selectively recognize biomolecules can lead to the miniaturization of biological sensors, the literature data showing that magnetic nanomaterials have been extensively studied for various applications in biomedicine, such as: moving, positioning, or separating cells, treating cancer, and so on.

Thus, a possible application of magnetic nanoparticles is the manipulation of living cells (transport—movement, positioning, or separation of cells). In their paper, Choi et al. developed a technique for transporting and positioning living cells with internalized Ni nanowires using magnetic field alignment. Thus, Ni nanowires were first internalized into neuroblastoma cells and then the latter were placed between two electrodes using a magnetic field. The study allows the creation of an interface between neurons and electronic devices [20].

Hultgren et al. [21] studied cell separation using Ni nanoparticles. The presented results showed a higher efficiency of cell separation when this was carried out using Ni than those obtained by using magnetic beads of comparable volume. Another conclusion of this study showed that the use of Ni nanoparticles allows the production of purer cell populations than the use of spherical nanoparticles. The increase in the efficiency of cell separation when magnetic nanowires are used for this can be attributed, according to the study, to the fact that Ni nanomaterials have a higher magnetic moment. The effectiveness of cell separation with Ni nanoparticles compared to that achieved with magnetic nanoparticles was also studied by Gao et al. [22]. The authors demonstrated that functionalized Ni nanoparticles in the shape of nanowires have a higher cell separation efficiency compared with magnetic nanoparticles with spherical shape, even at a concentration 60 times lower.

Another possible important application of magnetic nanomaterials is the use of these materials for cancer treatment. In the presence of a magnetic field, the magnetic nanomaterials can be heated or rotated so that they can transmit either heat, or a mechanical force to the cells they come in contact with, thus helping to eradicate cancer cells. Due to the shape anisotropy and the increased magnetic moment of materials in shape of nanowires, it is possible to perform hyperthermia at lower field strengths in order to minimize damage to healthy cells.

Another well-defined class of highly exploited metallic materials for biomedical applications is represented by noble metals, especially gold and silver nanoparticles. Currently, the most known route to quickly synthesize colloidal gold is the Turkevich method that uses sodium citrate both as a reduction agent of chloroaurate and stabilizer of the synthesized AuNPs. Moreover, green methods relying on different natural reduction agents, such as plant extracts, algae, and bacteria, are also used to produce stable and non-toxic colloidal gold for biomedical applications [23].

Due to their tunable size, shape, and specific optical properties, AuNPs have been used in biomedical imaging (e.g., X-ray-computed tomography, photo-acoustic imaging, dark field microscopic imaging, magnetic resonance imaging, and fluorescence imaging), immunoassay (ELISA, and lateral flow, which is the most used immunoassay method for SARS-CoV-2 detection), nanomedicine (e.g., drug-targeted drug and gene delivery), photothermal/photodynamic therapy (e.g., cancer treatment), and biosensing (e.g., electrochemical, colorimetric, or fluorescence-based sensing, but also surface plasmon resonance biosensors and surface enhanced Raman scattering based sensors) [23,24].

AuNPs are also the most commonly used in vaccinology. Once administered, they are quickly internalized by macrophages and dendritic cells, inducing their activation [25,26]. AuNPs have attracted huge interest in vaccinology due to their reliable surface functionalization, high biocompatibility, customized dimensions and shapes, and excellent optical properties. Moreover, owing to their inertness, AuNPs can be exploited both as delivery agents and adjuvants in vaccines, being able to boost the immune responses while assuring minimal toxicity [27]. AuNP, with their unique properties, represent ideal platforms towards a new era of vaccinology. Their high surface area and straightforward functionalization allow simultaneous and multivalent antigen presentation and make them excellent candidates for innovative nano-constructs in the field of vaccine development.

Silver nanoparticles are also of interest in medical sciences due to their anti-inflammatory, anti-angiogenesis, antiviral, antifungal, and antibacterial effects, being suitable for catheter modification, dental applications, or wound and bone healing. Moreover, AgNPs showed promising anticancer effects for several human cancerous cell lines [28].

The studies have revealed the importance of AgNPs as suitable means to treat different bacterial diseases, from Malaria leishmaniosis to Escherichia coli infections [29]. As a basic mechanism of action, by releasing silver ions, AgNPs increase the permeability of cell membranes and induce generation of reactive oxygen species, leading to interruption of the replication of the DNA [30].

In conclusion, this special issue aims to provides advances in biomaterials research field for applications in medical sciences. The use of the nanomaterials for biomedical applications has the potential to lead towards increasing life expectancy and improving quality of life.

Author Contributions: Conceptualization, O.-G.D.-P. and D.-D.H.; writing—original draft preparation, O.-G.D.-P., D.-D.H. and G.B.; writing—review and editing, O.-G.D.-P. and G.B.; supervision, O.-G.D.-P.; project administration, O.-G.D.-P.; funding acquisition, O.-G.D.-P. and D.-D.H. All authors have read and agreed to the published version of the manuscript.

Funding: This research was funded by the Romanian Executive Agency for Higher Education, Research, Development, and Innovation (UEFISCDI), contract number 110PCE/2022 (Project PN-III-P4-PCE-2021-1395/GreenEn), and contract number PCE75/2022 (Project PN-III-P4-PCE-2021-1081/Gold-SARS).

Conflicts of Interest: The authors declare no conflict of interest.

References

1. Alzahrani, A.S. Effect of TiO_2 on the sinter crystallization of nepheline glasses for dental application. *Int. J. Appl. Glass Sci.* **2022**. [CrossRef]
2. Wertz, M.; Fuchs, F.; Hoelzig, H.; Wertz, J.M.; Kloess, G.; Hahnel, S.; Rosentritt, M.; Koenig, A. The Influence of Surface Preparation, Chewing Simulation, and Thermal Cycling on the Phase Composition of Dental Zirconia. *Materials* **2021**, *14*, 2133. [CrossRef] [PubMed]
3. Catauro, M.; Ciprioti, S.V. Characterization of Hybrid Materials Prepared by Sol-Gel Method for Biomedical Implementations. A Critical Review. *Materials* **2021**, *14*, 1788. [CrossRef]
4. Ma, Z.; Li, J.; Cao, F.; Yang, J.; Liu, R.; Zhao, D. Porous silicon carbide coated with tantalum as potential material for bone implants. *Regen. Biomater.* **2020**, *7*, 453–459. [CrossRef] [PubMed]
5. Shen, C.; Witek, L.; Flores, R.L.; Tovar, N.; Torroni, A.; Coelho, P.G.; Kasper, F.K.; Wong, M.; Young, S. Three-Dimensional Printing for Craniofacial Bone Tissue Engineering. *Tissue Eng. Part A* **2020**, *26*, 1303–1311. [CrossRef]
6. Xu, F.; Ren, H.; Zheng, M.; Shao, X.; Dai, T.; Wu, Y.; Tian, L.; Liu, Y.; Liu, B.; Gunster, J.; et al. Development of biodegradable bioactive glass ceramics by DLP printed containing EPCs/BMSCs for bone tissue engineering of rabbit mandible defects. *J. Mech. Behav. Biomed. Mater.* **2020**, *103*, 103532. [CrossRef] [PubMed]

7. Zhang, Y.; Li, J.; Habibovic, P. Magnetically responsive nanofibrous ceramic scaffolds for on-demand motion and drug delivery. *Bioact. Mater.* **2022**, *15*, 372–381. [CrossRef]
8. Shumskaya, A.; Korolkov, I.; Rogachev, A.; Ignatovich, Z.; Kozlovskiy, A.; Zdorovets, M.; Anisovich, M.; Bashouti, M.; Shalabny, A.; Busool, R.; et al. Synthesis of Ni@Au core-shell magnetic nanotubes for bioapplication and SERS detection. *Colloids Surf. A Physicochem. Eng. Asp.* **2021**, *626*, 12707. [CrossRef]
9. Kargozar, S.; Mozafari, M.; Hamzehlou, S.; Kim, H.-W.; Baino, F. Mesoporous bioactive glasses (MBGs) in cancer therapy: Full of hope and promise. *Mater. Lett.* **2019**, *251*, 241–246. [CrossRef]
10. de Menezes, B.R.C.; Rodrigues, K.F.; Schatkoski, V.M.; Pereira, R.M.; Ribas, R.G.; Montanheiro, T.L.D.; Thim, G.P. Current advances in drug delivery of nanoparticles for respiratory disease treatment. *J. Mater. Chem. B* **2021**, *9*, 1745–1761. [CrossRef]
11. Yallapu, M.M.; Othman, S.F.; Curtis, E.T.; Gupta, B.K.; Jaggi, M.; Chauhan, S.C. Multi-functional magnetic nanoparticles for magnetic resonance imaging and cancer therapy. *Biomaterials* **2011**, *32*, 1890–1905. [CrossRef] [PubMed]
12. Kouhpanji, M.R.Z.; Zhang, Y.; Um, J.; Srinivasan, K.; Sharma, A.; Shore, D.; Gao, Z.; Chen, Y.; Harpel, A.; Porshokouh, Z.N.; et al. Bioapplications of magnetic nanowires:barcodes, biocomposites, heaters. *IEEE Trans. Magn.* **2022**. *In press*. [CrossRef]
13. Rajan, S.T.; Das, M.; Arockiarajan, A. Biocompatibility and corrosion evaluation of niobium oxide coated AZ31B alloy for biodegradable implants. *Colloids Surf. B-Biointerfaces* **2022**, *212*, 112342. [CrossRef]
14. Karaagac, O.; Kockar, H. Improvement of the saturation magnetization of PEG coated superparamagnetic iron oxide nanoparticles. *J. Magn. Magn. Mater.* **2022**, *551*, 169140. [CrossRef]
15. Awaad, A.; Elkady, E.F.; El-Mahdy, S.M. Time-dependent biodistribution profiles and reaction of polyethylene glycol-coated iron oxide nanoclusters in the spleen after intravenous injection in the mice. *Acta Histochem.* **2022**, *124*, 151907. [CrossRef] [PubMed]
16. Shen, L.Y.; Pan, V.; Li, H.F.; Zhang, Y.L.; Wang, P.F.; Ke, Y.G. Programmable assembly of gold nanoparticle nanoclusters and lattices. *J. Mater. Chem. B* **2020**, *8*, 6810–6813. [CrossRef]
17. Zhao, Y.; Shi, L.X.; Kuang, H.; Xu, C.L. DNA-Driven Nanoparticle Assemblies for Biosensing and Bioimaging. *Top. Curr. Chem.* **2020**, *378*, 18. [CrossRef]
18. Al-Terke, H.H.; Latikka, M.; Timonen, J.V.I.; Vekas, L.; Paananen, A.; Joensuu, J.; Ras, R.H.A. Functional Magnetic Microdroplets for Antibody Extraction. *Adv. Mater. Interfaces* **2021**, *9*, 2101317. [CrossRef]
19. Mahshid, S.S.; Mahshid, S.; Dolati, A.; Ghorbani, M.; Yang, L.; Luo, S.; Cai, Q. Electrodeposition and electrocatalytic properties of Pt/Ni—Co nanowires for non-enzymatic glucose detection. *J. Alloys Compd.* **2013**, *554*, 169–176. [CrossRef]
20. Choi, D.; Fung, A.; Moon, H.; Ho, D.; Chen, Y.; Kan, E.; Rheem, Y.; Yoo, B.; Myung, N. Transport of living cells with magnetically assembled nanowires. *Biomed. Microdevices* **2007**, *9*, 143–148. [CrossRef]
21. Hultgren, A.; Tanase, M.; Chen, C.S.; Meyer, G.J.; Reich, D.H. Cell manipulation using magnetic nanowires. *J. Appl. Phys.* **2003**, *93*, 7554. [CrossRef]
22. Gao, N.; Wang, H.; Yang, E.-H. An experimental study on ferromagnetic nickel nanowires functionalized with antibodies for cell separation. *Nanotechnology* **2010**, *21*, 105107. [CrossRef] [PubMed]
23. Bansal, S.A.; Kumar, V.; Karimi, J.; Singh, A.P.; Kumar, S. Role of gold nanoparticles in advanced biomedical applications. *Nanoscale Adv.* **2020**, *2*, 3764–3787. [CrossRef]
24. Nejati, K.; Dadashpour, M.; Gharibi, T.; Mellatyar, H.; Akbarzadeh, A. Biomedical Applications of Functionalized Gold Nanoparticles: A Review. *J. Clust. Sci.* **2022**, *33*, 1–16. [CrossRef]
25. Bastus, N.G.; Sanchez-Tillo, E.; Pujals, S.; Farrera, C.; Kogan, M.J.; Giralt, E.; Celada, A.; Lloberas, J.; Puntes, V. Peptides conjugated to gold nanoparticles induce macrophage activation. *Mol. Immunol.* **2009**, *46*, 743–748. [CrossRef]
26. Kang, S.; Ahn, S.; Lee, J.; Kim, J.Y.; Choi, M.; Gujrati, V.; Kim, H.; Kim, J.; Shin, E.C.; Jon, S. Effects of gold nanoparticle-based vaccine size on lymph node delivery and cytotoxic T-lymphocyte responses. *J. Control. Release* **2017**, *256*, 56–67. [CrossRef]
27. Ferrando, R.M.; Luigi Lay, L.; Polito, L. Gold nanoparticle-based platforms for vaccine development. *Drug Discov. Today Technol.* **2021**, *38*, 57–67. [CrossRef]
28. Burdușel, A.C.; Gherasim, O.; Grumezescu, A.M.; Mogoantă, L.; Ficai, A.; Andronescu, E. Biomedical Applications of Silver Nanoparticles: An Up-to-Date Overview. *Nanomaterials* **2018**, *8*, 681. [CrossRef]
29. Yaqoob, S.B.; Adnan, R.; Rameez Khan, R.M.; Rashid, M. Gold, Silver, and Palladium Nanoparticles: A Chemical Tool for Biomedical Applications. *Front. Chem.* **2020**, *3*, 376. [CrossRef]
30. Yin, I.X.; Zhang, J.; Zhao, I.S.; Mei, M.L.; Li, Q.; Chu, C.H. The Antibacterial Mechanism of Silver Nanoparticles and Its Application in Dentistry. *Int. J. Nanomed.* **2020**, *15*, 2555–2562. [CrossRef]

coatings

MDPI

Article

The Effect of Ti(C,N)-Based Coating Composition on Ni-Cr Alloys on the Initial Adhesion of *E. coli* Bacteria and *C. albicans* Fungi

Katarzyna Banaszek [1], Katarzyna Dąbrowska [2], Witold Jakubowski [3], Leszek Klimek [3] and Zofia Kula [4,*]

1. Restorative Dentistry, Department of General Dentistry, Medical University of Lodz, 251 Pomorska Street, 92-213 Lodz, Poland; katarzyna.banaszek@umed.lodz.pl
2. Conservative Dentistry and Endodontics, Department of Endodontics, Medical University of Lodz, 251 Pomorska Street, 92-213 Lodz, Poland; katarzyna.anna.dabrowska@umed.lodz.pl
3. Institute of Materials Science and Engineering, Lodz University of Technology, 1/15 Stefanowskiego Street, 90-934 Lodz, Poland; witold.jakubowski@p.lodz.pl (W.J.); leszek.klimek@p.lodz.pl (L.K.)
4. Department of Dental Technology, Medical University of Lodz, 251 Pomorska Street, 92-213 Lodz, Poland

* Correspondence: zofia.kula@umed.lodz.pl

Abstract: Under natural physiological conditions, the oral cavity is colonized by a diverse range of microorganisms, which inhabit its anatomical structures as well as prosthetic restorations and the supragingival surfaces of implants. The metabolic activity of these microorganisms can contribute to microbiological corrosion, leading to the degradation of metal prosthetic materials. No material used for prosthetic elements is entirely resistant to bacterial adhesion. However, the application of protective coatings, such as Ti(C,N) coatings, on prosthetic surfaces can significantly reduce microorganism adherence. This study aimed to evaluate the influence of carbon and nitrogen content in Ti(C,N) coatings on reducing microorganism adhesion. Tests were conducted on five groups of Ni-Cr alloy specimens, each coated with Ti(C,N) layers containing varying amounts of carbon and nitrogen. The adhesion of *E. coli* bacteria and *C. albicans* fungi was assessed under both stationary and dynamic flow conditions. Results demonstrated that all tested coatings significantly reduced microorganism adhesion compared to uncoated Ni-Cr alloy samples.

Keywords: bacterial adhesion; dentistry; Ni-Cr alloy; Ti(C,N) coatings

1. Introduction

The oral cavity, under natural physiological conditions, hosts diverse microbiota, integral to the functioning of the human organism. Bacteria and fungi are the most abundant, though protozoa and viruses have also been identified. These microorganisms colonize all anatomical regions of the oral cavity, with a particular affinity for adhering to hard surfaces such as teeth, prosthetic restorations, and implant components [1–4]. In the oral cavity, microorganisms classified as physiological flora can promote plaque accumulation in areas with increased bacterial retention, such as regions difficult to access during hygiene procedures. This process facilitates the permanent adhesion of microbial cells to surfaces, ultimately leading to the formation of biofilms, including multi-species biofilms. These biofilms may harbor bacteria associated with the onset and progression of marginal periodontal diseases [5,6]. Biofilm formation occurs on various surfaces within the oral cavity, including teeth (as dental plaque), gums, and dentures. The process of microbiological colonization begins with the adsorption of proteins onto the surface, followed by the adhesion of individual microbial cells. Subsequent stages involve microbial proliferation

https://doi.org/10.3390/coatings15020121

and the production of an exopolysaccharide matrix, which binds the cells together and facilitates the development of three-dimensional biofilm structures [7–11]. The composition of dental plaque and the extent of colonization are influenced by its specific location within the oral cavity, the composition of saliva, dietary habits, natural cleansing mechanisms, and compliance to oral hygiene practices.

Microorganisms present in the oral cavity can contribute to the corrosive degradation of metal prosthetic materials through microbiologically influenced corrosion (MIC) [12–19]. The accumulation of bacterial plaque on prosthetic surfaces is particularly conducive to promoting these corrosive processes. Even alloys regarded as corrosion-resistant are susceptible to colonization by bacteria or fungi capable of initiating corrosion [18–23]. The metabolic activity of microorganisms contributes to the initiation and/or intensification of electrochemical and chemical processes, ultimately leading to the corrosive degradation of materials [24]. Studies by Laurent et al. [25], involving *Actinomyces viscosus*, have demonstrated the impact of pathogenic bacteria on the corrosion resistance of dental alloys. Potentiodynamic measurements revealed a reduction in the corrosion resistance of alloys colonized by bacteria. Furthermore, microbial metabolic activity can alter the surrounding environment, promoting the development of physicochemical processes that drive material degradation [20–22]. Biocorrosion processes on metal surfaces in the oral cavity are driven by the metabolic activity of bacteria, as well as by enzymes, acids, and volatile substances such as hydrogen sulfide and ammonia released by microbial cells [26]. Additionally, fluctuations in the pH of the oral cavity during food consumption play a significant role in the progression of biocorrosion. These corrosion processes facilitate the release of metal ions, which can exacerbate oral diseases [27]. Therefore, the selection of alloys for dental applications must carefully consider their interactions with the biotic factors present in the oral environment.

Prosthetic restorations are constructed from a variety of materials, including metallic options such as base metal alloys, noble metals and titanium and its alloys, as well as non-metallic materials like ceramics and plastics. However, no material is entirely immune to bacterial adhesion, including gold, titanium, and their alloys [11]. When analyzing the bacterial colonization of prosthetic surfaces, several parameters of the material's surface layer must be considered, including surface roughness, wettability, surface free energy (SFE), chemical composition, surface charge, and the presence of proteins [4,19,28–30].

Base metal alloys, despite their numerous drawbacks, are likely to remain in use for an extended period, particularly in prosthetics and orthodontics, as a more cost-effective alternative to gold and platinum alloys. Their application in prosthetic restorations is further supported by their favorable mechanical and technological properties. However, the use of base metal alloys is associated with the risk of allergic reactions, including contact allergies, due to the cytotoxicity of certain constituent elements. In the search for innovative solutions for the reconstruction of missing teeth that meet health, strength, and economic criteria, various surface modifications of prosthetic elements can be considered. The toxicity of non-precious metal alloys can be mitigated through the application of protective layers and coatings [8,31,32]. These coatings, typically based on carbides, oxides, or nitrides of various metals [31–34], serve not only to protect against corrosion but also to enhance wear resistance and prevent the migration of harmful ions from the substrate into the oral cavity. Studies have demonstrated that such coatings can also reduce bacterial adhesion on the coated surfaces [5,7,27,34,35].

Among the various types of coatings, TiC and TiN coatings are relatively widely used. Research has demonstrated that these coatings possess notable mechanical and biological properties [36–38]. Their primary application has been as coatings for various types of implants [34,35,39]. However, their favorable properties also suggest potential for use as protective coatings on prosthetic elements made from non-precious alloys. These coatings enhance wear resistance, improve corrosion resistance, and act as a barrier, reducing the release of ions from the substrate into the surrounding environment [36,40,41]. A particularly advantageous feature of these coatings is their substantial reduction of the cytotoxic effects associated with chromium–cobalt and chromium–nickel prosthetic alloys [42,43].

Previous studies have demonstrated that surface modifications can effectively reduce microbial adhesion. Ti(C,N)-type coatings are hypothesized to exhibit similar properties. The objective of this study was to evaluate the potential of Ti(C,N)-type coatings in reducing the adhesion of fungi and bacteria to specimen surfaces and to identify which coating composition offers the greatest efficacy in minimizing microbial adhesion. The null hypothesis is that Ti (C,N) coatings do not influence the adhesion of bacteria and fungi.

2. Materials and Methods

Cylindrical specimens with a diameter of 8 mm and height of 10 mm, made from Ni-Cr alloy, were used as the research material. The initial composition of the alloy, determined by X-ray fluorescence analysis using a SIEMENS XRS300 spectrometer (SIEMENS, Karlsruhe, Germany), is presented in Table 1.

Table 1. Composition of Ni-Cr alloy.

Element Percentage by Weight [wt%]						
Cr	**Mo**	**Si**	**Fe**	**Co**	**Mn**	**Ni**
24.79	8.89	1.57	1.33	0.17	0.12	residue

The cylinders were divided into six groups, with 20 samples in each group. The control group (S0) consisted of cylinders without coating, while the remaining five groups were coated with Ti(C,N) layers of varying carbon and nitrogen contents. The deposition process conditions and chemical composition of coatings on Ni-Cr alloy (S0) were detailed in previously published studies [40]. Prior to the application of the coatings, the surfaces of the specimen substrates were polished. Following this treatment, the surface roughness (Ra) values ranged from 0.375 to 0.418 μm.

After coating, the specimens were cleaned with distilled water using an ultrasonic cleaner and subsequently sterilized in steam at 121 °C for 21 min using a Prestige Medical autoclave (Prestige Medical, Blackburn, UK).

Twenty samples from each group were divided into four subgroups, with 5 samples per subgroup. The subgroups were assigned to incubation with bacteria (two subgroups—one in stationary incubation and the other in flow incubation) and to incubation with fungi (two subgroups—one in stationary incubation and the other in flow incubation).

Samples from each group were placed in 100 mL of LB medium (for *E. coli*) or YPG medium (for *C. albicans*). A standardized number of cells (*E. coli* bacteria and *C. albicans* fungi), derived from cultures in the logarithmic growth phase, were introduced into the medium. A standardized laboratory method was developed to analyze specimens in suspensions of bacteria or fungi during their logarithmic growth phase. In this method, an inoculum containing 2×10^3 *E. coli* cells or 3.5×10^4 *C. albicans* cells was introduced into a sterile medium, resulting in a final bacterial or fungal cell density of $2–3 \times 10^8$ cells per milliliter. Cultures were incubated for 24 h at 37 °C. After incubation, the specimens were aseptically removed from the bioreactor and rinsed with sterile distilled water to remove non-adherent microorganisms. The sample surfaces were then examined to count adherent bacterial or fungal cells using a fluorescence microscope. Observations were performed with an Olympus GX71 inverted fluorescence microscope equipped with a DP 73 digital camera. Image acquisition was conducted using Stream software (Stream Essential 1.9), while cell counting was performed with the Image J software (version no. 1.54K), utilizing the 'cell counter' add-on.

The level of surface colonization was assessed using Live/Dead staining with two fluorescent dyes: bis-benzidine (Sigma-Aldrich, Saint Louis, MO, USA) and propidium iodide (Sigma-Aldrich, Saint Louis, MO, USA). Bis-benzidine penetrates bacterial cells and intercalates with DNA, emitting a white-blue fluorescence under UV light, thereby identifying viable cells. In contrast, propidium iodide binds to DNA but does not penetrate intact cell membranes, resulting in the visualization of dead cells in red. The tests were conducted under two independent conditions: stationary culture and flow culture. For flow culture, stationary samples were placed in a bioreactor, and medium circulation was maintained by the motion of an electromagnetic stirrer at a velocity of approximately 2 cm/s. Five images were captured of each specimen at different locations, resulting in 25 measurements per group.

The Ti-6Al-4V alloy specimen was used as the control group for comparisons.

Five images were captured of each specimen at different locations, resulting in 25 measurements per group.

The results were statistically analyzed using IBM SPSS Statistics, version 29 (IBM Corp., Armonk, NY, USA). The Shapiro–Wilk test for normality ($p < 0.05$) was applied to assess the normality of the variables. Due to a non-normal distribution, non-parametric Kruskal–Wallis tests followed by multiple comparisons with the Bonferroni correction were used to compare specimen types within the variables. A significance level of 5% ($\alpha = 0.05$) was adopted.

The analysis aimed to determine whether the chemical composition of Ti(C,N) coatings influences:

- the number of adhered bacterial and fungal cells;
- the percentage of viable bacterial and fungal cells.

The results of the statistical analysis for the individual groups are presented in Figures 1–8 and Tables 2–13. A significance level of $\alpha = 0.05$ was applied to all analyses.

3. Results

The results of the tests on the total number of adhered bacteria, including viable bacteria, under stationary conditions are presented in Table 2.

Table 2. Adhered bacteria count under stationary conditions.

Group	Adhered Bacteria Cell Count		Viable Bacteria Cell Count		Viable Bacteria Cell Percentage	
	Mean	SD	Mean	SD	Mean	SD
S0	281	14.2	171	18.2	61	13.1
S1	152	4.2	135	13.1	89	6.2
S2	100	5.1	83	11.0	83	5.9
S3	98	4.2	79	9.6	81	5.1
S4	58	4.6	44	15.1	76	10.2
S5	63	7.8	54	15.5	85	10.8

Table 3 shows the corresponding results for bacteria in flow conditions.

Table 3. Adhered bacteria count under flow conditions.

Group	Adhered Bacteria Cell Count		Viable Bacteria Cell Count		Viable Bacteria Cell Percentage	
	Mean	SD	Mean	SD	Mean	SD
S0	271	5.8	179	5.4	66	2.8
S1	79	1.7	66	13.9	84	7.9
S2	98	2.2	78	19.2	80	12.2
S3	84	2.5	66	15.1	78	8.4
S4	103	2.4	76	21.3	74	14.3
S5	97	1.9	70	20.1	72	11.4

Table 4 presents similar adhesion results for fungal cells under stationary conditions, while Table 5 provides the results for flow conditions.

Table 4. Adhered fungi count under stationary conditions.

Group	Adhered Fungi Cell Count		Viable Fungi Cell Count		Viable Fungi Cell Percentage	
	Mean	SD	Mean	SD	Mean	SD
S0	265	8.6	173	5.7	65	2.8
S1	160	5.1	111	5.6	70	2.6
S2	97	4.7	82	4.5	85	2.3
S3	89	5.5	68	3.9	76	2.1
S4	83	4.2	65	4.3	78	2.2
S5	51	3.2	43	2.3	84	1.8

Table 5. Adhered fungi count under flow conditions.

Group	Adhered Fungi Cell Count		Viable Fungi Cell Count		Viable Fungi Cell Percentage	
	Mean	SD	Mean	SD	Mean	SD
S0	198	6.7	107	6.3	54	3.1
S1	105	4.9	84	9.4	80	5.1
S2	67	5.2	44	12.6	66	8.5
S3	63	5.1	43	4.5	68	2.1
S4	58	3.6	36	19.1	62	16.4
S5	39	2.3	21	6.2	55	3.5

3.1. Bacteria—Stationary Conditions

The data presented in Figure 1 and Table 6 show that all Ti(C,N)-type coatings resulted in a decrease in the number of adhered bacteria on the surfaces of the coated samples. For all coated groups, the differences in the number of adhered bacteria compared to the uncoated group were statistically significant. However, when comparing the coated groups, statistically significant differences were observed between group S1 and groups S3 and S5, as well as between group S2 and group S5. In all other comparisons, the differences were not statistically significant.

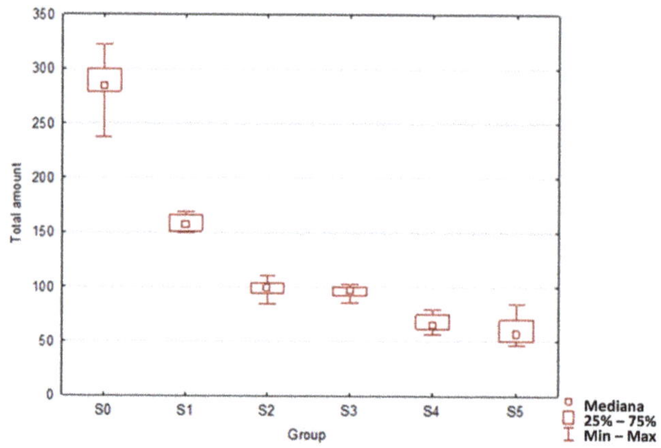

Figure 1. Adhered bacteria count on specimens from each group.

Table 6. Kruskal–Wallis test p-values for the number of adhered bacteria under stationary conditions.

Tested Groups	p-Value					
	S0	**S1**	**S2**	**S3**	**S4**	**S5**
S0		0.0475	0.0346	0.0117	$p < 0.0001$	$p < 0.0001$
S1	0.0475		1.0000	0.5621	0.0003	$p < 0.0001$
S2	0.0346	1.0000		1.0000	0.1985	0.0485
S3	0.0117	0.5621	1.0000		0.4572	0.1276
S4	$p < 0.0001$	0.0003	0.1985	0.4572		1.0000
S5	$p < 0.0001$	$p < 0.0001$	0.0485	0.1276	1.0000	

The data presented in Figure 2 and Table 7 show that all Ti(C,N)-type coatings resulted in a percentage increase in the number of viable bacteria adhering to the surfaces of the coated samples. For all coated groups, the differences in the number of adhered bacteria compared to the uncoated group were statistically significant. However, when comparing the coated groups, statistically significant differences were observed between group S1 and groups S4 and S5, as well as between group S2 and group S5. In all other comparisons, the differences were not statistically significant.

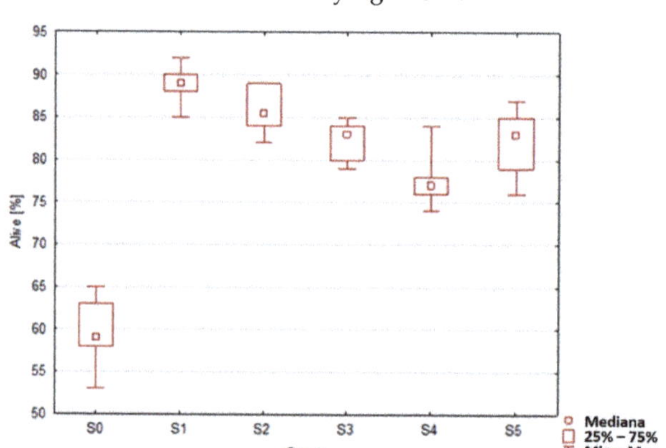

Figure 2. Viable bacteria percentage on specimens from each group.

Table 7. Kruskal–Wallis test *p*-values for the percentage of viable bacteria under stationary conditions.

Tested Groups	*p*-Value					
	S0	**S1**	**S2**	**S3**	**S4**	**S5**
S0		$p < 0.0001$	$p < 0.0001$	0.0117	0.0346	0.0297
S1	$p < 0.0001$		1.0000	0.5621	0.0003	$p < 0.0001$
S2	$p < 0.0001$	1.0000		1.0000	0.1985	0.0485
S3	0.0117	0.5621	1.0000		0.4572	0.1276
S4	0.0346	0.0003	0.1985	0.4572		1.0000
S5	0.0297	$p < 0.0001$	0.0485	0.1276	1.0000	

3.2. *Bacteria—Flow Conditions*

The data presented in Figure 3 and Table 8 show that all Ti(C,N)-type coatings resulted in a decrease in the number of adhered bacteria on the surfaces of the coated samples. For all coated groups, the differences in the number of adhered bacteria compared to the uncoated group were statistically significant. However, when comparing the coated groups, statistically significant differences were observed between group S1 and groups S2, S4, and S5, as well as between group S3 and group S4. In all other comparisons, the differences were not statistically significant.

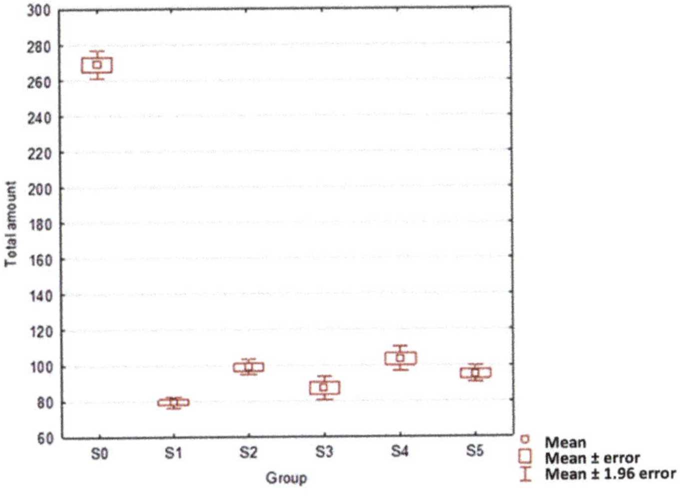

Figure 3. Adhered bacteria count on specimens from each group.

Table 8. Kruskal–Wallis test *p*-values for the number of adhered bacteria under flow conditions.

Tested Groups	*p*-Value					
	S0	**S1**	**S2**	**S3**	**S4**	**S5**
S0		0.0001	0.0001	0.0001	0.0001	0.0001
S1	0.0001		0.0004	0.4340	0.0001	0.0059
S2	0.0001	0.0004		0.0653	0.9063	0.9222
S3	0.0001	0.4340	0.0653		0.0039	0.4340
S4	0.0001	0.0001	0.9063	0.0039		0.3508
S5	0.0001	0.0059	0.9222	0.4340	0.3508	

The data presented in Figure 4 and Table 9 show that, compared to the uncoated group, statistically significant differences in the percentage of viable bacteria were observed in groups S1, S2, and S3. No statistically significant differences were found in groups S4 and S5. When comparing the coated groups, statistically significant differences were observed between group S1 and group S4, as well as between group S2 and group S4. In all other comparisons, the differences were not statistically significant.

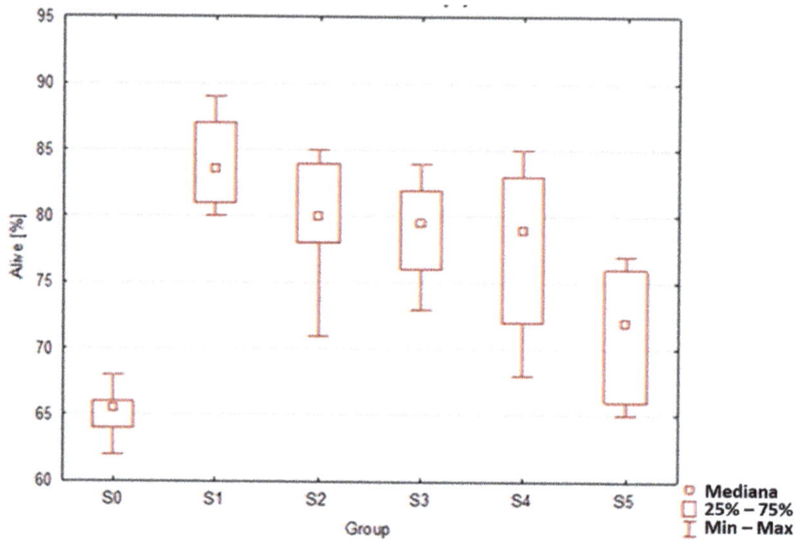

Figure 4. Viable bacteria percentage on specimens from each group.

Table 9. Kruskal–Wallis test p-values for the percentage of viable bacteria under flow conditions.

Tested Groups	*p*-Value					
	S0	**S1**	**S2**	**S3**	**S4**	**S5**
S0		$p < 0.0001$	$p < 0.0001$	0.0161	1.0000	1.0000
S1	$p < 0.0001$		1.0000	0.0658	0.0001	0.0728
S2	$p < 0.0001$	1.0000		1.0000	0.0286	1.0000
S3	0.0161	0.0658	1.0000		1.0000	1.0000
S4	1.0000	0.0001	0.0286	1.0000		1.0000
S5	1.0000	0.0728	1.0000	1.0000	1.0000	

3.3. *Fungi—Stationary Conditions*

The data presented in Figure 5 and Table 10 show that all Ti(C,N)-type coatings resulted in a decrease in the number of adhered fungal cells on the surfaces of the coated samples. For all coated groups, the differences in the number of adhered fungal cells compared to the uncoated group were statistically significant. However, when comparing the coated groups, no statistically significant differences were found between group S3 and group S4. In all other comparisons, the differences in the number of adhered fungal cells were statistically significant.

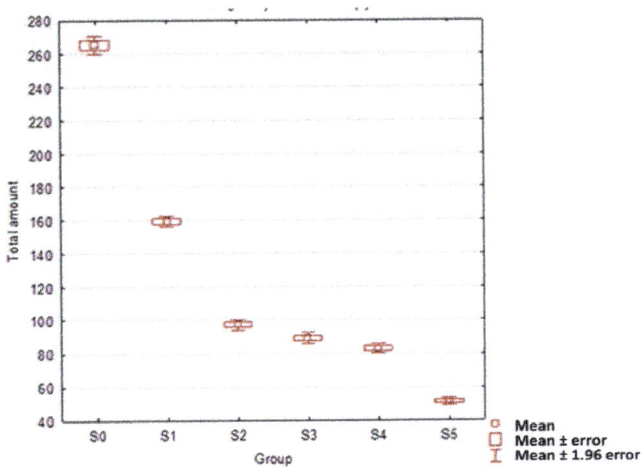

Figure 5. Adhered fungal cell count on specimens from each group.

Table 10. Kruskal–Wallis test *p*-values for the number of adhered fungal cells under stationary conditions.

Tested Groups	*p*-Value					
	S0	**S1**	**S2**	**S3**	**S4**	**S5**
S0		0.0001	0.0001	0.0001	0.0001	0.0001
S1	0.0001		0.0001	0.0001	0.0001	0.0001
S2	0.0001	0.0001		0.0348	0.0001	0.0001
S3	0.0001	0.0001	0.0348		0.1513	0.0001
S4	0.0001	0.0001	0.0001	0.1513		0.0001
S5	0.0001	0.0001	0.0001	0.0001	0.0001	

The data presented in Figure 6 and Table 11 indicate that, in comparison with the the uncoated group, statistically significant differences in the percentage of viable fungal cells were observed in groups S2, S4, and S5. No significant differences were found in groups S1 and S3. However, when comparing the coated groups, statistically significant differences were found between group S1 and groups S2 and S5, as well as between group S2 and group S3. In all other comparisons, the differences were not statistically significant.

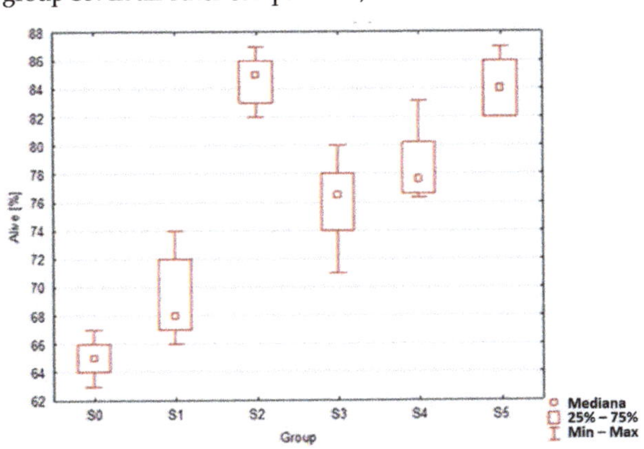

Figure 6. Viable fungal cell percentage on specimens from each group.

Table 11. Kruskal–Wallis test *p*-values for the percentage of viable fungal cells under stationary conditions.

Tested Groups	*p*-Value					
	S0	**S1**	**S2**	**S3**	**S4**	**S5**
S0		1.0000	$p < 0.0001$	0.0422	0.0063	$p < 0.0001$
S1	1.0000		0.0001	1.0000	0.3515	0.0003
S2	$p < 0.0001$	0.0001		0.0419	0.4147	1.0000
S3	0.0422	1.0000	0.0419		1.0000	0.0713
S4	0.0063	0.3515	0.4147	1.0000		0.6266
S5	$p < 0.0001$	0.0003	1.0000	0.0713	0.6266	

3.4. Fungi—Flow Conditions

The data presented in Figure 7 and Table 12 indicate that all Ti(C,N)-type coatings resulted in a decrease in the number of adhered fungal cells on the surfaces of the coated samples. For all coated groups, the differences in the number of adhered fungal cells compared to the uncoated group were statistically significant. However, when comparing the coated groups, statistically significant differences were observed between group S1 and groups S4 and S5, as well as between group S2 and group S5. In all other comparisons, the differences were not statistically significant.

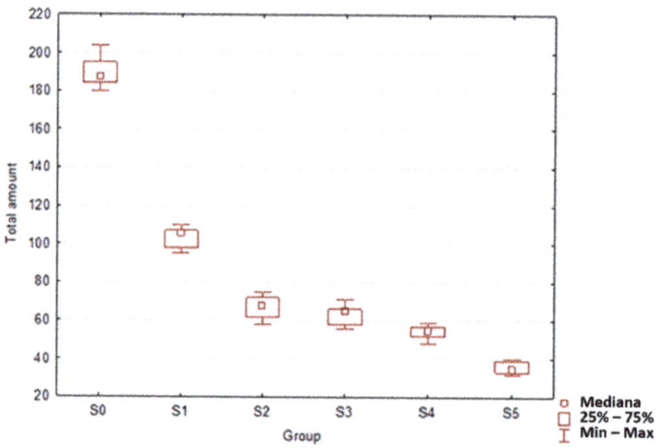

Figure 7. Adhered fungal cell count on specimens from each group.

Table 12. Kruskal–Wallis test *p*-values for the number of adhered fungal cells under flow conditions.

Tested Groups	*p*-Value					
	S0	**S1**	**S2**	**S3**	**S4**	**S5**
S0		0.0459	0.0255	0.0052	$p < 0.0001$	$p < 0.0001$
S1	0.0459		1.0000	0.3232	0.0032	$p < 0.0001$
S2	0.0255	1.0000		1.0000	0.6561	0.0088
S3	0.0052	0.3232	1.0000		1.0000	0.0713
S4	$p < 0.0001$	0.0032	0.6561	1.0000		1.0000
S5	$p < 0.0001$	$p < 0.0001$	0.0088	0.0713	1.0000	

The data presented in Figure 8 and Table 13 indicate that, compared to the group without coating, statistically significant differences in the percentage of viable fungal cells were observed only between the S1 and S3 groups. No statistically significant differences were found in the S2, S4, and S5 groups. However, among the coated groups, significant differences were observed only between the S1 group and the S4 and S5 groups. In all other comparisons, the differences were not statistically significant

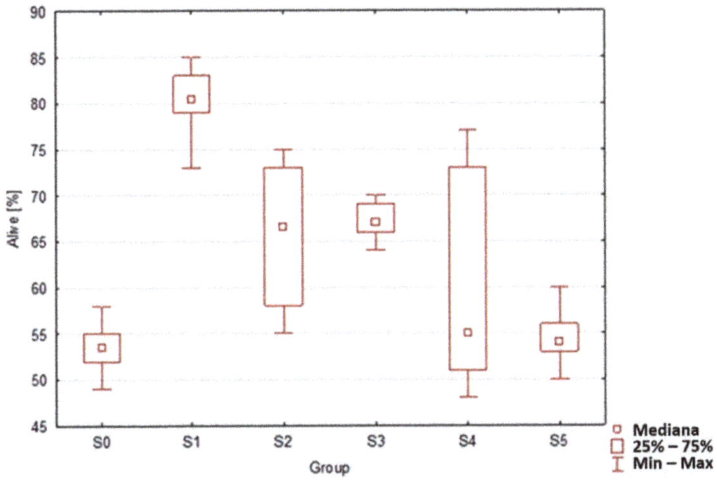

Figure 8. Viable fungal cell percentage on specimens from each group.

Table 13. Kruskal–Wallis test *p*-values for the percentage of viable fungal cells under flow conditions.

Tested Groups	*p*-Value					
	S0	**S1**	**S2**	**S3**	**S4**	**S5**
S0		$p < 0.0001$	0.0537	0.0428	1.0000	1.0000
S1	$p < 0.0001$		0.2678	0.3232	0.0015	$p < 0.0001$
S2	0.0537	0.2678		1.0000	1.0000	0.1538
S3	0.0428	0.3232	1.0000		1.0000	0.1253
S4	1.0000	0.0015	1.0000	1.0000		1.0000
S5	1.0000	$p < 0.0001$	0.1538	0.1253	1.0000	

4. Discussion

The microorganism adhesion studies were conducted in two variants: the stationary variant and the constant flow variant. This research design was chosen to reflect the conditions encountered in the oral cavity. During fluid intake, conditions similar to flow are present, as fluids wash over tissues and prosthetic elements. Conversely, during sleep, conditions akin to stationary environments prevail, with only minimal flow occurring due to saliva swallowing.

The presented studies on bacterial and fungal cell adhesion demonstrate that all Ti(C,N)-type coatings significantly reduce the adhesion of these microorganisms to the surfaces of the coated samples, compared to the uncoated Ni-Cr alloy surfaces. In each case, the differences in the number of adhered bacterial and fungal cells between the uncoated group (S0) and the coated groups were statistically significant. In the case of bacterial adhesion under stationary conditions, the number of adhered bacterial cells decreased compared to the uncoated samples (S0 group) by approximately 54% for the S1 group and up to about 22% for the S5 group. Under flow conditions, the number of adhered bacteria decreased from approximately 38% for group S4 to about 29% for group S1. When comparing the groups coated with Ti(C,N) coatings, it is evident that, both in stationary and flow conditions, all groups exhibited similar behavior. The adhesion of fungal cells under stationary conditions decreased, compared to the uncoated samples (group S0), from about 60% for group S1 to approximately 19% for group S5. Under flow conditions, the number of adhered fungal cells decreased from about 53% for group S1 to around 20% for group S5.

The obtained results appear to be highly beneficial in terms of preventing inflammatory complications associated with the introduction of prosthetic restorations into the oral cavity environment. It is important to note that the formation of a microbiological biofilm on prosthetic elements, as well as other surfaces, begins with the adhesion of cells. Therefore, a reduced number of adhered bacteria serves as a key factor in decreasing the likelihood of biofilm development. Upon analyzing the individual coatings, it appears that, from the perspective of bacterial adhesion, the coatings in the S5 group (TiN) are the most favorable, exhibiting the lowest numbers of adhered bacteria. In contrast, the coatings in the S1 group (TiC) were the least favorable, with the highest levels of bacterial adhesion. However, it is important to note that these differences were not statistically significant and should be considered as a trend rather than definitive results. A similar pattern was observed with fungal cell adhesion, where the coatings in the S5 group (TiN) performed the best, while the S1 group (TiC) exhibited the poorest performance. Given that the substrate preparation prior to coating application was consistent (with all samples being polished), the observed differences in adhesion are likely attributable to the composition of the coatings. It can be concluded that a higher carbon content in Ti(C,N) coatings appears to promote the initial adhesion of bacterial cells.

In terms of bacterial and fungal survival, all coatings demonstrated a higher number of viable microorganisms remaining on the surface of the coated samples compared to the uncoated samples. The percentage of viable bacteria increased, compared to the uncoated alloy, by approximately 125% for group S4 and up to 146% for group S1 under stationary conditions. It should be noted that, under stationary conditions, the differences between the groups were minimal and not statistically significant. Therefore, it can be concluded that all coated groups exhibited similar behavior. Under flow conditions, the percentage of viable bacteria increased from approximately 109% for group S5 to around 127% for group S1. However, as previously indicated, these differences were not statistically significant in any of the comparisons. Regarding fungal cell survival under stationary conditions, the percentage of viable cells increased from approximately 108% for group S1 to about 131% for group S2. The differences between the coated sample groups were small, and in several instances, statistically insignificant. In flow conditions, the percentage of viable fungal cells increased from 102% for group S5 to 148% for group S1. Similarly to stationary conditions, the differences in the number of viable cells were statistically significant not in all groups.

Considering the role of microorganisms in the processes occurring in the oral cavity, it seems most reasonable to seek solutions that reduce or eliminate bacterial adhesion to metal surfaces. One of the simplest approaches is the polishing of the metal surface, which, by reducing the available surface area for bacterial attachment, can decrease the number of adhered microorganisms. Such a relationship has been demonstrated in studies on Co-Cr alloys and ceramics, particularly in the context of Streptococcus mutans adhesion [44]. Similar conclusions were drawn by L.C.M. Dantas et al. in a study evaluating the effect of surface roughness on bacterial adhesion to various clinical biomaterials. The study concluded that manual polishing reduces both surface roughness and bacterial adhesion [45].

However, this method does not completely eliminate bacteria from the metal surface, making it necessary to incorporate metal surface coatings into clinical practice. By combining surface polishing of metal prosthetic elements with the application of protective coatings, the effectiveness can be significantly enhanced compared to polishing alone. As demonstrated by the presented research results, Ti(C,N)-type coatings are particularly well suited for this purpose. Similar conclusions were reached by S.E.A. Camargo and colleagues, who investigated the influence of titanium nitride (TiN) and silicon carbide (SiC) coatings on the adhesion of Porphyromonas gingivalis bacteria and the activity of osteoblasts on the surfaces of titanium implants. The results demonstrated that TiN and SiC coatings can reduce bacterial adhesion while promoting osteoblast adhesion, which is advantageous for dental implants [46]. Consistent with our findings, titanium-doped diamond-like carbon (Ti-DLC) coatings have been shown to reduce bacterial adhesion by up to 75% compared to stainless steel surfaces, highlighting their potential for biomedical applications [46].

Another crucial aspect is that surface modification through coating can help mitigate biocorrosion [43,47]. This is highly beneficial, as biocorrosion can lead to metal allergies, inflammation and potentially even cancer. Nickel ions, in particular, are highly allergenic and can cause pain, contact dermatitis and erythema in patients [5]. Eliminating these issues in patients will significantly enhance their comfort with prosthetic restorations. It is also important to note that saliva plays a key role in regulating biofilm growth [48]. With age, patients become more susceptible to various infections, inflammations and ulcerations. Additionally, older adults often experience reduced salivary secretion, with medications further contributing to this reduction [49]. Therefore, efforts to minimize bacterial and fungal biofilm on the surfaces of metal alloys used in prosthetic restorations are well founded.

The findings of this study allow for the identification of the coating type that most effectively reduces microorganism adhesion to surfaces. However, it is important to recognize that adhesion reduction is not the sole factor influencing the suitability of these coatings for prosthetic applications. Key considerations include corrosion resistance, wear resistance, coating adhesion to the substrate, and potential toxicity, as reported in earlier studies [39–42]. A comprehensive evaluation of all these factors is essential for selecting the optimal coating. Additionally, it should be noted that prosthetic elements are often veneered with ceramics or composite materials. Future research should focus on evaluating whether Ti(C,N) coatings affect the adhesion properties of these veneering materials.

5. Conclusions

The conducted studies have demonstrated that Ti(C,N)-type coatings significantly reduce the adhesion of bacteria and fungi to surfaces compared to uncoated alloy, thereby potentially minimizing biofilm formation on prosthetic elements covered with these coatings. While the exact mechanism of such interaction remains unclear, several factors likely contribute to the observed reduction in microbial adhesion. A key factor is the chemical inertness of Ti(C,N) carbonitrides, which possess low surface energy. This reduces interactions between the coating surface and microbial cells, thereby limiting adhesion [50]. Another critical factor is surface wettability. Reduced wettability correlates with lower microorganism adhesion. Essentially, microbial adhesion is generally less possible on hydrophobic surfaces, as the reduced affinity for water may hinder the initial adhesion of bacteria and fungi. As shown in our previous studies, Ti(C,N) coatings significantly reduce the wettability of the alloy surface with water [51]. For instance, the wetting angle of the uncoated sample surface was approximately 83° (hydrophilic by definition, angle < 90°), whereas coated samples exhibited angles as high as 115° for the TiN coating (hydrophobic by definition, angle > 90°). Additionally, the zeta potential, or the electrical potential at the interface between the adsorption and diffusion phases (the "shear plane") of a molecule situated close to the surface, may also play a role. In this case, it refers to the interaction between the charge on the specimen surface and the microorganisms. A reduction in zeta potential due to the presence of coatings could decrease electrostatic attraction, thereby further limiting microbial adhesion. Although this parameter was not investigated in the present study, it represents an important direction for future research.

Author Contributions: Conceptualization, K.B., L.K. and Z.K.; methodology, W.J. and L.K.; software, K.B. and W.J.; validation, K.B., K.D. and W.J.; formal analysis, Z.K. and L.K.; investigation, K.B.; resources, K.B.; data curation, K.B., W.J. and L.K.; writing—original draft preparation, K.B., K.D., W.J. and Z.K.; writing—review and editing, K.D., W.J. and Z.K.; visualization, K.B. and L.K.; supervision, W.J. and L.K.; project administration, K.B., Z.K. and L.K. All authors have read and agreed to the published version of the manuscript.

Funding: This research received no external funding.

Institutional Review Board Statement: Not applicable.

Informed Consent Statement: Not applicable.

Data Availability Statement: Data are contained within the article.

Conflicts of Interest: The authors declare that there are no conflicts of interest regarding the publication of this paper.

References

1. Giofinei-Schreibei, B.; Ghepentrog, M.; Haustein, I.; Miiller, W.D.; Lange, K.P.; Briedigkeit, H.; Gobel, U.B. Plaque formation on surface modified dental implants An in vitro study. *Ciln. Oral. Impl. Res.* **2001**, *12*, 543–551.
2. Jansen, V.K.; Conrads, G.; Richter, E.J. Microbial leakage and marginal fit of the implant-abutment interface. *Int. J. Oral. Maxillofac. Implants* **1997**, *12*, 527–540. [PubMed]
3. Wolinsky, L.E.; De Camargo, P.M.; Evard, C.; Newman, M.G. A study of in vitro attachment of *Streptococcus sanguis* and *Actinomyces viscosus* to saliva-treated titanium. *Int. J. Oral. Maxillofac. Implant.* **1989**, *4*, 17–31.
4. Silva, R.C.S.; Agrelli, A.; Andrade, A.N.; Mendes-Marques, C.L.; Arruda, I.R.S.; Santos, L.R.L.; Vasconcelos, N.F.; Machado, G. Titanium Dental Implants: An Overview of Applied Nanobiotechnology to Improve Biocompatibility and Prevent Infections. *Materials* **2022**, *15*, 3150. [CrossRef] [PubMed]
5. Mystkowska, J.; Niemirowicz-Laskowska, K.; Łysik, D.; Tokajuk, G.; Dąbrowski, J.R.; Bucki, R. The Role of Oral Cavity Biofilm on Metallic Biomaterial Surface Destruction–Corrosion and Friction Aspects. *Int. J. Mol. Sci.* **2018**, *19*, 743. [CrossRef] [PubMed]
6. Adya, N.; Alam, M.; Ravindranath, T.; Mubeen, A.; Saluja, B. Corrosion in titanium dental implants: Literature review. *J. Indian. Prosthodont. Soc.* **2005**, *53*, 126–131. [CrossRef]
7. Jakubowski, W.; Bartosz, G.; Niedzielski, P.; Szymanski, W.; Walkowiaka, B. Nanocrystalline diamond surface is resistant to bacterial colonization Diamond & Related. *Materials* **2004**, *13*, 1761–1763.
8. Zhang, X.; Liu, L.; Peng, W.; Dong, X.; Gu, Y.; Ma, H.; Gan, D.; Liu, P. Phosphonate/zwitterionic/cationic terpolymers as high-efficiency bactericidal and antifouling coatings for metallic substrates. *J. Mater. Chem. B* **2021**, *30*, 4169–4177. [CrossRef] [PubMed]
9. Ha-Young, K.; In-Sung, Y.; Jai-Bong, L.; Sung-Hun, K.; Dae-Joon, K.; Jung-Suk, H. Initial in vitro bacterial adhesion on dental restorative materials. *Int. J. Artif. Organs* **2012**, *35*, 773–779.
10. Kohavi, D.; Klinger, A.; Steinberg, D.; Selo, M.N. Adsorption of salivary proteins onto prosthetict itanium components. *J. Proslhetic Dent.* **1995**, *74*, 531–534. [CrossRef]
11. Jakubowski, W.; Walkowiak, B. Resistance of Oxidative Stress in Biofilm and Planktonic Cells. *Braz. Arch. Biol. Technol.* **2015**, *58*, 300–308. [CrossRef]
12. Asri, R.I.M.; Harun, W.S.W.; Samykano, M.; Lah, N.A.C.; Ghani, S.A.C.; Tarlochan, F.; Raza, M.R. Corrosion and surface modification on biocompatible metals: A review. *Mater. Sci. Eng.* **2017**, *77*, 1261–1274. [CrossRef]

13. Jamesh, M.I.; Li, P.; Bilek, M.M.; Boxman, R.L.; McKenzie, D.R.; Chu, P.K. Evaluation of corrosion resistance and cytocompatibility of graded metal carbon film on Ti and NiTi prepared by hybrid cathodic arc/glow discharge plasma-assisted chemical vapor deposition. *Corros. Sci.* **2015**, *97*, 126–138. [CrossRef]
14. Lian, K.Y.; Sałek, P.; Jin, M.; Ding, D. Density-functional studies of plasmons in small metal cluster. *Chem. Phys.* **2009**, *130*, 174701. [CrossRef]
15. Mercieca, S.; Conti, C.M.; Buhagiar, J.; Camilleri, J. Assessment of corrosion resistance of cast cobalt- and nickel-chromium dental alloys in acidic environments. *J. Appl. Biomater. Funct. Mater.* **2018**, *16*, 47–54. [CrossRef] [PubMed]
16. Bodunrin, M.O.; Chown, L.H.; Van Der Merwe, J.W.; Alaneme, K.K.; Oganbule, C.; Klenam, D.E.P.; Mphasha, N.P. Corrosion behavior of titanium alloys in acidic and saline media: Role of alloy design, passivation integrity, and electrolyte modification. *Corros. Rev.* **2020**, *38*, 25–47. [CrossRef]
17. Beech, I.B.; Gaylarde, C.C. Recent advances in the study of biocorrosion—An overview. *Rev. De Microbiol.* **1999**, *30*, 177–190. [CrossRef]
18. Chern Lin, J.H.; Lo, S.J.; Ju, C.P. Biocorrosion study of titanium-nickel alloys. *J. Oral. Rehabil.* **1996**, *23*, 129–134. [CrossRef]
19. Wilson, M.; Patel, H.; Kpendema, H.; Noar, H.J.; Hunt, P.N.; Mordan, N.J. Corrosion of intra-oralmagnets bymulti-species biofilms in the presence and absence of sucrose. *Biomaterials* **1997**, *18*, 53–57. [CrossRef]
20. Wylie, C.M.; Sheltonb, R.M.; Fleming, G.J.P.; Davenporta, A.J. Corrosion of nickel-based dental casting alloys. *Dent. Mater.* **2007**, *23*, 714–723. [CrossRef]
21. Upadhyaya, D.; Panchal, M.A.; Dubey, R.S.; Srivastava, V.K. Corrosion of alloys used in dentistry: A review. *Mater. Sci. Eng.* **2006**, *432*, 1–11. [CrossRef]
22. Manaranchea, C.; Hornbergerb, H. A proposal for the classification of dental alloys according to their resistance to corrosion. *Dent. Mater.* **2007**, *23*, 1428–1437. [CrossRef] [PubMed]
23. Viswanathan, S.; Saji, V.S.; Han-Cheol, C. Electrochemical behavior of Co-Cr and Ni-Cr dental cast alloys. *Trans. Nonferrous Met. Soc. China* **2009**, *19*, 785–790.
24. Tiller, A.K. *Electrochemical Aspects of Microbial Corrosion: An Overview, in Microbial Corrosion*; The Metals Society: London, UK, 1983; pp. 54–65.
25. Laurent, F.; Grosgogeat, B.; Reclaru, L.; Dalard, F.; Lissac, M. Comparison of corrosion behaviour in presence of oral bacteria. *Biomaterials* **2001**, *22*, 2273–2282. [CrossRef] [PubMed]
26. Suito, H.; Iwawaki, Y.; Goto, T.; Tomotake, Y.; Ichikawa, T. Oral Factors Affecting Titanium Elution and Corrosion: An In Vitro Study Using Simulated Body Fluid. *PLoS ONE* **2013**, *8*, e66052. [CrossRef]
27. Banaszek, K.; Szymanski, W.; Pietrzyk, B.; Klimek, L. Adhesion of *E. coli* Bacteria Cells to Prosthodontic Alloys Surfaces Modified by TiO2 Sol-Gel Coatings. *Adv. Mater. Sci. Eng.* **2013**, *2013*, 179241. [CrossRef]
28. Hiltbert, L.R.; Bagge−Ravn, D.; Kold, J.; Gram, L. Infuence of surface roughness of stainless steel on microbial adhesion and corrosion resistance. *Int. Biodeter. Biodegr.* **2003**, *52*, 175–185. [CrossRef]
29. Holban, A.M.; Farcasiu, C.; Andrei, O.C.; Grumezescu, A.M.; Farcasiu, A.T. Surface Modification to Modulate Microbial Biofilms-Applications in Dental Medicine. *Materials* **2021**, *14*, 6994. [CrossRef]
30. Sharma, P.K.; Rao, K.H. Analysis of different approaches for evaluation of surface energy of microbial cells by contact angle goniometry. *Adv. Colloid. Interface Sc.* **2002**, *98*, 341–463. [CrossRef] [PubMed]
31. Gvetadze, R.; Arutyunov, S.; Kryuchkov, S.; Antipov, M.; Bazhin, P.; Mustafaev, M.; Deshev, A.; Tsarev, V.; Andreev, M.; Katkov, I.; et al. Cermet coatings obtained by electric spark alloying to increase service life of dental instruments. *Ceram. Int.* **2024**, *50*, 52613–52621. [CrossRef]
32. Cao, Z.; Chen, C.; Li, R.; Shou, M.; Zheng, K.; Luo, R.; Wei, B.; Wang, T.; Wu, L. Mechanical properties and current-carrying tribological behaviors of sliver-titanium nitride composite ceramic coating. *Ceram. Int.* **2024**, *50*, 23419–23428. [CrossRef]
33. Batory, D.; Blaszczyk, T.; Clapa, M.; Mitura, S. Investigation of anti-corrosion properties of Ti:C gradient layers manufactured in hybrid deposition system. *J. Mater. Sci.* **2008**, *43*, 3385–3391. [CrossRef]
34. Szymanowski, H.; Sobczyk, A.; Gazicki-Lipman, M.; Jakubowski, W.; Klimek, L. Plasma enhanced CVD deposition of titanium oxide for biomedical applications. *Surf. Coat. Tech.* **2005**, *200*, 1036–1040. [CrossRef]
35. Sáenz deViteri, V.; Barandika, M.G.; Bayon, R.; Fernandeza, X.; Ciarsolo, I.; Igartua, A.; Tanoira, R.P.; Moreno, J.E.; Peremarch, C.P.-J. Development of Ti–C–N coatings with improved tribological behavior and antibacterial properties. *J. Mech. Behav. Biomed. Mat.* **2016**, *55*, 75–86. [CrossRef] [PubMed]
36. Balázsi, K.; Lukács, I.E.; Gurbán, S.; Menyhárd, M.; Bacáková, L.; Vandrovcová, M.; Balázsi, C. Structural, mechanical and biological comparison of TiC and TiCN nanocomposites films. *J. Eur. Ceram. Soc.* **2013**, *33*, 2217–2221. [CrossRef]
37. De Viteri, V.S.; Barandika, M.G.; De Gopegui, U.R.; Bayón, R.; Zubizarreta, C.; Fernández, X.; Igartua, A.; Agullo-Rueda, F. Characterization of Ti-C-N coatings deposited on Ti6Al4V for biomedical applications. *J. Inorg. Biochem.* **2012**, *117*, 359–366. [CrossRef]

38. Bramaa, M.; Rhodese, N.; Hunte, J.; Riccia, A.; Teghild, R.; Migliacciob, S.; Della Roccac, C.; Leccisottia, S.; Lioia, A.; Scandurraa, M.; et al. Effect of titanium carbide coating on the osseointegration response in vitro and in vivo. *Biomaterials* **2007**, *28*, 595–608. [CrossRef]
39. Banaszek, K.; Maślanka, M.; Semenov, M.; Klimek, L. Corrosive Studies of a Prosthetic Ni-Cr Alloy Coated with Ti(C,N) Type Layers. *Materials* **2022**, *15*, 2471. [CrossRef]
40. Banaszek, K.; Klimek, L. Ti(C, N) as Barrier Coatings. *Coatings* **2019**, *9*, 432. [CrossRef]
41. Banaszek, K.; Wiktorowska-Owczarek, A.; Kowalczyk, E.; Klimek, L. Possibilities of applying Ti (C,N) coatings on prosthetic elements—Research with the use of human endothelial cells. *Acta Bioeng. Biomater.* **2016**, *18*, 119–126.
42. Banaszek, K.; Klimek, L.; Zgorzynska, E.; Swarzynska, A.; Walczewska, A. Cytotoxicity of titanium carbonitride coatings for prostodontic alloys with different amounts of carbon and nitro gen. *Biomed. Mater.* **2018**, *13*, 045003. [CrossRef] [PubMed]
43. Kozmos, M.; Virant, P.; Rojko, F.; Abram, A.; Rudolf, R.; Raspor, P.; Zore, A.; Bohinc, K. Bacterial Adhesion of Streptococcus mutans to Dental Material Surfaces. *Molecules.* **2021**, *26*, 1152. [CrossRef] [PubMed]
44. Dantas, L.C.; Da Silva-Neto, J.P.; Dantas, T.S.; Naves, L.Z.; das Neves, F.D.; da Mota, A.S. Bacterial Adhesion and Surface Roughness for Different Clinical Techniques for Acrylic Polymethyl Methacrylate. *Int. J. Dent.* **2016**, *2016*, 8685796. [CrossRef] [PubMed]
45. Camargo, S.E.A.; Roy, T.; Carey IV, P.H.; Fares, C.; Ren, F.; Clark, A.E.; Esquivel-Upshaw, J.F. Novel Coatings to Minimize Bacterial Adhesion and Promote Osteoblast Activity for Titanium Implants. *J. Funct. Biomater.* **2020**, *11*, 42. [CrossRef]
46. Zhao, Y.Y.; Zhao, B.; Su, X.; Zhang, S.; Wang, S.; Keatch, R.; Zhao, Q. Reduction of bacterial adhesion on titanium-doped diamond-like carbon coatings. *Biofouling* **2018**, *34*, 26–33. [CrossRef] [PubMed]
47. Shokeen, B.; Zamani, L.; Zadmehr, S.; Pouraghaie, S.; Ozawa, R.; Yilmaz, B.; Lilak, S.; Sharma, S.; Ogawa, T.; Moshaverinia, A.; et al. Surface Characterization and Assessment of Biofilm Formation on Two Titanium-Based Implant Coating Materials. *Front. Dent. Med.* **2021**, *2*, 695417. [CrossRef]
48. Zhang, R.; Han, B.; Liu, X. Functional Surface Coatings on Orthodontic Appliances: Reviews of Friction Reduction, Antibacterial Properties, and Corrosion Resistance. *Int. J. Mol. Sci.* **2023**, *24*, 6919. [CrossRef] [PubMed]
49. Woods, D.L.; Mentes, C.J. Spit: Saliva in nursing research, uses and methodological considerations in older adults. *Biol. Res. Nurs.* **2011**, *13*, 320–327. [CrossRef] [PubMed]
50. Yu, J.; Zhou, M.; Zhang, L.; Wei, H. Antibacterial Adhesion Strategy for Dental Titanium Implant Surfaces: From Mechanisms to Application. *J. Funkc. Mater. Biomater.* **2022**, *13*, 169. [CrossRef]
51. Banaszek, K.; Klimek, L. Wettability and surface free energy of Ti(C,N) coatings on nickel-based casting prosthetic alloys. *Arch. Foundry Eng.* **2015**, *15*, 11–16. [CrossRef]

coatings

MDPI

Article

Sol-Gel SiO_2 Coatings with Curcumin and Thymol on 3D Printouts Manufactured from Ti6Al4V ELI

Natalia Słabęcka, Karolina Czajkowska, Paulina Łataś, Patryk Śniarowski, Laura Ciosek, Krzysztof Jastrzębski and Bożena Pietrzyk *

Institute of Materials Science and Engineering, Faculty of Mechanical Engineering, Lodz University of Technology, Stefanowskiego 1/15, 90-537 Lodz, Poland; nataliaslabecka@gmail.com (N.S.); kczajkowska2701@gmail.com (K.C.); paulinalatas7@gmail.com (P.Ł.); patryk.sniarowski@wp.pl (P.Ś.); laura.ciosek@tlen.pl (L.C.); krzysztof.jastrzebski@p.lodz.pl (K.J.)

* Correspondence: bozena.pietrzyk@p.lodz.pl

Abstract: Bacterial biofilm on implants may cause inflammation, which disturbs the process of the implant's integration with the surrounding tissues. Such problems are becoming critical for patients' health, especially in connection with the presence of antibiotic-resistant bacterial strains. Among the existing alternatives for drug treatments are natural-based substances. This study focused on the examination of silica coatings with curcumin and thymol, which were deposited using the sol-gel method on 3D printouts made of Ti6Al4V ELI. This substrate material is commonly used in medicine. The selective laser melting technique used for the manufacturing of samples was in line with the existing procedures applied for individual orthopedic implants. The examination involved the assessment of the coatings' morphology, chemical composition, and biological effect. The antibacterial properties were tested using a flow cytometer using *Escherichia coli*, and the cytotoxicity on Saos-2 cells was assessed using the LIVE/DEAD test. The obtained results showed that it is possible to produce silica sol-gel coatings with the addition of specific natural substances in concentrations assuring a bacteriostatic effect. The produced coatings did not show any cytotoxic effect, which confirms the possibility of using both curcumin and thymol as additives to coatings used in medicine, e.g., for orthopedic implants.

Keywords: sol-gel coatings; natural antibacterials agents; curcumin; thymol; 3D printing

1. Introduction

Antibiotics, in their almost one hundred years of conscious usage, have become an integral part of medicine. However, their excessive use (both as medical therapeutic sand in the food industry) has resulted in a more frequent occurrence of microorganisms resistant to their effects and causes problems with the treatment of patients exposed to action of those microbes. The problem of antibiotic resistance has been recognized by the WHO as a key threat to public health, affecting all countries regardless of their level of development [1]. The limited capability to straightforwardly fight infection takes humanity back to the beginning of the 20th century and has resulted in a rising number of medical complications and even the deaths of patients due to antibiotic-resistant microorganisms. In addition, there are also economic aspects of antibiotic resistance, including increasing costs of care and an extended length of hospitalization [1,2].

Approximately 60%–70% of antibiotics have poor permeability resulting from their insufficient intracellular retention, which means that these substances are not effective enough for any type of infection [3]. In addition, microorganisms have developed mechanisms to cope with the detrimental effects of antibiotics, such as cell wall synthesis, DNA replication, and protein translation machinery. Resistance mechanisms cover various approaches from the expression of enzymes that are capable of degrading, modifying, or inactivating given antibiotics, to changing the cell wall composition or the activity of ion pumps, or to directly influencing regions most susceptible to antibiotic attack, for example by post-translational modification [4].

Among the most dangerous bacterial species under the careful attention of medical doctors are methicillin-resistant *Staphylococcus aureus* (MRSA) [5–10] and cephalosporin-resistant *E. coli* [11–13]. According to the 2019 Antibiotic Resistance Report, over 2.8 million infections with antibiotic-resistant bacteria occur each year in the USA, resulting in over 35,000 deaths [14].

Among the difficult-to-treat infections are those related to the inhabitation of the surface of implants with microbes and the formation of biofilms. When bacteria attach to a given surface using adhesion proteins, they begin to not only multiply on it but also produce exopolysaccharides. This film protects them against external factors, e.g., antibiotics, and creates a firmer attachment to the surface. If peri-implant infection occurs, the osseointegration process may be impaired. Bacterial infection causes inflammation, which initiates the process of bone tissue resorption around the implant. This further leads to the loosening of the implant as a result of micro-movements. Peri-implant infections may also cause systemic infections, which prolong hospitalization and contribute to the length of the patient's recovery [5].

One of the alternatives to antibiotics is nanoparticles, or metal ions such as silver, zinc, copper, gold, and iron [15]. However, their safety regarding human cells and the natural environment is increasingly being questioned [16–18]. Research shows that nanoparticles can accumulate in mammalian cells and have a toxic effect on them. However, the cellular response strongly depends on the shape, size, and surface properties of metal ions and nanoparticles [16,17]. Due to their dimensions, nanoparticles can enter the circulatory system and even penetrate the blood–brain barrier [16]. The common usage of products with metal nanoparticles obviously results in their release into the natural environment, resulting in further interactions with eukaryotic and prokaryotic cells [18]. Additionally, the widespread use of silver compounds as antibacterial agents has caused bacteria to develop resistance to silver [19].

The risk associated with the excessive use of antibiotics and nanoparticles as antibacterial agents motivates the search for alternative solutions that prevent bacterial infections. One of the possible solutions is the use of enzymes like lysozyme (N-acetyl-muramin hydrolase, also known as muramidase). The antibacterial effect of this naturally occurring compound is multidirectional: it causes the breakdown of polysaccharides in cell walls and disrupts the quorum-sensing systems [20]. Among the artificially synthetized compounds is parylene F, which shows antibacterial and anti-adhesive properties against Gram-negative bacteria [21].

A common approach is introducing protective antimicrobial coatings. These act on the basis of releasing active substances, e.g., metal ions, nanoparticles, chemical compounds, etc., killing microbes on contact, for example, in the case of chitosan or synthetic cationic polymers [22,23]. An interesting solution is the use of galvanic coatings produced using plasma-assisted implantation or electrospraying [24,25]. The potential difference in the metal constituents of such a coating causes the creation of electric microcells in the physiological solution, affecting bacterial health. In this case, there is no threat of gradual washing out of the antibacterial substance from such a coating [26]. More traditional approaches involve doped diamond-like carbon coatings [27] or silica sol-gel coatings with additives [28]. Sol-gel coatings in particular can be easily applied to elements of even complex shapes. The coating matrix (silica) itself does not have a bactericidal effect and is safe for human tissues, but the released substances may have a detrimental effect on bacteria [28].

Among the natural antibacterial substances, the most frequently discussed are turmeric, thyme, and oregano. The action of such a group of substances can be both active and passive. First, their action may be related to affecting the bacterial cell membrane or block metabolic processes. On the other hand, their passive effect results from the modification of environmental conditions or surface properties. They make it difficult for microorganisms to adhere or reproduce, e.g., due to changing the pH. A positive aspect of using natural substances is related to their synergistic effects with other drugs [29,30]. Natural substances can be used directly, in the form of oils or extracts, and their key active ingredients can be isolated from them [29,30]. Regarding curcumin (found in turmeric), it does not directly kill bacteria but rather inhibits the biofilm formation and prevents microbial adhesion via affecting the quorum-sensing systems [31]. Thymol, present in thyme, causes destabilization of the bacterial cell membrane. Small thymol molecules cross the lipid barrier, leading to changes in the conformation and activity of the internal and membrane proteins of bacteria. Thymol has a biofilm-reducing effect in the case of infections such as *Escherichia coli, Listeria monocytogenes, Pseudomonas putida, Staphylococcus aureus* and even methicillin-resistant *Staphylococcus aureus* [32,33]. The use of natural substances is also potentially safer for organisms and the natural environment, which makes them promising candidates as antibacterial additives for coatings.

The following paper focuses on the evaluation of the antibacterial potential of SiO_2 coatings manufactured by the sol-gel method on 3D-printed surfaces obtained via the selective laser sintering (SLS) technique. The antibacterial effect was exerted by the use of the natural compounds curcumin and thymol. The metal substrate (Ti6Al4V ELI) and printing technique (SLS) were aligned for direct application in printouts used for personalized orthopedical implants.

2. Materials and Methods

2.1. Substrate Material

Titanium alloy Ti6Al4V ELI samples in the shape of cylinders (Figure 1) with a diameter of 16 mm and a height of 4 mm, manufactured by Medgal Ltd. (Białystok, Poland), were used as the substrate for sol-gel coatings. For this purpose, a ProX 320 A printer (3D Systems Corporation, Rock Hill SC, USA) was used, equipped with a working platform with dimensions of 250 × 250 mm. Printouts with an accuracy of ±0.2% were obtained using a laser power of 500 W. In order to remove the stresses resulting from the printing process, the samples were heat-treated in a high-vacuum oven before being detached from the substrate. The samples were made from powder for which 95% of the average grain size was in the range of 28–36 μm, and the maximum size did not exceed 40 μm. Samples made of Ti6Al4V rod cut to the same dimensions as mentioned above were mechanically ground, polished to a mirror-like state, and used as reference samples for the morphology examination. For the evaluation of the coatings' thickness, monocrystalline silicon wafers (100) were used as substrates.

Figure 1. Ti6Al4V ELI titanium alloy substrate prepared using the SLS method.

2.2. Coatings

2.2.1. Sol Preparation

The SiO_2 sol was prepared by mixing 5 mL of tetraethoxysilane (TEOS) (Aldrich, Steinheim Germany), 2 mL of 36% hydrochloric acid (Chempur, Piekary Śląskie, Poland), and 50 mL of ethanol (Chempur, Piekary Śląskie, Poland). The sol was stirred using a magnetic stirrer for approximately 2 h and left for 24 h before coating deposition. The sol with 0.25% (*w*/*v*) curcumin (Sigma-Aldrich, St. Louis, MO, USA) and with 3% (*w*/*v*) thymol (Warchem S.A., Zakręt, Poland) was prepared by adding an appropriate amount of the natural agent to the measured volume of sol and stirring it for at least 10 min until fully dissolved in the sol. The choice of curcumin and thymol concentrations was based on studies by other researchers, where materials with these substances were tested in terms of antibacterial activity [31,32,34,35]. Raduly et al. tested antibacterial properties of siloxane host matrices with 0.27% curcumin dyes [34]. Irizar et al. tested bio-based hybrid nanocomposites for stone conservation and for this purpose examined the addition of 3% thymol and 3% thymol-SiO_2 mesoporous nanoparticles [35].

2.2.2. Coating Deposition

Before the deposition of coatings, substrate materials were cleaned in a series of sonicated baths. First they were cleaned in tap water with a small amount of detergent for 3 min, then with distilled water for 1 min, and finally in acetone for 3 min. A laboratory dip-coater, model TLO 0.1 (MTI Corporation, Richmond, CA, USA), was used for the deposition of coatings. The coatings were applied by immersing the substrate in a given sol and withdrawing it at a constant speed of 0.5 mm/s. After deposition, samples were left for 96 h at room temperature to allow the coatings to develop their final mechanical properties.

2.2.3. Sample Marking

The samples are abbreviated in the following manner: Ti—substrate without coating, Si—substrate with SiO_2 coating, Cur—substrate with SiO_2 coating and 0.25% curcumin, Tym—substrate with SiO_2 coating and 3% thymol. Mirror-polished Ti6Al4V was used for better visualization of coatings in the microscopic examination. To easily distinguish the samples, "3D printed" or "polished" was added to each abbreviation.

2.3. Microscopic Examination

Coating morphologies were assessed using a VHX950F (Keyence, Mechelen, Belgium) digital microscope via comparison with uncoated samples. Surface analysis was performed at 100× and 300× magnification. Further observations were performed with the use of the scanning electron microscopy (SEM) technique with a JSM-6610LV microscope (JEOL, Tokyo Japan). Observations were made using the secondary electron mode at an accelerating voltage of 20 kV. Additionally, energy-dispersive X-ray spectroscopy (EDS) was used for the verification of the chemical composition of the coatings. This study was performed in 2 repetitions.

2.4. Surface Roughness and Thickness of Coatings

The evaluation of the surface roughness was performed with the use of an optical profilometer (Sensofar, Barcelona, Metrology, Spain) on samples before and after the coating deposition. The arithmetic deviation from the mean line (Ra) and the highest roughness height (Rz) in the tested regions were determined. This study was performed in 2 repetitions. The thicknesses of coatings deposited on silicon substrates were determined with use of a reflectance spectrometer (Thin Film Analyzer UV20, Filmetrics, San Diego, CA, USA).

2.5. Assessment of Antibacterial Properties

Flow cytometry was used to assess the samples' antibacterial properties. The test was performed using a suspension of *Escherichia coli* O157:H7 (ATCC® 43890) with a density of 2×10^7 CFU/mL. On four 3D-printed samples (Ti, Si, Cur, Tym), 20 µL of bacterial suspension was applied and left for two different time intervals: 10 or 60 min. After the given exposure time, the suspension was collected from the examined surfaces and mixed with 2 µL of LIVE/DEAD kit and 200 µL of PBS buffer. The positive and negative controls were prepared in a similar manner but with 20 µL of untreated *E. coli* suspension. To kill the bacteria (positive control), an additional 20 µL of ethyl alcohol was added. This study was performed in 2 repetitions.

For visualization, the results are presented in graphs as mean ± standard deviation. Additionally, a one-way ANOVA test with a significance level of $p < 0.05$ was used for the statistical verification of the data set.

2.6. Assessment of Cytotoxicity

The cytotoxicity test using the LIVE/DEAD method was carried out in accordance with standard [36]. Bone-forming osteoblast cells, Saos-2 line cultured in consecutive McCoy's medium (with 15% of fetal bovine serum), were used for this examination. Four 3D printed samples (Ti, Si, Cur, Tym) were incubated in 12-well plates each with 2000 µL of culture medium with approximately 12×10^4 cells for the period of 24 h (in standard conditions). After that time, the medium was removed, cells were rinsed with a phosphate buffer, and a mixture of ethidium bromide (20 µL), calcein (3 µL) was added. The incubation with dyes lasted 30 min. After that time, the surfaces of the samples were observed with the use of fluorescence microscopy (Olympus GX71, Tokyo, Japan). For each sample, 5 photos were taken at various random locations. The number of cells (green: live; red: dead) was calculated. This study was performed in 2 repetitions.

3. Results

3.1. Microscopic Examination

Figure 2 shows the surfaces of the samples produced by 3D printing with and without modifications. In all cases, the applied coating was visible despite significant surface irregularities. The presence of curcumin in sample Cur was easily observed by the yellow color of the coating. The change in the intensity of that color showed an uneven thickness of the film; thicker groves formed after printing.

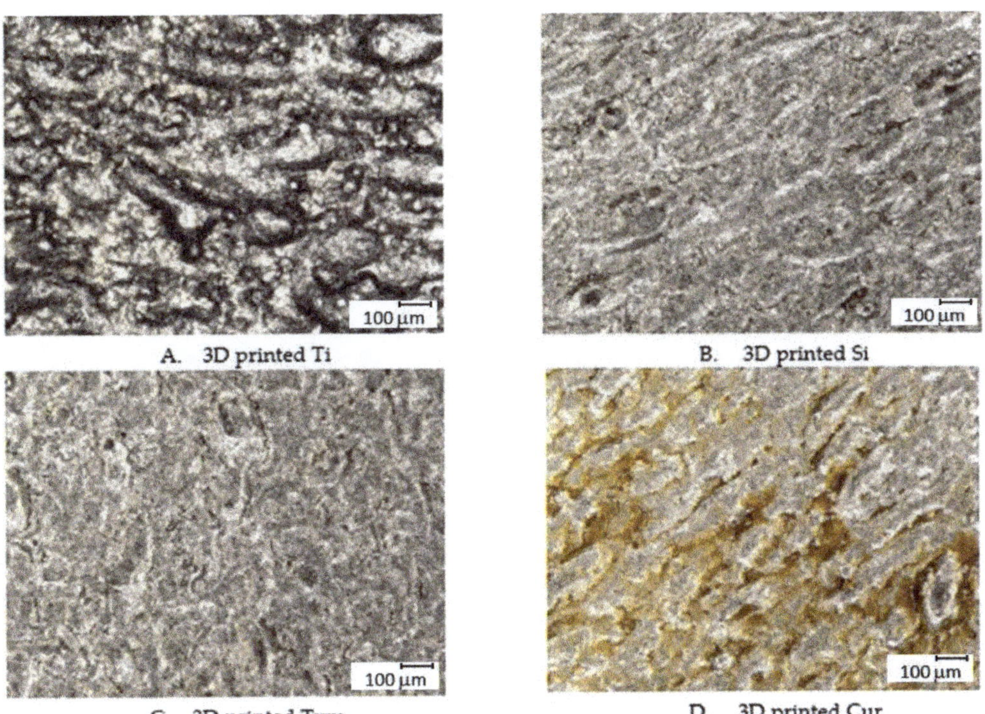

Figure 2. Surfaces of examined 3D-printed samples visualized by means of optical microscopy.

For better visualization of the coatings, the deposition was also performed on mirror-like polished samples (see Figure 3). No cracks or points of delamination were visible. Nevertheless, for the samples with coatings, there were some visible points of distortions and inclusions. In the case of the SiO_2-coated samples (Si), these may have originated from the partial gelation of the sol before its application. For other coatings, this may have been the result of processes common for SiO_2-coated samples and the incomplete dissolution of the natural additives.

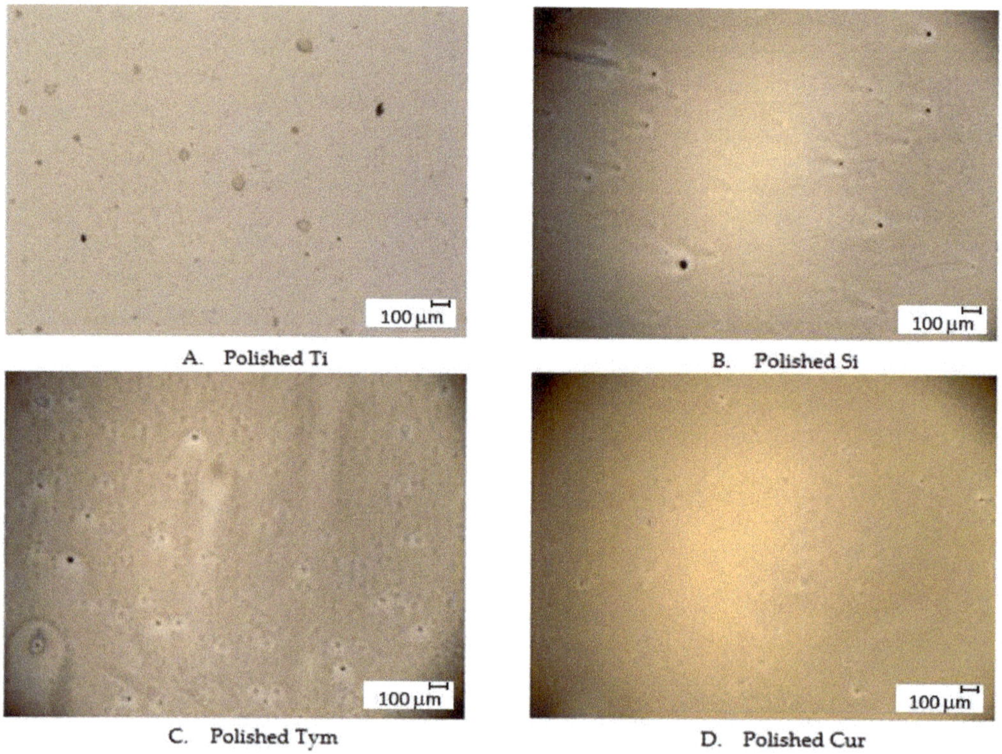

Figure 3. Surfaces of examined mirror-like polished samples visualized by means of optical microscopy.

SEM images of the coatings on printouts are shown in Figure 4. Especially on the surface of the untreated samples, there were visible surface patterning and powder residues after the SLS process. A portion of the used powder was loosely attached and visible as spherical particles on the surface. However, in the case of samples with coatings, those structures were covered with deposited films.

The presence of coatings was also evidenced by the chemical composition of the analyzed surfaces. Table 1 shows the percentage of the silicon content of the examined samples. As expected, the Ti sample did not contain silicon. The contents of this element were different for the Si, Cur, and Tym samples, probably due to differences in the thickness of the coatings.

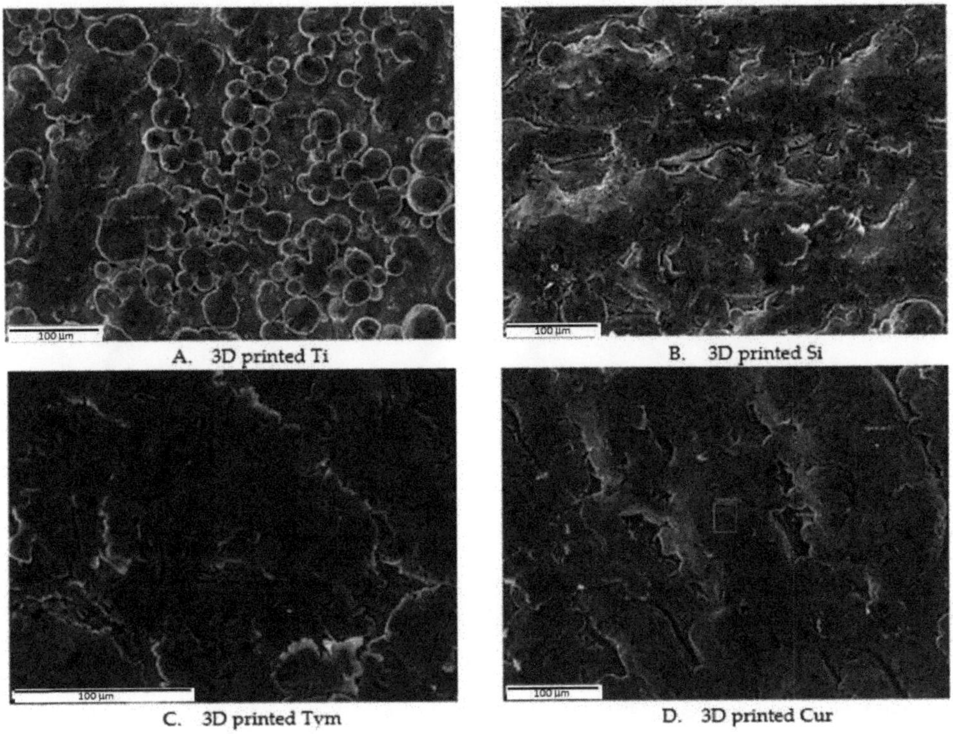

Figure 4. Surfaces of examined 3D-printed samples visualized by SEM.

Table 1. Silicon content in examined printouts obtained with the use of SEM-EDS.

Sample Abbreviation	Ti	Si	Cur	Tym
Silicon content [%]	0.00 ± 0.01	1.36 ± 0.20	1.72 ± 0.17	2.82 ± 0.35

3.2. *Surface Roughness and Thickness of Coatings*

An increase in roughness was visible for the coatings with added thymol (see Figure 5). That modification led to an over 2.6 rise in the Rz value and a two-times-higher Ra. In the case of Cur, a small increase in the value of Rz was visible but it was on the margin of the marked measurement error. The roughness of Ti and Si was similar.

In order to make an estimate of the coating thickness and determine the effect of additives, the thickness of the coatings deposited on the silicon wafers was measured. The thicknesses of the coatings deposited on the silicon wafers were similar for all produced coatings and amounted to 46 ± 5 nm. It was assumed that the thickness of the coatings produced on 3D printouts was higher due to the roughness of the substrate, but the addition of active agents did not affect their thickness.

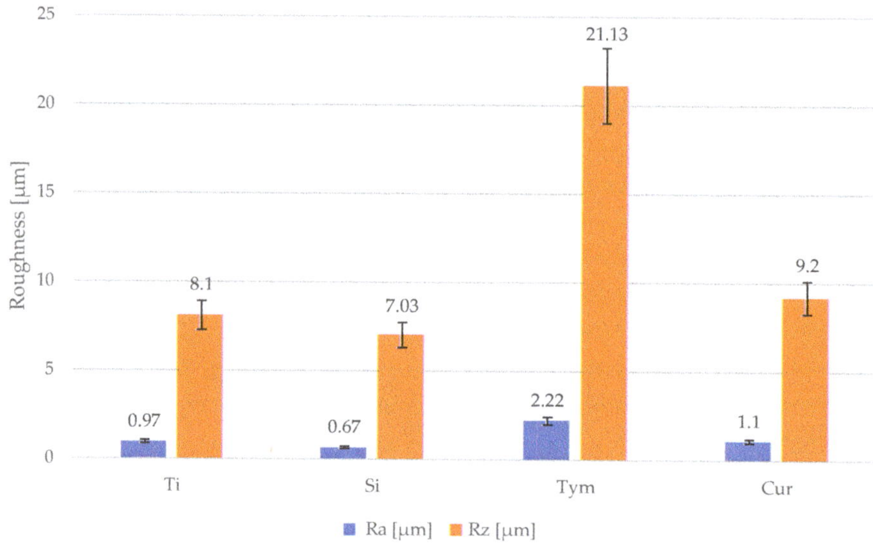

Figure 5. Summary of the roughness of examined 3D-printed surfaces.

3.3. *Assessment of Antibacterial Properties*

Data on the percentage of live and dead bacterial cells after 10 min and 60 min of incubation are presented in Figures 6 and 7, respectively. Regardless of the coatings' composition, the percentage of dead bacteria increased after 60 min of exposure in comparison to that of the untreated sample. For the 10 min exposure, such a trend was not visible for Tym. In fact, that modification in the mentioned conditions led to almost no dead bacteria. With the increase in the exposure time, the number of dead cells also rose. That change was not visible only for the Si sample. The highest limitation of living microbes was visible for the Cur sample after 60 min of exposure. In that case, almost one-third of the bacteria were dead.

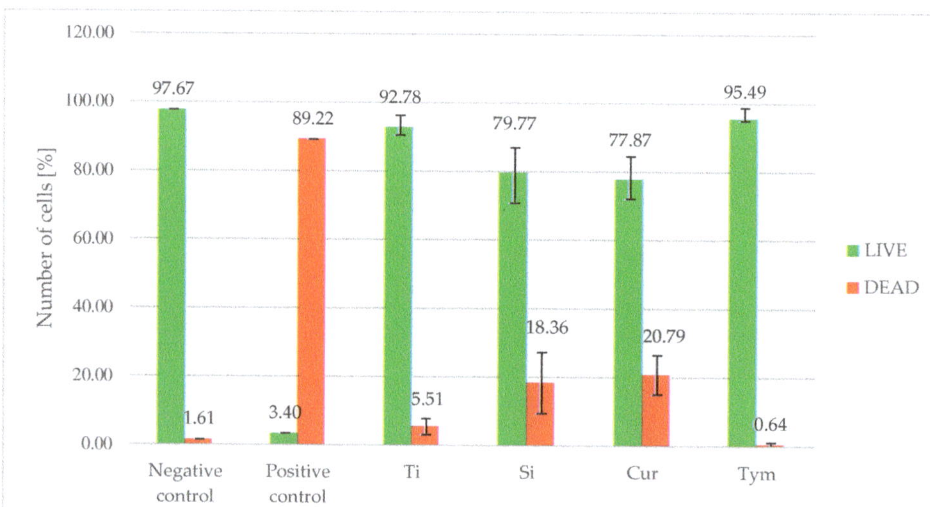

Figure 6. Percentage of live and dead bacterial cells after 10 min of exposure to the tested surfaces.

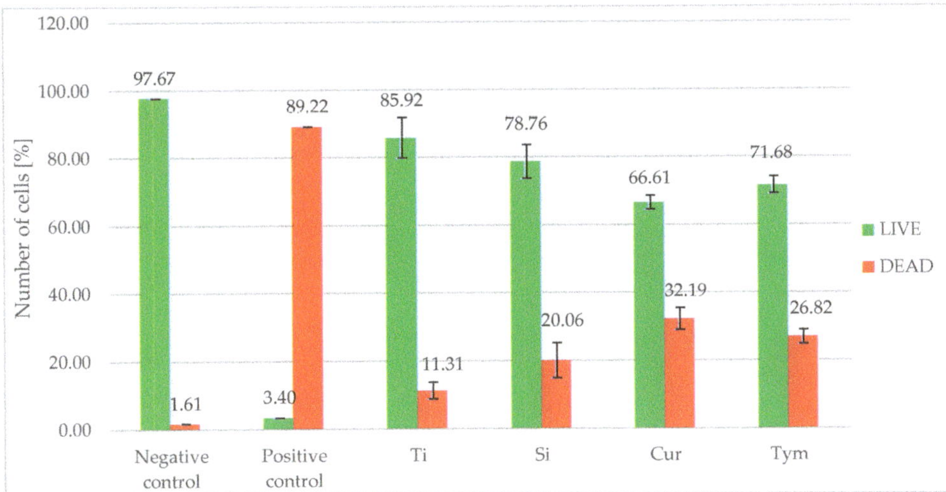

Figure 7. Percentage of live and dead bacterial cells after 60 min of exposure to the tested surfaces.

3.4. *Assessment of Cytotoxicity*

Figure 8 shows the percentages of the proliferation of Saos-2 cells on the examined surfaces. For all the examined materials, the mean proliferation was no lower than 90%. However, it should be noted that proliferation shows the tendency and ability of cells to reproduce and does not directly provide their amount. Among the modified materials, the largest number of cells was present on the Tym sample (see Figure 9). The total mean value (the sum of live and dead cells) for that type of sample was only slightly lower than the number of cells on the Ti sample (219.2 and 222.6, respectively), but the number of dead cells was almost three times higher (reaching 22.1 and 7.6, respectively). In the case of the printout with the sol-gel coating without any additive, the mean total value of cells was about 20% lower. On the surface of the Cur sample, there was the lowest number of cells, over 50% less than in any of the other cases. The number of dead cells on the Ti, Si, and Cur samples was, in each case, close to 10.

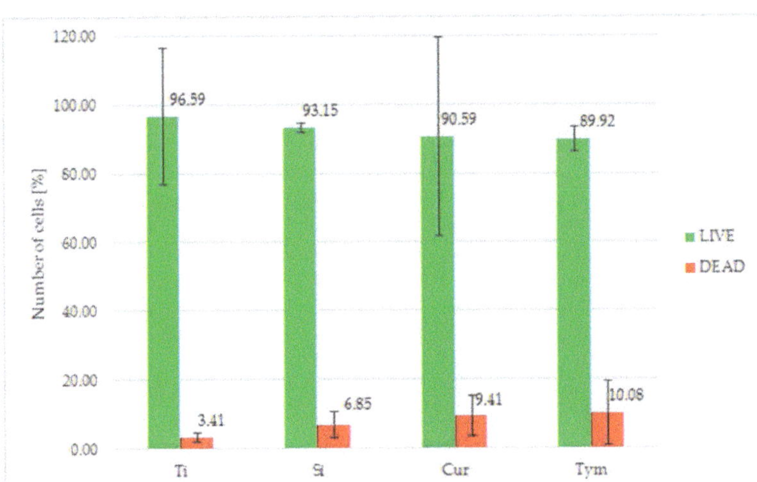

Figure 8. Evaluation of cell proliferation of Saos-2 cell line on examined surfaces.

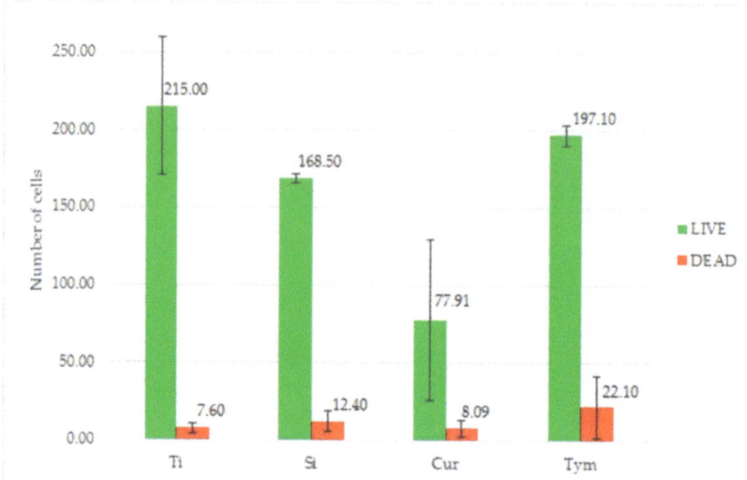

Figure 9. Number of Sos-2 cells on examined surfaces.

4. Discussion

Although only the Cur coating on the 3D printouts allowed the visual assessment of its uneven thickness, it can be assumed that a similar situation also occurs for Si and Tym. Nevertheless, the lack of cracks and signs of delamination proves the successful formation of sol-gel coatings on the samples manufactured by the SLS method. The main application of the proposed coating is not related to increasing corrosion resistance but rather to affecting microbial colonization. It means that even in the case of the discontinuation of the film, the release of natural substances to the surrounding environment is secured, and this reduces the amount of bacteria, as shown by the results of flow cytometry. The 3D printouts did not undergo any postprocessing to remove the loosely bound particles of powder after SLS. Nevertheless, due to the deposition of coatings, the remaining spheres were firmly covered, which may, to some extent, have prevented their detachment from the surface. The presented increase in surface roughness, especially for Tym, was not observed on either optical or SEM, which provided information about the smoothening of the surface. The Ra and Rz parameters may have been affected by the previously mentioned inclusions related to the incomplete dissolution of the additives or sol gelation prior to deposition. An increase in the surface roughness may promote bacterial adhesion to the surface of a material [23]. However, a threshold value that would clearly correspond to the increase in biofilm formation has not yet been established. In the case of the Tym coating, an increase in dead bacteria was still visible.

The results of the bactericidal assessment using flow cytometry did not show a clear bactericidal effect since the portion of dead bacterial cells was not higher than that of the living ones. However, the introduction of thymol and curcumin had a bacteriostatic effect, proved by the significant reduction in living bacteria. Even in the case of the SiO_2 coating without additives, some positive effect on the reduction in bacterial colonization was present. The literature has already described antibacterial properties due to the addition of curcumin [31,34] or thymol [32,33] but with higher contents of these additives. In this study, a low content of the natural agents was selected so as not to affect gelation in the process of the coating deposition. Eventually, the additives of natural agents may have had some effect on the coatings' morphology since their inclusion was present in the Tym and Cur samples. The increase in the number of dead *E. coli* after prolonging the exposure suggests that the antibacterial effect requires the release of the active substance from the coating over time, especially since this effect was stronger for doped films.

The results of the LIVE/DEAD cytotoxicity assessment prove that sol-gel SiO_2-based coatings do not have a cytotoxic effect on Saos-2 cells regardless of their composition. This means that the orthopedic application of such films has no drawbacks concerns surrounding the bone tissue response.

5. Conclusions

Silica sol-gel coatings with natural additives were successfully deposited on 3D printouts made of Ti6Al4V ELI. The addition of both curcumin and thymol led to a bactericidal effect of the synthetized films that is promising for the medical application of such materials and the further use of natural compounds to fight microbial infections. The potential of sol-gel coatings on 3D printouts manufactured in a manner resembling the production of individual implants also had no adverse effect on osteoblasts. The additional advantage of the proposed solution is the covering of loosely attached powders remaining after printing, which may be treated as postprocessing of such printouts. Further studies concerning both a wider range of curcumin and thymol concentrations as well as the evaluation of the durability of the coating are required.

Author Contributions: Conceptualization, N.S., K.C., P.Ł., P.Ś., and L.C; investigation, N.S., K.C., P.Ł., P.Ś., and L.C.; methodology N.S., K.C., P.Ł., P.Ś., and L.C.; supervision, B.P. and K.J.; writing—original draft, N.S., K.C., P.Ł., P.Ś., and L.C.; writing—review and editing, K.J. and B.P. All authors have read and agreed to the published version of the manuscript.

Funding: This research received no external funding.

Institutional Review Board Statement: Not applicable.

Informed Consent Statement: Not applicable.

Data Availability Statement: Data are contained within this article.

Acknowledgments: The authors are thankful for the involvement and support of the members of the Institute of Materials Science and Engineering at the Lodz University of Technology, Witold Jakubowski, Marta Kamińska, Anna Jędrzejczak, and Marcin Makówka in conducting this study. Gratitude is also expressed to Medgal Ltd. for preparing the samples.

Conflicts of Interest: The authors declare no conflicts of interest.

References

1. World Health Organization. Antimicrobial Resistance. 2023. Available online: https://www.who.int/news-room/fact-sheets/detail/antimicrobial-resistance (accessed on 3 May 2024).
2. Jonas, O.B.; Irwin, A.; Berthe, F.C.J.; Le Gall, F.G.; Marquez, P.V. *Drug-Resistant Infections: A Threat to Our Economic Future*; World Bank: Washington, DC, USA, 2017; Volume 2.
3. Slavin, Y.N.; Asnis, J.; Häfeli, U.O.; Bach, H. Metal Nanoparticles: Understanding the Mechanisms behind Antibacterial Activity. *J. Nanobiotechnol.* **2017**, *15*, 65. [CrossRef] [PubMed]
4. Godoy-Gallardo, M.; Eckhard, U.; Delgado, L.M.; de Roo Puente, Y.J.D.; Hoyos-Nogués, M.; Gil, F.J.; Perez, R.A. Antibacterial Approaches in Tissue Engineering Using Metal Ions and Nanoparticles: From Mechanisms to Applications. *Bioact. Mater.* **2021**, *6*, 4470–4490. [CrossRef] [PubMed]
5. Pokrowiecki, R.; Tyski, S.; Zaleska, M. Problematyka Zakażeń Okołowszczepowych. *Postep. Mikrobiol.* **2014**, *53*, 123–134.
6. Cornejo-Juárez, P.; Vilar-Compte, D.; Pérez-Jiménez, C.; Ñamendys-Silva, S.A.; Sandoval-Hernández, S.; Volkow-Fernández, P. The Impact of Hospital-Acquired Infections with Multidrug-Resistant Bacteria in an Oncology Intensive Care Unit. *Int. J. Infect. Dis.* **2015**, *31*, 31–34. [CrossRef]
7. Martina, P.F.; Martinez, M.; Centeno, C.K.; Von Specht, M.; Ferreras, J. Dangerous Passengers: Multidrug-Resistant Bacteria on Hands and Mobile Phones. *J. Prev. Med. Hyg.* **2019**, *60*, E293–E299. [CrossRef] [PubMed]
8. Glajzner, P. Infections Associated with Healthcare—A Still Serious Problem of Healthcare in Poland. *Farm. Pol.* **2019**, *75*, 148–156. [CrossRef] [PubMed]
9. Caselli, E. Hygiene: Microbial Strategies to Reduce Pathogens and Drug Resistance in Clinical Settings. *Microb. Biotechnol.* **2017**, *10*, 1079–1083. [CrossRef]
10. Silver, S.; Phung, L.T.; Silver, G. Silver as Biocides in Burn and Wound Dressings and Bacterial Resistance to Silver Compounds. *J. Ind. Microbiol. Biotechnol.* **2006**, *33*, 627–634. [CrossRef]

11. Lin, W.P.; Huang, Y.S.; Wang, J.T.; Chen, Y.C.; Chang, S.C. Prevalence of and Risk Factor for Community-Onset Third-Generation Cephalosporin-Resistant Escherichia Coli Bacteremia at a Medical Center in Taiwan. *BMC Infect. Dis.* **2019**, *19*, 245. [CrossRef]
12. Gramundi, I.; Albornoz, E.; Boutureira, M.; Rapoport, M.; Gomez, S.; Corso, A.; Castro, G.; Faccone, D. Characterization of Third Generation Cephalosporin-Resistant Escherichia Coli Clinical Isolates from Ushuaia, Argentina. *Rev. Argent. Microbiol.* **2023**, *55*, 43–48. [CrossRef]
13. Kourtis, A.P.; Sheriff, E.A.; Weiner-Lastinger, L.M.; Elmore, K.; Preston, L.E.; Dudeck, M.; McDonald, L.C. Antibiotic Multidrug Resistance of Escherichia Coli Causing Device- and Procedure-Related Infections in the United States Reported to the National Healthcare Safety Network, 2013–2017. Clin. *Infect. Dis.* **2021**, *73*, e4552–e4559. [CrossRef]
14. Centers for Disease Control and Prevention. *Antibiotic Resistance Threats in the United States*; U.S. Department of Health and Human Services, CDC: Atlanta, GA, USA, 2019.
15. Rai, M.; Deshmukh, S.D.; Ingle, A.P.; Gupta, I.R.; Galdiero, M.; Galdiero, S. Metal Nanoparticles: The Protective Nanoshield Against Virus Infection. *Crit. Rev. Microbiol.* **2016**, *42*, 46–56. [CrossRef]
16. Kumah, E.A.; Fopa, R.D.; Harati, S.; Boadu, P.; Zohoori, F.V.; Pak, T. Human and Environmental Impacts of Nanoparticles: A Scoping Review of the Current Literature. *BMC Public. Health* **2023**, *23*, 1059. [CrossRef] [PubMed]
17. Walkowiak, B.; Walkowiak-Przybyło, M.; Komorowski, P. *Biological Evaluation of Materials*; IOP Publishing: Bristol, UK, 2022.
18. Taghavi, S.M.; Momenpour, M.; Azarian, M.; Ahmadian, M.; Souri, F.; Taghavi, S.A.; Sadeghain, M.; Karchani, M. Effects of Nanoparticles on the Environment and Outdoor Workplaces. *Electron. Physician* **2013**, *5*, 706–712. [CrossRef]
19. Hosny, A.E.D.M.S.; Rasmy, S.A.; Aboul-Magd, D.S.; Kashef, M.T.; El-Bazza, Z.E. The Increasing Threat of Silver-Resistance in Clinical Isolates from Wounds and Burns. *Infect. Drug Resist.* **2019**, *12*, 1985–2001. [CrossRef] [PubMed]
20. Cegielska-Radziejewska, R.; Leśnierowski, G.; Kijowski, J. Properties and Application of Egg White Lysozyme and Its Modified Preparations—A Review. *Pol. J. Food Nutr. Sci.* **2008**, *58*, 5–10.
21. Bing, W.; Cai, Y.; Jin, H.; Tian, L.; Tian, L.; Yin, Y.; Teng, Y.; Wang, P.; Hou, Z.; Bai, X. An Antiadhesion and Antibacterial Platform Based on Parylene F Coatings. *Prog. Org. Coat.* **2021**, *151*, 106021. [CrossRef]
22. Ao, H.; Yang, S.; Nie, B.; Fan, Q.; Zhang, Q.; Zong, J.; Guo, S.; Zheng, X.; Tang, T. Improved Antibacterial Properties of Collagen I/Hyaluronic Acid/Quaternized Chitosan Multilayer Modified Titanium Coatings with Both Contact-Killing and Release-Killing Functions. *J. Mater. Chem. B* **2019**, *7*, 1951–1961. [CrossRef]
23. Mu, M.; Wang, X.; Taylor, M.; Castillo, A.; Cisneros-Zevallos, L.; Akbulut, M.; Min, Y.; Akbulut, Y. Min, Multifunctional coatings for mitigating bac-terial fouling and contamination. *Colloid Interface Sci. Commun.* **2023**, *55*, 100717. [CrossRef]
24. Jin, G.; Qin, H.; Cao, H.; Qian, S.; Zhao, Y.; Peng, X.; Zhang, X.; Liu, X.; Chu, P.K. Synergistic Effects of Dual Zn/Ag Ion Implantation in Osteogenic Activity and Antibacterial Ability of Titanium. *Biomaterials* **2014**, *35*, 7699–7713. [CrossRef]
25. Cao, H.; Liu, X.; Meng, F.; Chu, P.K. Biological Actions of Silver Nanoparticles Embedded in Titanium Controlled by Micro-Galvanic Effects. *Biomaterials* **2011**, *32*, 693–705. [CrossRef] [PubMed]
26. Haufe, H.; Muschter, K.; Siegert, J.; Böttcher, H. Bioactive Textiles by Sol–Gel Immobilised Natural Active Agents. *J. Sol-Gel Sci. Technol.* **2008**, *45*, 97–101. [CrossRef]
27. Birkett, M.; Zia, A.W.; Devarajan, D.K.; Soni; Panayiotidis, M.I.; Joyce, T.J.; Tambuwala, M.M.; Serrano-Aroca, A. Multi-Functional Bioactive Silver- and Copper-Doped Diamond-Like Carbon Coatings for Medical Implants. *Acta Biomater.* **2023**, *167*, 54–68. [CrossRef]
28. Molina, J.; Valero-Gómez, A.; Pocoví-Martínez, S.; Ibiza-Palacios, M.S.; Bosch, F. Simple and Effective Sol-Gel Methodology to Obtain a Bactericidal Coating for Prostheses. *J. Sol-Gel Sci. Technol.* **2023**, *108*, 809–826. [CrossRef]
29. Stan, D.; Enciu, A.M.; Mateescu, A.L.; Ion, A.C.; Brezeanu, A.C.; Stan, D.; Tanase, C. Natural Compounds With Antimicrobial and Antiviral Effect and Nanocarriers Used for Their Transportation. *Front. Pharmacol.* **2021**, *12*, 723233. [CrossRef]
30. Chouhan, S.; Sharma, K.; Guleria, S. Antimicrobial Activity of Some Essential Oils—Present Status and Future Perspectives. *Medicines* **2017**, *4*, 58. [CrossRef] [PubMed]
31. Zheng, D.; Huang, C.; Huang, H.; Zhao, Y.; Khan, M.R.U.; Zhao, H.; Huang, L. Antibacterial Mechanism of Curcumin: A Review. *Chem. Biodivers.* **2020**, *17*, e2000171. [CrossRef]
32. Nagoor Meeran, M.F.; Javed, H.; Al Taee, H.; Azimullah, S.; Ojha, S.K. Pharmacological Properties and Molecular Mechanisms of Thymol: Prospects for Its Therapeutic Potential and Pharmaceutical Development. *Front. Pharmacol.* **2017**, *8*, 380. [CrossRef]
33. Kowalczyk, A.; Przychodna, M.; Sopata, S.; Bodalska, A.; Fecka, I. Thymol and Thyme Essential Oil—New Insights into Selected Therapeutic Applications. *Molecules* **2020**, *25*, 4125. [CrossRef]
34. Raduly, F.M.; Rădiţoiu, V.; Rădiţoiu, A.; Frone, A.N.; Nicolae, C.A.; Purcar, V.; Ispas, G.; Constantin, M.; Răut, I. Modeling the properties of curcumin derivatives in relation to the architecture of the siloxane host matrices. *Materials* **2021**, *15*, 267. [CrossRef]
35. Irizar, P.; Pintor-Rial, A.; Martinez-Arkarazo, I.; Olazabal, M.A.; Ruiz-Rubio, L.; Cardiano, P.; Gomez-Laserna, O. Bio-based hybrid nanocomposites as multifunctional sustainable materials for stone conservation. *Prog. Org. Coat.* **2023**, *185*, 107899. [CrossRef]
36. *ISO PN-EN 10993-5*; Biological evaluation of medical devices—Part 5: In vitro cytotoxicity tests. International Standard Organization: Geneva, Switzerland, 2009.

coatings

MDPI

Article

Effect of Anodic Oxidation Pulse Voltage on Antibacterial Properties and Biocompatibility of Ti-Ag Alloy

Zhen Ma [1,2,3], Yudong Yan [1], Chang Shi [2], Kexin Di [1], Jianwei Xu [1], Qicong Liu [1], Liting Mu [1,4,*], Jianming Zheng [3], Jiali Hu [2] and Erlin Zhang [2,*]

1 Heilongjiang Provincial Key Laboratory of Oral Biomedical Materials and Clinical Applications, Jiamusi University, Jiamusi 154007, China; jmsdxmz@163.com (Z.M.); yanyudongzg@163.com (Y.Y.); jmsdxdikexin@163.com (K.D.); jmsdxxjw@126.com (J.X.); liuqicong1211@163.com (Q.L.)

2 School of Materials Science and Engineering, Northeastern University, Shenyang 110819, China; sc15004572083@163.com (C.S.); hujiali523@163.com (J.H.)

3 Zhejiang Wanfeng Technology Development Co., Ltd., Shaoxing 312499, China; 252930179@163.com

4 College of Pharmacy, Jiamusi University, Jiamusi 154007, China

* Correspondence: muliting@163.com (L.M.); zhangel@atm.neu.edu.cn (E.Z.)

Abstract: For the application of titanium and titanium alloys in orthopedic implant materials, the antibacterial properties and cell biocompatibility determine whether the implant surgery is successful. In this study, a functional anodic oxidation (AO) coating was successfully prepared to modify the surface of Ti-Ag alloy. The surface characteristics of the anodized Ti-Ag alloy were analyzed using techniques such as X-ray diffraction (XRD), scanning electron microscopy (SEM), X-ray photoelectron spectroscopy (XPS), and contact angle measurements. The corrosion characteristics of Ti-Ag samples were tested by an electrochemical workstation. In addition, the antibacterial properties and cell activity were studied by the plate count method and MC3T3-E1 cell staining. The results indicate that the AO process can generate a multi-functional TiO_2/Ag_2O coating with a large number of block and flower-like structures on the surface of a Ti-Ag alloy. When the AO voltage of the sample is 120 V, the maximum roughness is 0.73 μm and the minimum wetting degree is 23°, which improves the biocompatibility. The corrosion test results show that AO treatment can improve the corrosion resistance of a Ti-Ag alloy. The oxidation voltage is 20 V and the coating has the best corrosion resistance. The corrosion open circuit potential (Eocp) is 107.621 mV and the corrosion current density (icorr) is 2.241×10^{-8} A·cm^{-2}. This coating can promote ion release and show more than 99% of a strong antibacterial ability against *S. aureus*. The results of the compatibility evaluation by cultured cells showed that the multifunctional coating formed by the anodic oxidation process did not cause cytotoxicity and promoted the adhesion of MC3T3-E1 cells.

Keywords: Ti-Ag alloy; anodic oxidation; antibacterial property; biocompatibility; corrosion resistance

1. Introduction

Titanium (Ti) and Ti alloys are commonly utilized in orthopedic implant materials due to their exceptional mechanical properties, favorable biocompatibility, low toxicity, and cost-effectiveness [1,2]. However, traditional Ti alloy implants do not have antibacterial properties and have infection problems during and after implantation. In addition, the inadequate disinfection of implants and surgical instruments and postoperative wound contamination can also aggravate infection [3]. The above factors can lead to implantation failure and even cause fatal damage to the human body, increasing medical costs and patient suffering.

The main reason for the failure of implantation surgery is that bacterial strains adhere to the surface of Ti and Ti alloy implants, proliferate, and form bacterial biofilms, which can lead to infection [4–6]. In addition, the long-term use of high-dose antibiotics by some patients can lead to bacterial mutations, strong drug resistance, and the formation of superbugs, exacerbating the infection process [7–11]. If the surface of the implant has excellent antibacterial properties, it will prevent bacteria from adhering to the surface of the implant, thereby preventing the formation and expansion of the biofilm [12,13]. Therefore, creating a stable and long-lasting antibacterial environment around the implant is of great significance for inhibiting bacterial attachment and biofilm formation [14].

To improve the antibacterial properties of Ti and Ti alloys, material scientists have tried to change the antibacterial properties of the surface and the antibacterial properties of the material itself to prevent bacterial adhesion and inhibit the formation of biofilms. At present, the antibacterial and antiviral properties of organic and inorganic antibacterial agents have been widely recognized [15]. Among them, organic antibacterial agents mainly include antibacterial proteins, polysaccharides, and antibacterial peptides, which inactivate bacteria by inhibiting bacterial adhesion and destroying extracellular structures [16]. However, organic antibacterial agents have the disadvantages of a short duration and poor stability [17]. Inorganic antibacterial agents are mainly metal element antibacterial agents, such as Ag, Cu, Zn, etc., [18]. They can kill bacteria by binding to bacterial proteins and changing the state of bacterial membranes [19,20]. They have a wide antibacterial spectrum, high antibacterial efficiency, and good safety [21]. Therefore, considering the complexity of the human implantation environment and the safety of antibacterial agents, the development of Ti alloys containing inorganic antibacterial agents is considered a suitable method for Ti alloys to achieve antibacterial properties and inhibit biofilm formation [22].

The methods of developing Ti alloys composite inorganic antibacterial implants mainly include the surface modification technology and the manufacture of antibacterial Ti alloys [23]. Surface modification is mainly to prepare coatings doped with inorganic antibacterial agents on the surface of Ti alloys by electrophoretic deposition (EPD), anodic oxidation (AO), micro-arc oxidation (MAO), and other technologies to enhance the antibacterial properties and biocompatibility [24]. M. H et al. [25] deposited a Cu-containing hydroxyapatite (HAp) coating on the TC4 alloy by the EPD process. But, the coating has a poor inhibitory effect on Staphylococcus aureus (*S. aureus*). Hou et al. [26] prepared Ag-doped TiO_2 nanotubes by the anodic oxidation method and the implanted samples had good antibacterial properties. K. M et al. [27] deposited nano-silver on the anodic oxidation surface of pure Ti for modification and observed that the osteoblast adhesion and proliferation were enhanced by the coating, showing higher biocompatibility. However, the antibacterial coating prepared by surface modification technology has defects such as coating shedding and dissolution, and it will be difficult for it to maintain continuous antibacterial properties and biological activity [28].

Due to their excellent mechanical processing properties and long-lasting antibacterial properties, antibacterial metal materials have been used as antibacterial alloy materials in many studies [23]. Antibacterial alloy materials are usually made by adding inorganic antibacterial agents containing metallic elements and through processes such as powder metallurgy, vacuum sintering, and arc melting [29,30]. Zhang et al. [31] utilized arc melting technology to fabricate titanium alloys with varying copper contents. The antibacterial efficacy of the as-cast titanium–copper alloys against *S. aureus* was found to be 37%. In a separate study, Chen et al. [23] employed powder metallurgy technology to prepare a Ti-Ag alloy. The results indicated that alloys with a 5 wt% Ag content exhibited an antibacterial rate ranging from 60% to 80% against *S. aureus*. The observed antibacterial activity was highly dependent on the Ag content, as well as the presence and distribution of the Ti_2Ag phase. It has been proved that Ti alloys containing inorganic antibacterial agents have certain antibacterial properties. However, the antibacterial properties of antibacterial metal materials still need to be further improved [32]. In addition, a Ti alloy is a bio-inert material that requires high biocompatibility and corrosion resistance when implanted into the human body [33].

Generally, surface modification techniques are used to improve antibacterial properties and biocompatibility. Hu et al. [34] utilized ultrasonic-MAO technology to create a coating on the Ti-3Cu alloy, exhibiting robust long-term antibacterial properties and promoting early cell adhesion. Schwartz et al. [35] prepared TiO_2 and HAp bioactive coatings by plasma electrolytic oxidation in aqueous electrolytes and molten salts, and cleaned the coatings with autoclave (A), ultraviolet light (UV), radio frequency (RF), air plasma (P), and UV-ozone cleaner (O), providing an effective and mild surface modification treatment method for cleaning and disinfecting organic pollutants on the surface of bioactive coatings. Through the research of Liu et al. [10], a multifunctional TiO_2/Cu_2O coating was prepared on the Ti-5Cu alloy by the AO process. The coating improvement significantly enhances the antibacterial properties of the alloy, has no cytotoxicity, and is beneficial to accelerate cell adhesion. Cao et al. [8], by anodizing Ti-5Cu, found that the double-layer-structure oxide coating formed on the Ti-5Cu improves the corrosion resistance by ten times, and the antibacterial rate is improved. In addition, studies have shown that anatase TiO_2 formed on the Ti alloys by AO treatment is consistent with HAp, showing higher bone conductivity [36]. To date, the majority of prior investigations have consistently demonstrated that surface modification techniques applied to antibacterial Ti alloys can concurrently improve both the antibacterial properties and biocompatibility of the alloys. This dual enhancement holds significant promise for the broader utilization of antibacterial Ti alloys.

For many antibacterial metal ions, the antibacterial and antiviral properties of Ag have been widely considered to be one of the best and most effective antibacterial agents [13]. Silver interacts with sulfhydryl groups in enzymes and proteins to form reactive oxygen species and destroys the regulatory system of reactive oxygen species in bacteria, resulting in high oxidative stress in bacteria and the inactivation of strains. Ti alloys containing a metal Ag antibacterial agent will provide antibacterial properties for its implants. However, the antimicrobial properties and bioactivities of the antimicrobial Ti-Ag alloy have been rarely reported.

In this paper, the anodic oxidation technology was innovatively applied to a Ti-Ag antibacterial titanium alloy. Scanning electron microscopy (SEM), X-ray photoelectron spectroscopy (XPS), X-ray diffraction (XRD), an electrochemical analysis workstation, ion release, contact angle, plate counting method, and cell culture were used to detect. The effects of an anodizing voltage on the corrosion resistance, antibacterial properties, and biocompatibility of the Ti-Ag alloy were studied. The results indicate that the anodic oxidation process can significantly enhance the antibacterial properties of the Ti-Ag alloy, TiO_2/Ag_2O's, multi-functional coating, and the coating exhibits higher corrosion resistance. No cytotoxicity was found, and cell adhesion and proliferation were accelerated. In addition, the antibacterial mechanism of the film was also discussed. Therefore, anodic oxidation may be an effective surface modification method.

2. Materials and Methods

2.1. Preparation and AO Treatment of Ti-Ag Samples

Ti-5 wt% Ag (referred to as Ti-Ag) rods were synthesized using high-purity Ti (99.9 wt%) and Ag (99.9 wt%) in a high vacuum arc furnace. Additional information about the alloy is available in Ref. [37]. A sample measuring Φ15 mm × 2 mm was obtained from the bar using a wire-cutting machine. Ti-Ag samples were polished with 80, 400, 600, 1200, and 2000 particle sizes of SiC sandpaper (Yingpai, Foshan, Guangdong, China) in turn, and polished to the surface mirror state with polishing liquid. For the next step, the samples were cleaned and dried with ultrapure water and anhydrous ethanol. The sample was soaked in 120 g/L NaOH (Merck KGaA, Shanghai, China) solution for 10 min to remove the oil on the sample. Activation treatment was performed in a mixed solution with a volume ratio of HF (Merck KGaA, Shanghai, China): HNO_3 (Kermel, Tianjin, China): H_2O = 1:2:10 for 10 s. Finally, the samples were washed with high-purity water and dried.

The AO treatment was carried out in a solution of 0.1 mol/L H_3PO_4 (Kermel, Tianjin, China) for 5 min. Various voltage values, specifically 20 V, 40 V, 60 V, 80 V, and 120 V, were applied sequentially. The experimental procedure is illustrated in Figure 1. The sample names were AO-20 V, AO-40 V, AO-60 V, AO-80 V, and AO-120 V, and compared with the untreated sample (AO-0 V).

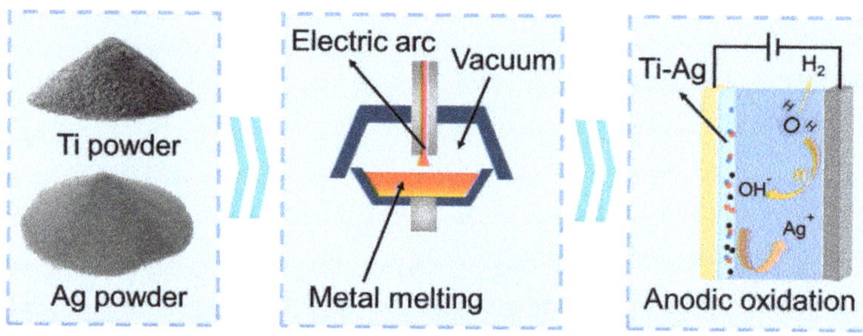

Figure 1. Preparation process of anodic oxidation coating.

2.2. XRD and Microstructure Characterization

The phase composition on the surface of Ti-Ag alloy subjected to various AO voltages was analyzed using XRD (D8 Advance, Bruker AXS, Karlsruhe, Germany) with copper Kα radiation. MDI Jade 6.5 software's database was employed for data interpretation and analysis. The scanning range was $30° < 2\theta < 90°$ and step scanning mode was used. The surface morphology of Ti-Ag samples subjected to various AO voltages was examined using cold-field-emission scanning electron microscopy (SEM). Additionally, the elemental composition was analyzed through energy dispersive spectrometry (EDS). The coating thickness gauge (LS225+N1500, Linshang, Shenzhen, China) is used to measure the coating at five different positions, and the relationship between coating thickness and voltage is measured.

2.3. XPS Analysis

The elemental species and chemical states on the coating surface following anodic oxidation of the Ti-Ag samples were investigated using X-ray photoelectron spectroscopy (XPS, Axis Supra, Kratos, Stretford, UK). To eliminate surface impurities, the samples were pre-treated before the analysis. Full spectrum scans and high-resolution scans for Ti, Ag, and O were conducted. The binding energy of C 1s was calibrated at 284.6 eV using standard data, and peak fitting was carried out using XPS Peak 41 software.

2.4. Corrosion Resistance

The electrochemical properties of Ti-Ag samples with different AO voltages in 0.9% NaCl (Kermel, Tianjin, China) solution were tested by an electrochemical workstation (Versa STAT 300, AMETEK, CA, USA) at 37 ± 1 °C. The standard three-electrode apparatus was used. The reference electrode, counter electrode, and working electrode were saturated calomel electrode (SCE, INESA, Shanghai, China), platinum electrode (INESA, Shanghai, China), and sample to be tested, respectively. The open circuit potential measurement time was 3600 s. Electrochemical impedance spectroscopy (EIS) tests were performed at frequencies of $0.01–10^5$ Hz. Then, the potentiodynamic polarization test was carried out within the range of -0.5~0.5 mV.

2.5. Surface Roughness and Water Wetting Angle

The surface morphology of the sample was examined using laser confocal microscopy (OLS4100, OLYMPUS, Tokyo, Japan), capturing both two-dimensional (2D) and three-dimensional (3D) images. Five different 100 μm × 100 μm regions were selected for measurement. Surface roughness (Ra) data for the samples were acquired using LEXT OLS4000 software. The contact angle measuring instrument (FCA2000A3E, Aifeisi, Shanghai, China) was employed to assess the hydrophilicity of the samples with different AO voltages. Using the hanging drop method, 1.5 μL of laboratory-grade deionized water was deposited on the sample surface, and measurements were taken at a minimum of three different positions, documented through photographs. Surface contact angles were calculated using software, and mean values along with standard deviations were determined for analysis.

2.6. Ion Release

The specimens labeled as AO-0 V, AO-20 V, AO-40 V, AO-60 V, AO-80 V, and AO-120 V were immersed in 5.5 mL of NaCl solution (surface area/volume ratio 2 cm^2/mL) for 24 h at a constant temperature of 37 °C. The concentration of Ag^+ in the solution was determined using inductively coupled plasma mass spectrometry (Optima 7300DV, Perkin Elmer, MA, USA).

2.7. Antibacterial Property

2.7.1. Plate Count Method

Staphylococcus aureus (*S. aureus*, ATCC 6538) is a commonly used bacterium for evaluating the antibacterial properties of implant materials. According to GB/T 31402-2015 [34], the antibacterial properties of the samples were determined by plate counting method. Before the experiment, all necessary samples and utensils underwent sterilization at 120 °C/0.15 MPa for 30 min. In each experimental group, three sets of parallel samples were selected for the experiment. According to the standard procedure, nutrient broth (NB, Chinese Medicine Reagents, Shanghai, China) and physiological agar (Xiya Reagent, Linyi, China) were used to culture bacteria, and cp-Ti was used as the control group. To ensure the sterile state of the samples, all samples were exposed to ultraviolet light for 1 h for sterilization. Subsequently, 100 μL of bacterial suspension with a concentration of 10^4 cfu/mL was uniformly coated on the sample to avoid the bacterial liquid overflowing the sample range during the dropping process. The samples were then placed in an incubator for 24 h. After incubation, 2 mL of normal saline and 100 μL of trypsin were added to each well and shaken for 5 min to ensure that all the bacteria fell into the solution. From each well, 100 μL of bacterial droplets were taken on the nutrient agar plate and evenly smeared using a triangular glass rod. Excess liquid was removed from the surface until no obvious liquid remained, followed by another 24 h incubation period. An automatic colony counter (Shineso V3, China) was employed to enumerate the viable bacteria following the guidelines of the Chinese national standard GB/T 4789.2-2010 [34]. The antibacterial rate was calculated using Equation (1):

$$AR = (N_{control} - N_{sample})/N_{control} \times 100\% \quad (1)$$

where $N_{control}$ and N_{sample} represent the count of colonies on both the control and test samples. As per the Chinese national standard GB/T 4789.2-2010, an antibacterial rate (AR) of $\geq$90% suggests antibacterial properties, while an AR of $\geq$99% indicates strong antibacterial properties.

2.7.2. Live/Dead Staining

Following a 24 h incubation period on the sample surface, the bacteria were stained using the LIVE/DEAD® BacLight™ bacterial activity kit (Thermo Fisher, MA, USA) in accordance with the provided instructions. The dye was applied to the sample surface and allowed to incubate for 15 min at room temperature in the dark. Subsequently, the bacteria were washed twice with normal saline to eliminate the influence of excess dye. The stained bacteria were then examined using a fluorescence microscope (Olympus BX51, Olympus, Tokyo, Japan). Living bacteria emitted green fluorescence, while the fluorescent signal from dead bacteria appeared red.

2.8. Cell Compatibility In Vitro

2.8.1. Cell Viability and Cytotoxicity

In in vitro cytotoxicity evaluation experiment, MC3T3-E1 cells were used for evaluation according to ISO 10993-5 standard [10], and blank control and control groups (cp-Ti) were set up. All samples underwent sterilization at 121 °C/0.15 MPa. Following sterilization, the samples were placed in a 24-well plate and 500 μL of cell suspension with a concentration of 2×10^4 cells/mL was added to each well. After that, the plates were incubated at 37 °C in a humid environment of 5% CO_2 for 1 day, 3 days, and 5 days, respectively. At each designated time point, 200 μL of MTT solution (Pricella, Wuhan, China) was added to each well and incubated at 37 °C for 4 h. Afterward, the MTT solution was removed and 300 μL of DMSO solution (Pricella, Wuhan, China) was added, followed by transfer to a 96-well plate for measurement. The optical density (OD) values were measured using an enzyme-labeled instrument (iMark, Bio-Rad, CA, USA). The relative cell growth rate (RGR) was calculated according to Equation (2):

$$RGR = (OD_{sample} - OD_{control})/(OD_{negative} - OD_{control}) \times 100\% \quad (2)$$

where OD_{sample}, $OD_{control}$, and $OD_{negative}$ represent the optical density values of the test sample, the cp-Ti, and the blank control group (negative sample), respectively. Cytotoxicity was deemed non-existent when the RGR was equal to or exceeded 75%.

2.8.2. Cell Adhesion and Morphology

As described in Section 2.8.1, MC3T3-E1 cells were seeded onto the surface of the test samples and incubated for 1, 4, and 24 h, followed by PBS (SCIENTIFIC, Shanghai, China) washing. To fix the cells and permeabilize the cell membrane, a 4% paraformaldehyde (PFA, SCIENTIFIC, Shanghai, China) solution and a 0.5% Triton X-100 solution (SCIENTIFIC, Shanghai, China) were applied to the samples at 37 °C. After another round of PBS washing, the samples were treated with the fluorescently labeled cyclic peptide. Subsequently, the cells were incubated in the dark for 30 min. Following this, DAPI (4,6-diamino-2-phenylindole, G-CLONE, Beijing, China) treatment was applied for 2 min. Lastly, the cell samples underwent blocking with an anti-fluorescence attenuation blocker. For additional information, please refer to other available sources [8]. The samples were placed under a fluorescence microscope to observe and record cell morphology.

2.9. Statistical Analysis

The biological outcomes were replicated three times, and the results are presented as the mean ± standard deviation. Representative results underwent quantitative statistical analysis to determine statistical significance, with $p < 0.05$ considered as statistically significant.

3. Results

3.1. Surface Morphology

Figure 2 presents the appearance of anodized Ti-Ag samples with different AO voltages. After AO treatment, the surface coating is uniform. When the AO voltage is 20 V, the color of the coating is dark blue. Upon increasing the voltage to 40 V, the color of the coating is khaki. As the voltage continues to rise, the color of the AO-60 V becomes magenta. When the voltage increases to 80 V, the color of the coating is dark cyan. Finally, upon increasing the voltage to 120 V, the color of the coating is light purple. The anodization voltage exerts a notable influence on the coloration of the resulting anodic oxide film.

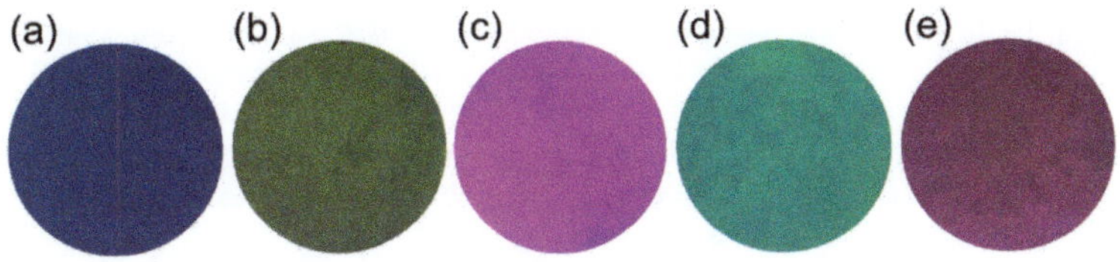

Figure 2. The coloring effect of Ti-Ag samples with different AO voltages: (**a**) AO-20 V; (**b**) AO-40 V; (**c**) AO-60 V; (**d**) AO-80 V; (**e**) AO-120 V.

3.2. Phase Identification

Figure 3 presents the phase patterns of anodized Ti-Ag samples with different AO voltages. According to Figure 3, AO-0 V represents the Ti-Ag sample without AO treatment, which is mainly composed of α-Ti and Ti_2Ag phases. The diffraction peak of the Ti_2Ag phase overlaps with that of α-Ti [23]. The phase composition of samples treated by AO with different voltages was semi-quantitatively analyzed using the Rietveld Refinement method; the results are shown in Table 1. Ti and Ag oxides were not detected on the AO-treated samples and the oxide film was thin.

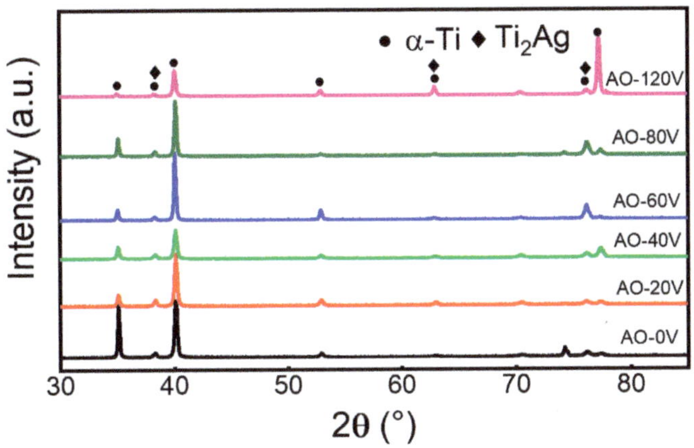

Figure 3. XRD patterns of Ti-Ag samples with different AO voltages.

Table 1. Semi-quantitative calculation results of coating phase composition.

Sample	α-Ti/wt%	Ti_2Ag/wt%	Rwp/%
AO-0 V	96.6 ± 1.2	3.4 ± 0.2	8.26
AO-20 V	97.3 ± 0.9	2.7 ± 0.4	9.12
AO-40 V	96.1 ± 1.4	3.9 ± 0.3	9.36
AO-60 V	96.3 ± 1.1	3.7 ± 0.6	10.26
AO-80 V	97.6 ± 0.8	2.4 ± 0.4	12.42
AO-120 V	98.5 ± 0.6	1.5 ± 0.5	8.37

3.3. Microstructure and EDS

Figure 4 presents the microstructure of Ti-Ag samples with different AO voltages. In Figure 4a, the coating surface is uniform when the oxidation voltage is 20 V. As the voltage increases, when the voltage reaches 40 V, as shown in Figure 4b, the surface is smoother and denser. As shown in Figure 4c, a large number of white block structures appear on the surface of AO-60 V. After local amplification, the block structure is aggregated, the shape is regular, and the size is <1 μm. Upon continuously increasing the voltage to 80 V, as shown in Figure 4d, the block structure is partially converted into a flower-like structure with a small amount of a petal structure. With the further increase in the voltage to 120 V, as shown in Figure 4e, there are many grooves on the surface, and the bulk structure is completely transformed into a flower-like structure with a size ranging from approximately 1~2 μm. Figure 4f shows the coating thickness of AO-treated samples with different voltages. With the increase in the AO treatment voltage, the oxide film on the surface of the sample becomes thicker. The thickness reaches the maximum at 120 V and the size is about 2.7 μm.

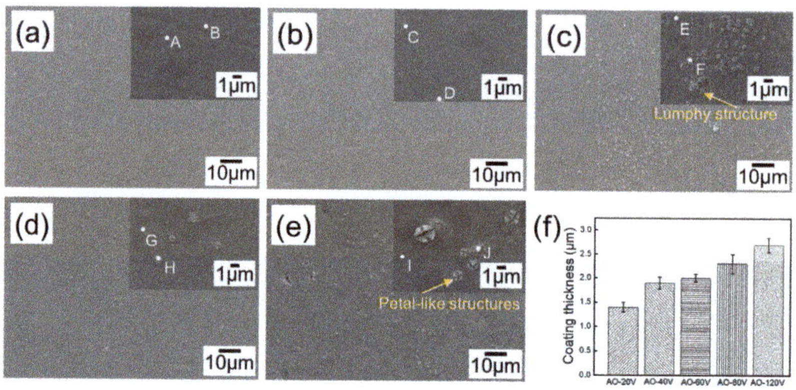

Figure 4. SEM morphology of (**a**) AO-20 V, (**b**) AO-40 V, (**c**) AO-60 V, (**d**) AO-80 V, (**e**) AO-120 V; (**f**) Coating thickness; Points A–I represent sampling points for EDS analysis.

Table 2 shows the EDS analysis results of Ti-Ag samples with different AO voltages. The EDS analysis results of point A–I in Figure 4, revealing the presence of Ti and O elements, and the content of Ag is low. With the increase in the voltage, the content of the Ti element in the coating decreases from the highest value of 79.0% of AO-20 V to 52.4% of AO-120 V. However, the content of the O element in the coating increased with the increase in the oxidation voltage, up to 45.0%. The bulk structure and flower-like structure in the coating were analyzed by EDS. The bulk structure shown in F point in Figure 4c was compared with the matrix E point. The content of the Ag element in the bulk structure reached 7.9% and the enrichment of the Ag element appeared. However, when the anodizing voltage increased to 120 V, the flower-like structure shown at the J point in Figure 4e was tested by EDS, and the content of the Ag element was only 2.6%. The research shows that excessive voltage will cause the burning of Ag elements and the content of Ag in the coating will decrease.

Table 2. EDS analysis results of point A–I in Figure 4.

Oxidation Potential	Point	Element (wt%)			σ		
		Ti	O	Ag	Ti	O	Ag
AO-20 V	A	79.0 ± 0.1	16.4 ± 0.1	4.6 ± 0.2	0.5	0.5	0.2
	B	73.4 ± 0.2	22.4 ± 0.3	4.2 ± 0.1	0.2	0.2	0.1
AO-40 V	C	75.5 ± 0.1	19.9 ± 0.1	4.6 ± 0.3	0.4	0.4	0.2
	D	68.8 ± 0.3	22.0 ± 0.1	9.1 ± 0.1	0.4	0.5	0.2
AO-60 V	E	73.0 ± 0.2	22.7 ± 0.2	4.3 ± 0.2	0.4	0.4	0.1
	F	63.9 ± 0.3	28.3 ± 0.1	7.8 ± 0.1	0.3	0.4	0.3
AO-80 V	G	64.9 ± 0.2	26.5 ± 0.3	4.1 ± 0.1	0.4	0.2	0.1
	H	64.9 ± 0.1	27.2 ± 0.1	7.9 ± 0.3	0.4	0.4	0.2
AO-120 V	I	64.0 ± 0.2	32.1 ± 0.2	3.8 ± 0.1	0.4	0.4	0.1
	J	52.4 ± 0.1	45.0 ± 0.1	2.6 ± 0.1	0.3	0.3	0.1

3.4. XPS

Figure 5 depicts the XPS full spectrum of Ti-Ag samples along with the high-resolution spectra of Ti, O, and Ag under the AO-20 V, AO-60 V, and AO-120 V processes. In the comprehensive spectrum (Figure 5(a1–c3)), the presence of Ti, Ag, and O elements was identified in the anodic oxidation coating of Ti-Ag samples, aligning with the EDS element detection results detailed in Table 2. In the high-resolution spectrum of the Ti element, as shown in Figure 5(a2–c2), the peaks of Ti 2p1/2 are detected at 464.0 eV and 464.2 eV, and the peaks of Ti 2p3/2 are detected at 458.4 eV and 458.7 eV, corresponding to the Ti-O bond in TiO_2 [38]. Simultaneously, the peak energy of Ti 2p1/2 is about 5.9 eV higher than that of Ti 2p3/2, indicating that the element Ti mainly exists in the oxidation coating in the form of TiO_2 crystal. The binding energies of Ti 2p1/2 at 461.2 eV and 461.9 eV and Ti 2p3/2 at 457.0 eV, 456.7 eV, and 456.9 eV correspond to the Ti-O bond in Ti_2O_3. It can be seen from the diffraction intensity that Ti mainly exists in TiO_2 and a small amount of Ti_2O_3 oxide colloid exists [8,23].

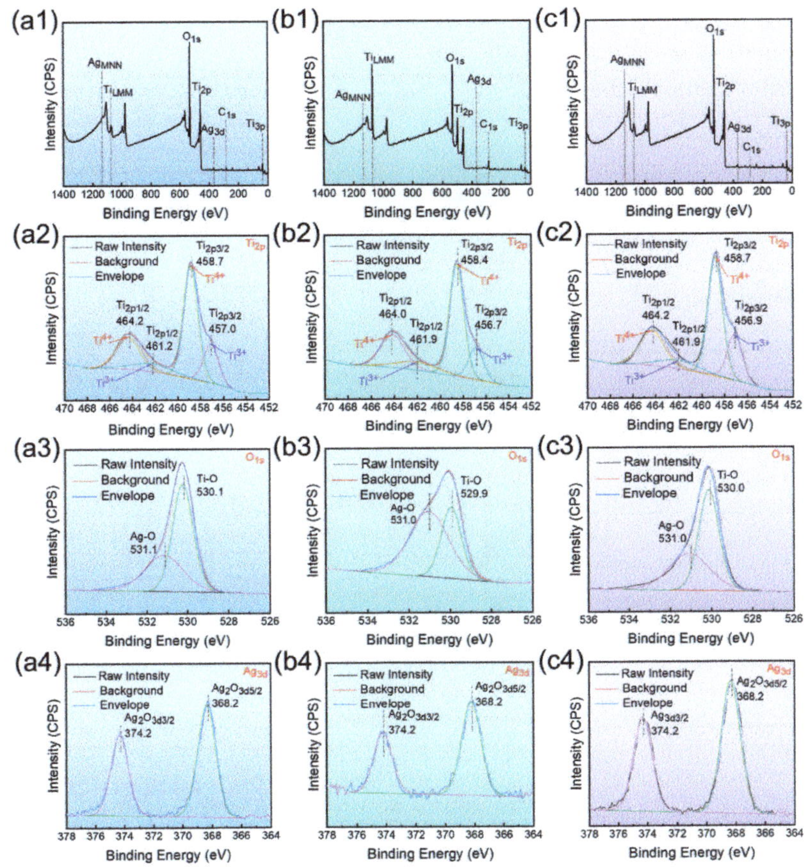

Figure 5. XPS analysis spectra of survey spectrum, Ti 2p, O 1s, Ag 3d: (**a1–a4**) AO-20 V; (**b1–b4**) AO-60 V; (**c1–c4**) AO-120 V.

In the high-resolution O 1s spectrum presented in Figure 5(a3–c3), it becomes evident that the bonding of Ti, Ag, and O elements within the coating is further confirmed. The O 1s spectrum is composed of Ti-O bonds with peaks of 530.1 eV, 529.9 eV, and 530.0 eV and Ag–O bonds with peaks of 531.1 eV and 531.0 eV. In Figure 5(a4–c4), the high-resolution spectra of Ag 3d are presented. The peaks corresponding to Ag 3d3/2 and Ag 3d5/2 are observed at 374.2 eV and 368.2 eV, respectively, indicating the existence of an Ag^{+} oxidation state. With the increase in the oxidation voltage, no significant variation is observed in the spectrum. Based on the above analysis, the coating is mainly composed of TiO_2 and Ag_2O.

3.5. Corrosion Resisting Property

Figure 6 shows the corrosion-resisting property of the Ti-Ag samples with different AO voltages, where Figure 6a–e is the electrochemical analysis and Figure 6f is the equivalent circuit diagram. From the corrosion open circuit potential (Eocp) curve shown in Figure 6a, the potential of the Ti-Ag alloy after AO treatment is significantly higher than AO-0 V. It can be seen from the Eocp data in Table 3 that when the oxidation voltage is 20 V, Eocp reaches a maximum of 107.621 mV. Compared with AO-0 V, Eocp increases by 351.759 mV. The test results show that AO treatment reduces the corrosion failure tendency of the Ti-Ag alloy, and the oxide film formed on the surface has high thermodynamic stability. Hence, at an oxidation voltage of 20 V, the Ti-Ag sample exhibits the least susceptibility to corrosion. The corrosion tendencies of the Ti-Ag alloy can be deduced from the Tafel curve in Figure 6b, wherein the Tafel curve shifts towards positive potential and a lower corrosion current density. The Tafel curve was fitted and calculated and the self-corrosion potential (Ecorr) and self-corrosion current density (icorr) of the sample were obtained. The polarization resistance (Rp_0) of the sample is calculated by the Stern–Geary Equation. The calculation formula is shown in Equation (3):

$$Rp_0 = \beta_a\beta_c/(2.303(\beta_a + \beta_c))\ i_{corr} \quad (3)$$

Table 3. Electrochemical data on OCP and Tafel curves of Ti-Ag samples.

Samples	Eocp/mV (vs. SCE)	Ecorr/mV (vs. SCE)	icorr/ (10^{-8} A·cm^{-2})	$R_{p0}/10^4\ \Omega\cdot cm^2$
AO-0 V	−244.138	−220.091 ± 0.012	21.477 ± 0.031	1.55 ± 0.23
AO-20 V	107.621	17.035 ± 0.016	2.241 ± 0.028	33.3 ± 0.16
AO-40 V	−82.112	−115.281 ± 0.014	3.661 ± 0.015	19.8 ± 0.19
AO-60 V	−99.991	−94.963 ± 0.019	2.969 ± 0.032	18.2 ± 0.13
AO-80 V	−115.687	−136.163 ± 0.013	2.378 ± 0.016	12.6 ± 0.24
AO-120 V	−106.042	−164.884 ± 0.017	9.411 ± 0.022	20.2 ± 0.15

Among them, Rp_0, β_a, β_c, and icorr are polarization resistance, anode slope, cathode slope, and corrosion current density, respectively. The test results are shown in Table 3. Through AO treatment, the Ecorr of the Ti-Ag samples was greater than −220.091 mV of AO-0 V. Among them, the Ecorr reaches a maximum of 17.035 mV at a voltage of 20 V. It can be seen from the icorr data in Table 3 that compared with 21.477×10^{-8} A/cm^2 of AO-0 V, the Ti-Ag samples treated by AO showed a lower corrosion current density. When the oxidation voltage is 20 V, the icorr of the Ti-Ag sample is the smallest, which is 2.241×10^{-8} A/cm^2. The data indicate that AO treatment can significantly increase the polarization resistance of the sample. At the same time, the AO treatment markedly lowers the corrosion current density or raises the corrosion potential, thereby enhancing the corrosion resistance of the Ti-Ag samples. However, when the anodizing voltage reaches a maximum of 120 V, the icorr of the sample increases, which may be related to the surface state of the sample and the size of the surface petal-like structure.

Figure 6c presents the Nyquist curves for Ti-Ag samples at varying AO voltages. The AO-treated sample has a larger capacitive arc radius than the AO-0 V sample. The polarization resistance of the passivation film can be inferred from the diameter of the semicircle, indicating that the AO-treated coating has a higher anti-polarization ability at the interface. Figure 6d,e shows the Bode curves of Ti-Ag samples, including the relationship between frequency and phase (Figure 6d) and the relationship between frequency and impedance amplitude (|Z|) (Figure 6e). In Figure 6d, the frequency corresponding to the maximum phase angle is only one, which means that there is only one obvious time constant in the electrochemical process. As the oxidation voltage of AO treatment increases, the frequency corresponding to the maximum phase angle moves to the high-frequency region, indicating that the charge transfer reaction in the electrochemical process becomes faster. In the intermediate frequency region (0.1~100 Hz) of Figure 6e, only a linear slope of about −1 is observed, which is a typical EIS monolayer structure. Therefore, a simple equivalent circuit $R_S(Q_PR_P)$ is used to fit the model of the passivation film. In this representation, R_S denotes the solution resistance, Q_P signifies the capacitance, and R_P represents the resistance of the passivation film. Table 4 presents the electrochemical parameters computed using the equivalent circuit.

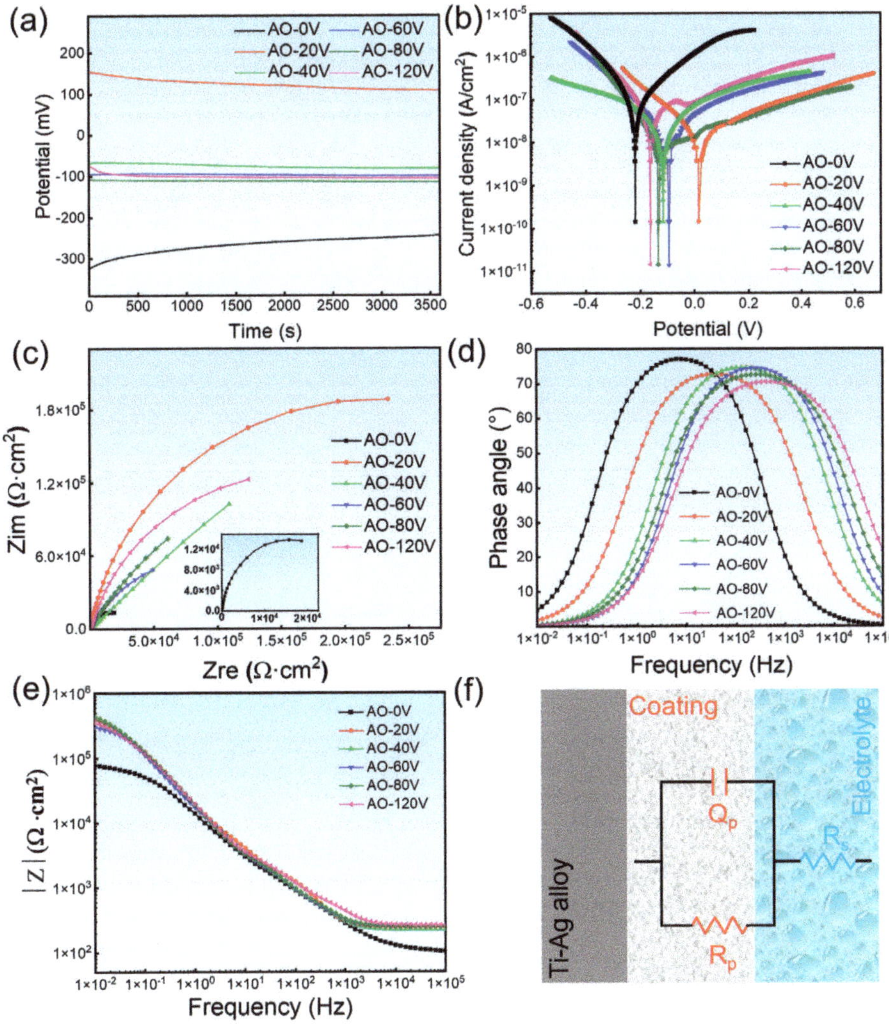

Figure 6. Electrochemical correlation curves of Ti-Ag samples at various AO voltages: (**a**) OCP curve, (**b**) Tafel curve, (**c**) Nyquist curve, (**d**,**e**) Bode plots, (**f**) Equivalent circuit model of samples.

Table 4. Electrochemical data of Ti-Ag samples in 0.9% NaCl solution from Nyquist and Bode.

Samples	$R_S/\Omega\cdot cm^2$	$Q_P/10^{-5}$ $F\cdot cm^{-2}$	n	$R_P/10^4$ $\Omega\cdot cm^2$
AO-0 V	105.7 ± 0.6	32.4 ± 0.12	0.915 ± 0.005	7.81 ± 0.12
AO-20 V	158.2 ± 0.9	2.48 ± 0.07	0.889 ± 0.002	44.82 ± 0.09
AO-40 V	112.3 ± 0.5	2.59 ± 0.03	0.828 ± 0.003	30.12 ± 0.17
AO-60 V	145.4 ± 0.8	7.95 ± 0.03	0.700 ± 0.006	29.28 ± 0.15
AO-80 V	115.6 ± 0.9	5.91 ± 0.05	0.669 ± 0.004	28.73 ± 0.17
AO-120 V	149.9 ± 0.7	7.71 ± 0.06	0.649 ± 0.005	32.67 ± 0.16

3.6. *Surface Roughness and Wetting Angle*

Figure 7 illustrates the surface roughness and water contact angle of Ti-Ag samples at various AO voltages. The impact of the anodic oxidation process on the biocompatibility and hydrophilicity of the samples was assessed. In Figure 7a, the surface roughness of the sample is approximately 0.10 μm after sandpaper grinding and polishing. With the increase in the oxidation voltages, the roughness increases obviously. When the oxidation voltage of AO treatment is 20 V, the surface roughness is 0.31 μm, which is 0.21 μm higher than that of AO-0 V. The surface roughness of AO-40 V is 0.25 μm. With the further increase in the voltages to 60 V and 80 V, the roughness of the sample is 0.42 μm and 0.51 μm. The surface roughness value reached a maximum of about 0.73 μm at an oxidation voltage of 120 V, and the roughness value continued to rise, increasing the possibility of cell adhesion during implantation.

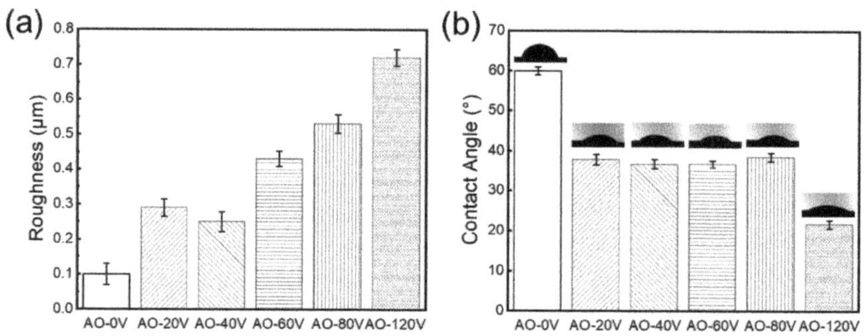

Figure 7. The surface roughness of Ti-Ag samples with different AO voltages: (**a**) surface roughness, (**b**) surface water contact angle.

Figure 7b illustrates the surface water contact angle of Ti-Ag samples at different AO voltages. The contact angle of the sample without AO treatment is approximately 60°. When the AO treatment voltage is between 20 and 80 V, the water contact angle is significantly reduced, but it does not change with the voltage, which is about 36~38°. When the oxidation voltage is 120 V, the lowest water contact angle is about 23°, indicating that the wetting angle decreases with the increase in voltage and AO treatment can improve the hydrophilicity of Ti-Ag samples.

3.7. *Ion Release*

Figure 8 shows the dissolution amount of Ag^+ with different AO voltages. The concentration of Ag^+ ions in the untreated sample was 3.48 μg/L. When the oxidation voltage was 20 V, the dissolution of Ag^+ reached 9.21 μg/L. Upon increasing the voltage to 40 V, the ion release was 10.94 μg/L. Upon continuously increasing the voltage to 60 V and 80 V, the ion release was 19.6 μg/L and 22.6 μg/L. Finally, upon increasing the maximum voltage to 120 V, the release of Ag^+ reaches a maximum of 35.9 μg/L. The bactericidal behavior of Ag-containing alloys is mainly due to the contribution of the long-term stable release of Ag^+. The concentration of Ag^+ increased with the increase in the oxidation voltage, indicating that AO treatment had a significant effect on the dissolution of Ag^{+}. Therefore, AO treatment can enhance the antibacterial properties of the Ti-Ag alloy.

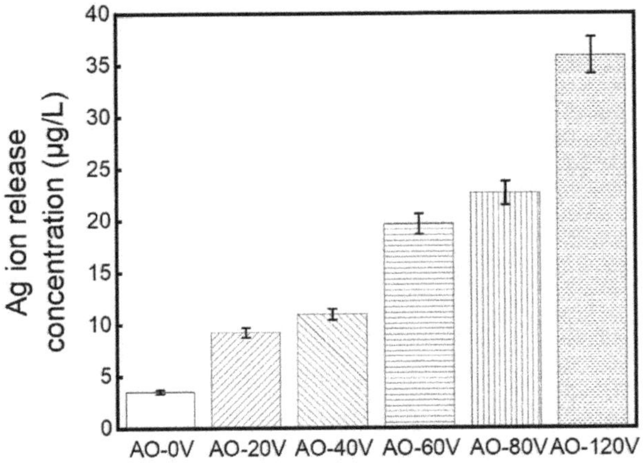

Figure 8. The release of Ag^{+} from Ti-Ag alloy with different AO voltages immersed in 0.9% NaCl solution for 24 h.

3.8. Antibacterial Property

3.8.1. Antibacterial Rate

Figure 9 displays the bacterial colonies and antibacterial rates of Ti-Ag samples at various AO voltages, determined through plate counting. A large number of *S. aureus* colonies were distributed on the cp-Ti as the blank control sample, as shown in Figure 9a, which confirmed that cp-Ti did not have antibacterial activity. However, the Ti-Ag control sample exhibited a reduction in the number of colonies compared to the number observed on the cp-Ti colonies, resulting in an antibacterial rate of approximately 68%. The antibacterial properties of the Ti-Ag alloy need further enhancement. Notably, only a few colonies were observed on the Ti-Ag samples after AO treatment, signifying that AO can substantially improve antibacterial properties. Ti-Ag samples subjected to anodic oxidation, as depicted in Figure 9c–f, all demonstrated an antibacterial rate exceeding 99% against *S. aureus*, highlighting the significant improvement in the antibacterial capability achieved through anodic oxidation. However, when the oxidation voltage reached 120 V, the number of colonies increased compared with other oxidation voltages. As shown in Figure 9g, the antibacterial rate was about 97%.

3.8.2. Live/Dead Fluorescence Staining

Figure 10 presents the live/dead fluorescence staining of *S. aureus* performed after 24 h of culture on Ti-Ag samples subjected to different AO voltages. On the surfaces of cp-Ti and Ti-Ag control samples, a substantial amount of green fluorescence and a limited amount of red fluorescence were observed. The green fluorescence represents live bacteria, while the red fluorescence indicates dead bacteria. This suggests the presence of numerous live bacteria on the sample surfaces, indicative of a weak antibacterial ability. In contrast, the surface of the Ti-Ag samples treated with AO exhibited a significant amount of red fluorescence, confirming the improved antibacterial properties achieved through AO treatment. These results are consistent with those calculated in the 3.8.1 plate counting experiment.

3.9. Cell Experiment

Figure 11 illustrates the chart depicting the cell proliferation and relative growth rate (RGR) of MC3T3-E1 cells when cultured with cp-Ti, AO-20 V, and AO-40 V sample extracts for 1, 3, and 5 days. In Figure 11a, it is evident that the OD value increased with the extension of the culture time, indicating healthy cell growth on the sample surface with an increasing cell count. At 1 day of culture, there was no significant difference in the OD value between AO-treated samples and cp-Ti. After 3 days of culture, the OD value of the AO-120 V sample was noticeably higher than that of the cp-Ti sample, while the OD values of AO-20 V and cp-Ti were not significantly different. Following 5 days of culture, the OD value of the AO-treated samples consistently exceeded that of cp-Ti. The calculated RGR results are shown in Figure 11b. After 3 and 5 days of culture, the RGR for the samples exceeded 100%, indicating Grade 0 cytotoxicity. These results suggest that AO-treated Ti-Ag samples exhibit no cytotoxicity to MC3T3-E1 cells.

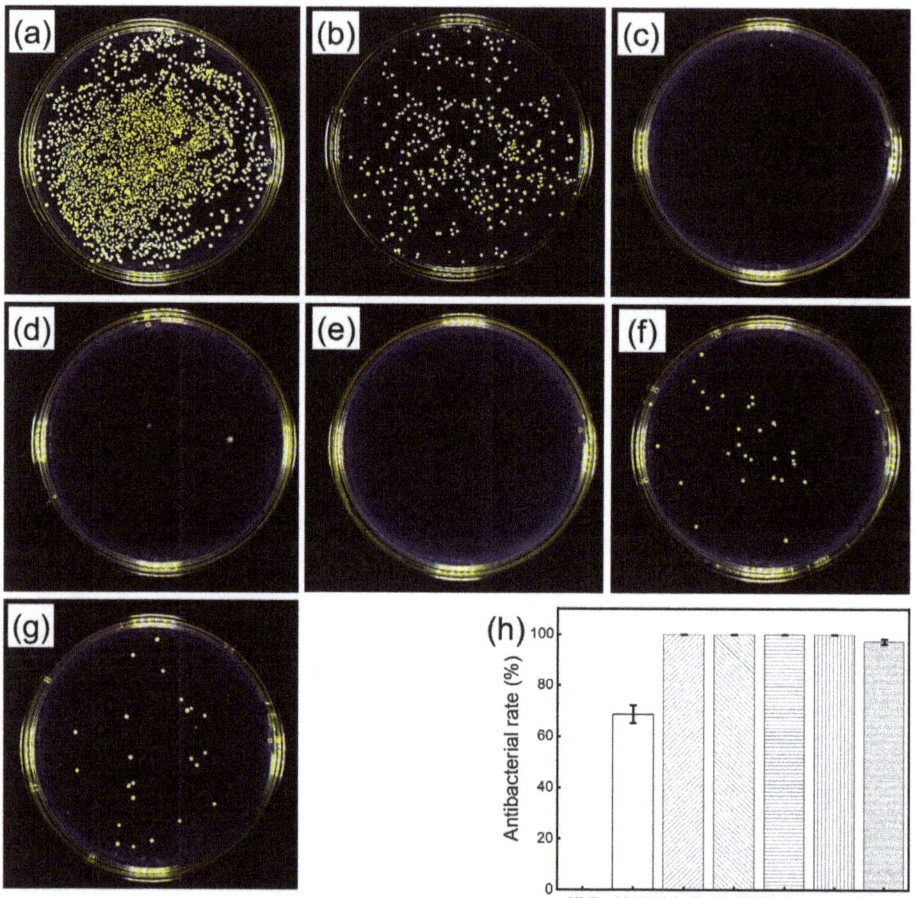

Figure 9. The number of colonies and antibacterial rate of S.aureus cultured for 24 h on Ti-Ag samples with different AO voltages: (**a**) cp-Ti; (**b**) AO-0 V; (**c**) AO-20 V; (**d**) AO-40 V; (**e**) AO-60 V; (**f**) AO-80 V; (**g**) AO-120 V; (**h**) antibacterial rates.

Figure 12 shows the adhesion state of MC3T3-E1 cells obtained by staining on cp-Ti, AO-20 V, and AO-120 V specimens for 4 h and 24 h. After 4 h of culture, the cell size on the AO-120 V sample was larger than that on the AO-20 V sample, and the cells showed a flat spherical shape. When the culture time was increased to 24 h, the cells grew pseudopods and cytoskeleton, and the cells were distributed on the sample in a large area.

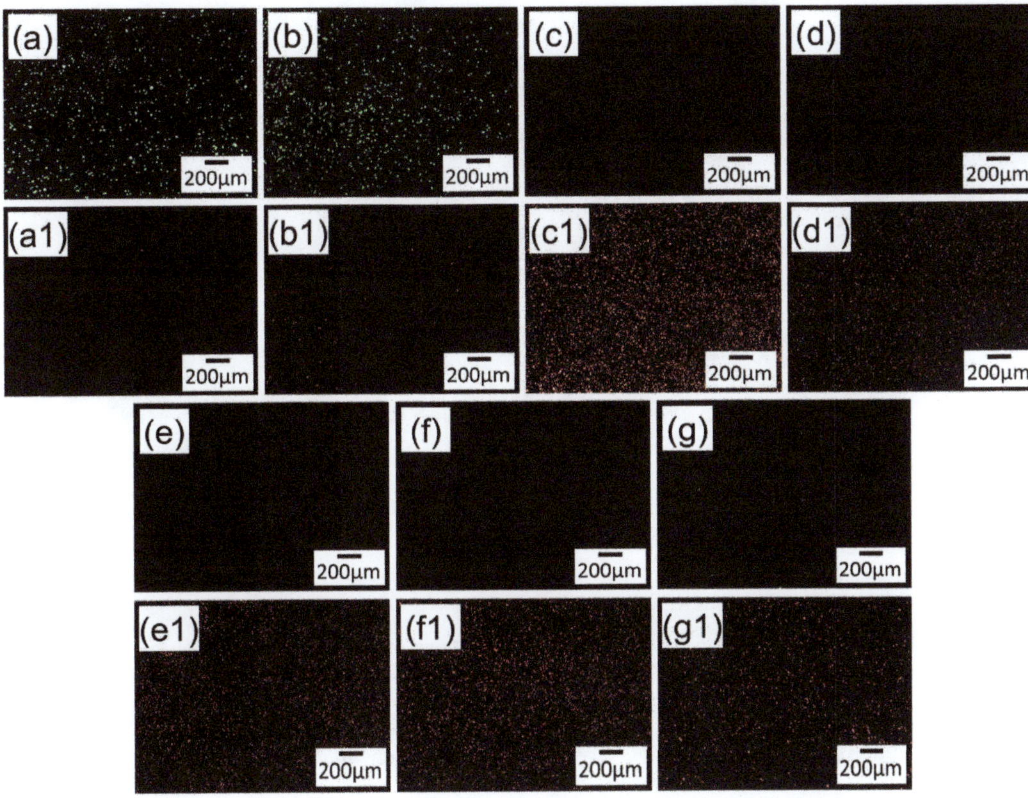

Figure 10. Live/dead fluorescence staining of *S. aureus* colonies on Ti-Ag samples: (**a**,**a1**) cp-Ti; (**b**,**b1**) AO-0 V; (**c**,**c1**) AO-20 V; (**d**,**d1**) AO-40 V; (**e**,**e1**) AO-60 V; (**f**,**f1**) AO-80 V; (**g**,**g1**) AO-120 V.

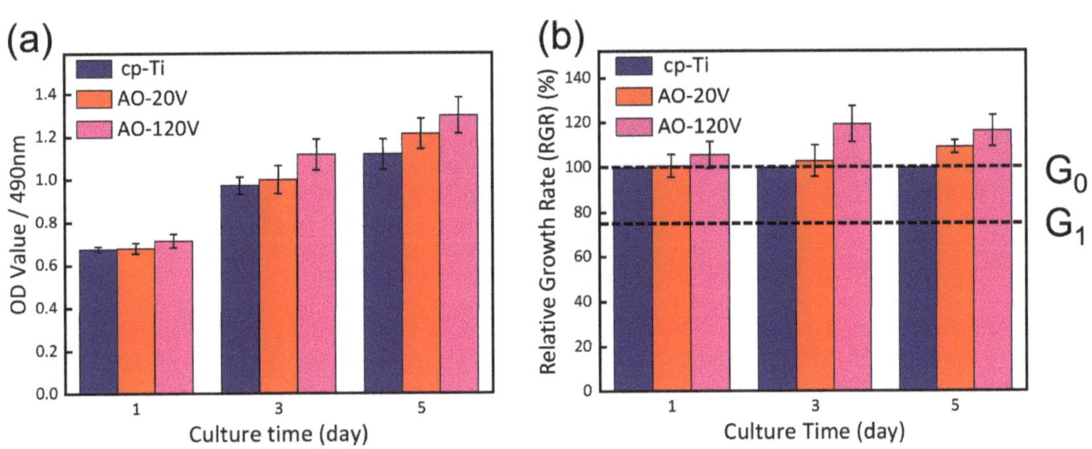

Figure 11. Cell viabilities of MC3T3-E1 cells after 1, 3, and 5 days of incubation on Ti-Ag samples: (**a**) OD Values at 490 nm, (**b**) RGR of cells on different samples.

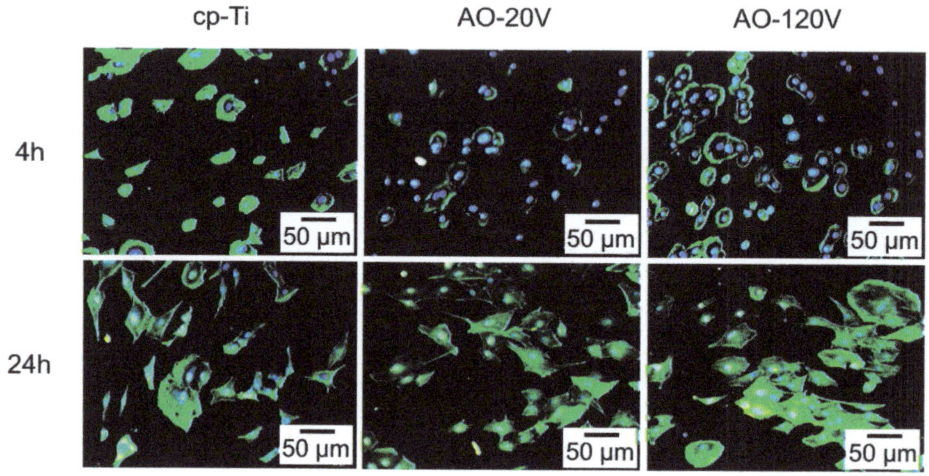

Figure 12. Cytoskeleton of MC3T3-E1 cells cultured on cp-Ti, AO-20 V, and AO-120 V specimens for 4 h and 24 h.

4. Discussion

4.1. Surface Physicochemical Performance

The anodic oxidation reaction induces alterations in the surface morphology and composition of Ti alloys [39]. The formation of anodic oxide film is formed by extracting O^{2-} and matrix metal from the plating solution under the action of the electrode [8]. As depicted in Figure 2a–e, the anodizing voltage markedly influences the staining of the AO film. The color of the oxide is largely determined by the thickness of the oxide [40]. The thickness of Ti oxide formed under natural conditions is only 6 nm [41]. The mixed oxide film formed by anodic oxidation is more uniform and has good adhesion to the alloy surface [42]. The change in the oxidation voltage has no obvious effect on the surface phase composition of the film, as shown in Figure 3. However, the scanning results in Figure 4a–e show that when the oxidation voltage is 60 V, a large number of block structures appear on the surface. EDS analysis shows that the Ag content of the block structure is higher than that of the matrix, and AO treatment can accelerate the enrichment of Ag-containing phases on the surface of the alloy [43]. When the voltage increases to 120 V, the 'flower-like structure' of the oxide coating is the manifestation of the dielectric breakdown of the coating [44]. However, the dielectric breakdown phenomenon caused by too high voltage will cause the burning loss of Ag element, resulting in the decrease of the Ag element content in the coating. Ultimately, the XPS results in both Figures 2 and 5 indicate the formation of the AO coating on the α-Ti and a small quantity of Ti_2Ag substrates. The coating is comprised of TiO_2 and Ag_2O. Consequently, it can be concluded that the oxidation voltage has a limited impact on the chemical composition and phase composition of the film.

In the initial stage of anodic oxidation, the surface of the sample undergoes a violent hydrolysis reaction under a constant applied voltage, which increases the pH value around the sample and produces a large amount of free OH^-. The reaction is shown in Equation (4):

$$2H_2O + 2e^- \rightarrow H_2\uparrow + 2OH^- \quad (4)$$

$$Ti \rightarrow Ti^{4+} + 4e^- \quad (5)$$

As the hydrolysis reaction proceeds, when the OH^- in the solution reaches a certain concentration, under the action of the electric field, OH^- reacts with the α-Ti and Ti_2Ag substrates, as shown in the Equations (6) and (7):

$$Ti_2Ag + 10OH^- \rightarrow Ti(OH)_4 + Ag(OH)_2{}^- + 9e^- \quad (6)$$

$$Ti^{4+} + 4OH^- \rightarrow Ti(OH)_4 \quad (7)$$

Subsequently, the reactions in Equations (8) and (9) occur on the surface, resulting in a large amount of Ag^+ migrating outward to form oxides. When the voltage is large, a white block structure is formed on the surface, as shown in Figure 4c.

$$2Ag(OH)^{2-} \rightarrow Ag_2O + H_2O + 2OH^- \quad (8)$$

$$2Ti(OH)_4 + 3e^- \rightarrow Ti_2O_3 + 2H_2O + 3OH^- \quad (9)$$

As the oxidation proceeds further, a new film will be formed outside [8]. When the voltage reaches the critical breakdown voltage, the white block structure appears as a gap and transforms into a flower-like structure. With the increase in the voltage and time, the coating size continues to thicken, and the size of the flower-like structure increases, as shown in Figure 4e,f, and with the progress of the reactions in Equations (10) and (11), the burning loss of the Ag element occurs. Finally, the formation rate and dissolution rate of the oxide film are balanced, and the surface is composed of oxides of TiO_2 and Ag_2O.

$$Ag_2O + 2H_2O \rightarrow 2Ag^+ + 2H^+ + 2e^- \quad (10)$$

$$2Ti_2O_3 + O_2 \rightarrow 4TiO_2 \quad (11)$$

The roughness and wetting angle are important criteria for evaluating the biocompatibility of Ti alloy implant materials [45]. Osteoblasts are more likely to adhere to the rougher Ti alloys' surface [46]. As depicted in Figure 7a, AO treatment markedly enhances the surface roughness of Ti-Ag samples. This manifestation indicates that the flower-like structure formed on the sample surface due to electron migration and bonding has a significant impact on cell adhesion and proliferation. In Figure 7b, the reduction in the wetting angle suggests that the alloy possesses high surface energy. At the same time, the ceramic oxide layer formed by AO treatment, especially TiO_2, has high ionic properties and shows increased wettability [47]. The above results indicate that the coating formed by AO treatment promotes cell adhesion and proliferation.

4.2. Corrosion Resistance and Ion Release

The corrosion characteristics and ion release of implant materials play a pivotal role in the durability, efficacy, and biocompatibility of these materials. The results of Figure 6, Tables 2 and 3 show that the TiO_2/Ag_2O coating formed on the Ti-Ag alloy by AO treatment enhances its corrosion resistance. The dense oxide layer formed by AO treatment can effectively prevent the erosion of the solution [48]. With the increase in the AO treatment voltage, the polarization resistance of the passivation film increases first and then decreases, which further reflects the protection of the coating for the substrate, as shown in Figure 6c. The phase angle and impedance values are higher than those of the Ti-Ag substrate in the whole frequency range, as shown in Figure 6d,e. The peak phase angle at a high frequency describes the dielectric properties of the TiO_2/Ag_2O coating grown by AO treatment, while the peak phase angle at a medium frequency represents a dense oxide film [49]. Enhancing corrosion resistance can mitigate the occurrence of pitting or stress corrosion in implant materials under the combined influence of bodily fluids and stress. This, in turn, reduces the risk of material fracture failure and enhances the safety of internal implant materials during use [34]. However, from the potentiodynamic polarization curve, with the increase in the AO treatment voltage, the AO-120 V sample showed a higher corrosion current density, which may be related to the formation of a surface 'flower-like structure', and the formation of protrusions provides the possibility of corrosion [50].

Although the coating formed on the surface after AO treatment can significantly enhance the corrosion resistance of the alloy, it also increases the surface content of Ag. The unstable Ag_2O formed on the surface is more likely to accelerate the accelerated release of Ag^+ and ionize in aqueous solution. The form is as in Equation (10), where the NaCl solution will accelerate the reaction. At the same time, the surface microstructure of the alloy is a critical factor influencing the release behavior of metal ions [51]. With the increase in the oxidation voltage, the flower-like structure formed on the surface increases the contact area between the film and the NaCl solution, which is the reason for promoting the release of Ag^+.

4.3. Antibacterial Property

The strong antibacterial property of the Ti-Ag alloy against *S. aureus* is mainly attributed to the release of Ag^+ [52]. At present, the research on the antibacterial mechanism of silver shows that silver can bind to bacteria [53], penetrate bacterial cell membranes [54], produce a large amount of reactive oxygen species (ROS) [55], and act as a regulator of microbial signal transduction pathways [56], and so it has received extensive attention in the study of the new generation of antibacterial drugs [57]. The above antibacterial properties of Ag are attributed to the oligodynamic effect of Ag, which enables Ag to target the microbial membrane and bind to it, thereby destroying its structural morphology and integrity [58–60]. Secondly, the abnormal structure of the microbial membrane can block the microbial outflow/inflow pump [61], which hurts the transport of microbial cells and leads to the loss of cell viability [62]. It can also target the mitochondria in the cytoplasm to produce ROS [63] until some serious DNA damage is induced [64], which then leads to mutations and genotoxic effects, thereby killing bacteria. At the same time, the precipitation of the Ti_2Ag phase and Ag-rich phase in Ti-Ag can also lead to the rupture of bacterial cell membranes and the death of bacteria [36,65]. The anodic oxidation coating on the Ti-Ag sample isolated the direct contact between the Ti_2Ag phase and the bacteria, thereby reducing the antibacterial ability. However, as shown in Figure 9, the coating treated with AO showed an antibacterial rate of up to 99.99%, manifesting that AO treatment restrained the growth of bacteria.

It has been reported that a Ag_2O coating can effectively inhibit the growth of *S. aureus* and E. coli [66]. Ali [67] et al. prepared Ag_2O nanoparticles by the green synthesis method and showed an excellent antibacterial ability against *E. coli*, *S. aureus*, and *P. aeruginosa*. It can be seen that the Ag_2O coating on the sample after AO treatment can inhibit bacterial proliferation and thus achieve an antibacterial effect. Based on the above analysis, the antibacterial mechanism of Ti-Ag alloy anodic oxidation is illustrated in Figure 13. Although the maximum release of Ag^+ in this experiment remains below the minimum antibacterial concentration of Ag^+ at 5.4 ppm [68], AO treatment significantly enhances the release of Ag^+, contributing to the antibacterial effect of free Ag^+ and nano-Ag_2O particles. However, when the oxidation voltage reaches 120 V, the release of Ag^+ is at its peak, but the antibacterial rate slightly decreases, possibly due to the reduction of the Ag content in the Ti-Ag matrix. Therefore, the antibacterial properties of AO-treated Ti-Ag samples are primarily attributed to the synergistic action of the presence of Ag_2O and the release of Ag^+.

4.4. Cytotoxicity

Low cytotoxicity and good biocompatibility are the basic requirements to ensure the clinical application of biomedical implant materials and surface modification. Figure 10 illustrates the outcomes of the interaction between MC3T3-E1 cells and the samples. The AO-treated samples exhibited higher cell viability compared to the control samples, and the viability of all AO-treated Ti-Ag samples exceeded 75%, indicating the absence of cytotoxicity compared to the cp-Ti samples. The cytoskeleton staining shown in Figure 11 showed that the cell morphology after AO treatment showed a typical polygon. In comparison to cp-Ti, cells on AO samples exhibited more filamentous pseudopods, with interconnected ends forming a network structure. The cell morphology of the AO-120 V samples was better than that of AO-20 V samples. The cells displayed a robust spreading and adhesion state on the sample, indicating a healthy and thriving growth condition.

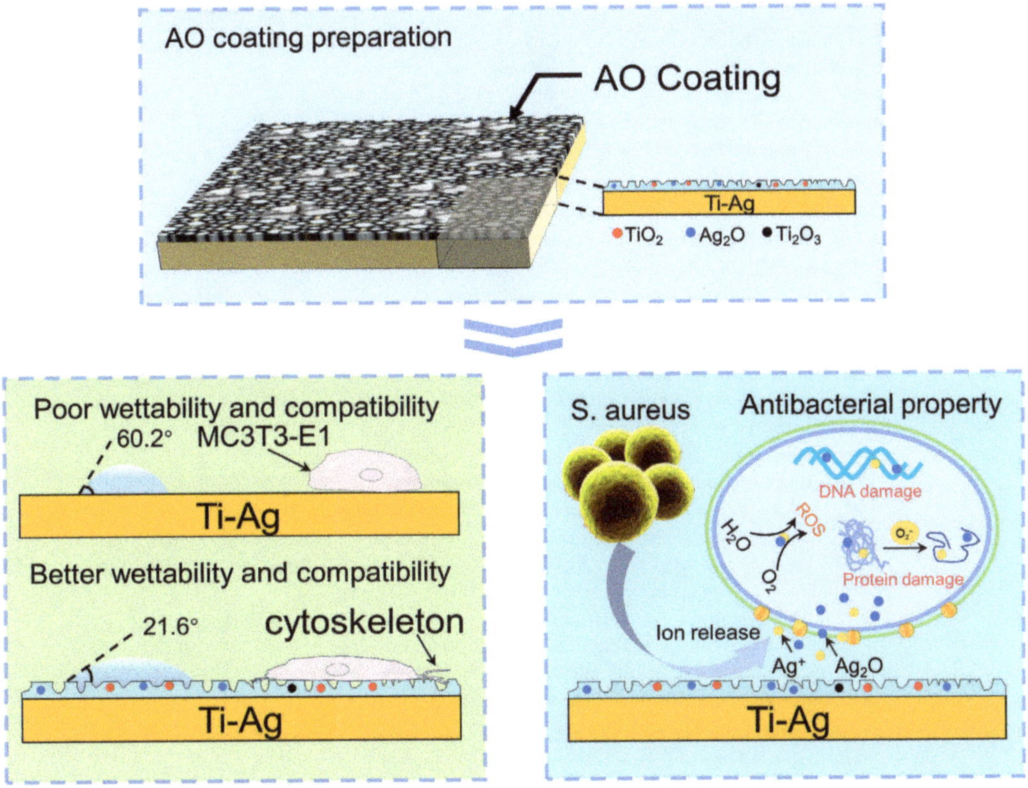

Figure 13. Formation and antibacterial mechanism of anodic oxidation of Ti-Ag alloy.

Silver is an effective antibacterial agent and is generally considered to be cytocompatible at low concentrations [69,70]. When the concentration of Ag^+ exceeds 5.4 ppm, it will produce cytotoxicity [71]. AgNPs themselves or Ag^+ can produce ROS in cells, resulting in DNA damage, protein denaturation, and thus apoptosis [72,73]. At the same time, Ag accumulates in the mitochondria of the cell, resulting in the decline of mitochondrial membrane potential (MMP), promoting the production of ROS, and causing mitochondrial dysfunction [74]. These pathways can lead to the cytotoxicity of the Ag^+ released from Ti-Ag samples. Even though AO treatment can increase the release of silver ions, the maximum release concentration of Ti-Ag samples in 0.9% NaCl solution for 24 h is still far less than the minimum concentration of cytotoxicity.

The surface microstructure, physicochemical state of the coating surface, and the composition of the oxide coating have different effects on the adhesion and growth of cells on the Ti alloys. The block and flower-like structures formed on the surface after AO treatment made a great contribution to the improvement of the roughness and hydrophilicity of the TiO_2/Ag_2O coating. At the same time, both anatase and rutile TiO_2 contained in the coating have good bioactivity and osteoinductivity [75].

In summary, AO treatment can obtain lower cytotoxicity and higher biocompatibility on the surface of Ti-Ag samples. The surface structure and composition of the TiO_2/Ag_2O coating are beneficial to antibacterial and cell proliferation. However, the antibacterial properties of Ag, the controllability of ion release, and in vivo biocompatibility still need more research in the future.

5. Conclusions

The impact of the AO process on the corrosion resistance, antibacterial properties, and biocompatibility of a Ti-Ag alloy was investigated by manipulating the surface composition and structure of the alloy through variations in the anodic oxidation voltage. This is of great significance for the application of the Ti-Ag alloy in the field of antibacterial titanium alloy bone implant materials. The following experimental conclusions can be drawn:

(1) Different voltage AO treatments can change the microstructure of the Ti-Ag alloy and form a multi-functional coating with an irregular flower-like structure with a size of 1~2 μm on the surface. The coating is mainly composed of Ag_2O and TiO_2. The change in voltage has no obvious effect on the XRD test results of the samples, but too high a voltage will cause the burning loss of Ag.

(2) After the AO treatment of the Ti-Ag alloy, the oxide film formed on the surface of the alloy improves the corrosion resistance and the corrosion resistance is higher than that of untreated samples. However, with the increase in the AO voltage, the corrosion resistance decreases. When the AO voltage is 20 V, the maximum Eocp is 107.621 mV, the maximum Ecorr is 17.035 mV, the minimum icorr is 2.241×10^{-8} $A{\cdot}cm^{-2}$, and the corrosion tendency and corrosion rate are the smallest.

(3) AO treatment enhanced the antibacterial properties of the Ti-Ag alloy and the antibacterial rate was more than 99%. The increase in the AO voltage can promote the release of Ag^+ and the improvement of the antibacterial performance is mainly attributed to the release of the Ag_2O and Ag^+ formed on the surface. However, the change in voltage has no obvious effect on the antibacterial performance.

(4) With the increase in the AO treatment voltage, the surface roughness can be increased, the wetting angle can be reduced, and the biocompatibility of the Ti-Ag alloy can be improved. When the AO voltage is 120 V, the maximum roughness is 0.73 μm and the minimum wetting angle is 23°, showing better cell compatibility and no cytotoxicity.

Therefore, in the face of the complex physiological environment of the human body, higher requirements are put forward for the performance of implant materials such as a Ti-Ag alloy. In the future, it is of great significance to study the long-term stability of implant materials, investigate a wider range of alloy systems, and carry out a comprehensive evaluation of the biological properties of implant materials in vivo.

Author Contributions: Z.M.: Software, Methodology, Validation, Investigation, Writing—original draft. Y.Y.: Software, Methodology, Formal analysis, Writing—original draft, Investigation. C.S.: Visualization, Investigation, Software. K.D.: Visualization, Investigation. J.X.: Visualization, Investigation. Q.L.: Visualization, Investigation. L.M.: Project administration, Supervision. J.Z.: Visualization, Investigation. J.H.: Resources, Software, Data curation. E.Z.: Software, Conceptualization, Supervision, Funding acquisition, Writing—review and editing. All authors have read and agreed to the published version of the manuscript.

Funding: This research was funded by the Heilongjiang Province Natural Science Foundation of China (No. YQ2022E042); the Postdoctoral Funding Project in the Heilongjiang Province (No. LBH-Z23297); the Heilongjiang Provincial Department of the Education Basic Research Business Fee Basic Research Project (No. 2023-KYYWF-0585); the Heilongjiang Provincial Key Laboratory of Oral Biomedical Materials and Clinical Applications Open Project (No. KQSW202302); and the Doctoral Special Research Fund Project of Jiamusi University (No. JMSUBZ2021-09).

Institutional Review Board Statement: Not applicable.

Informed Consent Statement: Not applicable.

Data Availability Statement: Data are unavailable due to privacy or ethical restrictions.

Conflicts of Interest: Authors Zhen Ma and Jianming Zheng were employed by the company Zhejiang Wanfeng Technology Development Co., Ltd. The remaining authors declare that the research was conducted in the absence of any commercial or financial relationships that could be construed as a potential conflict of interest.

References

1. Alves, A.C.; Sendao, I.; Ariza, E.; Toptan, F.; Ponthiaux, P.; Pinto, P.M.A. Corrosion behaviour of porous Ti intended for biomedical applications. *J. Porous Mater.* **2016**, *23*, 1261–1268. [CrossRef]
2. Huang, Y.; Song, G.; Chang, X.; Wang, Z.; Zhang, X.; Han, S.; Su, Z.; Yang, H.; Yang, D.; Zhang, X. Nanostructured Ag^+-substituted fluorhydroxyapatite-TiO_2 coatings for enhanced bactericidal effects and osteoinductivity of Ti for biomedical applications. *Int. J. Nanomed.* **2018**, *13*, 2665–2684. [CrossRef]
3. Cai, D.G.; Zhao, X.T.; Yang, L.; Wang, R.X.; Qin, G.W.; Chen, D.F.; Zhang, E.L. A novel biomedical titanium alloy with high antibacterial property and low elastic modulus. *J. Mater. Sci. Technol.* **2021**, *81*, 13–25. [CrossRef]
4. Gallo, G.; Schillaci, D. Bacterial metal nanoparticles to develop new weapons against bacterial biofilms and infections. *Appl. Microbiol. Biotechnol.* **2021**, *105*, 5357–5366. [CrossRef] [PubMed]
5. Zhang, W.; Zhang, S.Y.; Liu, H.; Ren, L.; Wang, Q.; Zhang, Y. Effects of surface roughening on antibacterial and osteogenic properties of Ti-Cu alloys with different Cu contents. *J. Mater. Sci. Technol.* **2019**, *48*, 278–284. [CrossRef]
6. Campos, P.A.D.; Sabrina, R.; Batistao, D.W.D.F.; Bruuna, F.A.; Queiroz, L.L.; Brito, S.S.; Paulo, P.G.F.; Ribas, R.M. Multidrug resistance related to biofilm formation in Acinetobacter baumannii and klebsiella pneumoniae clinical strains from different pulsotypes. *Curr. Microbiol* **2016**, *72*, 617–627. [CrossRef] [PubMed]
7. Wang, R.; Shi, M.S.; Xu, F.Y.; Qiu, Y.; Zhang, P.; Shen, K.L.; Zhao, Q.; Yu, J.G.; Zhang, Y.F. Graphdiyne-modified TiO_2 nanofibers with osteoinductive and enhanced photocatalytic antibacterial activities to prevent implant infection. *Nat. Commun.* **2020**, *11*, 4465. [CrossRef] [PubMed]
8. Cao, S.; Zhang, Z.-M.; Zhang, J.Q.; Wang, R.X.; Yang, L.; Chen, D.F.; Qin, G.W.; Zhang, E.L. Improvement in antibacterial ability and cell cytotoxicity of Ti-Cu alloy by anodic oxidation. *Rare Met.* **2022**, *41*, 594–609. [CrossRef]
9. Maharubin, S.; Hu, Y.B.; Sooriyaarachchi, D.; Cong, W.L.; Tan, G.Z. Laser engineered net shaping of antimicrobial and biocompatible titanium-silver alloys. *Mater. Sci. Eng. C* **2019**, *105*, 110059. [CrossRef] [PubMed]
10. Liu, S.; Zhang, Z.M.; Zhang, J.Q.; Qin, G.W.; Zhang, E.L. Construction of a TiO_2/Cu_2O multifunctional coating on Ti-Cu alloy and its influence on the cell compatibility and antibacterial properties. *Surf. Coat. Technol.* **2021**, *421*, 127438. [CrossRef]
11. Sadowska, J.M.; Genoud, K.J.; Kelly, D.J.; Brien, F.J. Bone biomaterials for overcoming antimicrobial resistance: Advances in non-antibiotic antimicrobial approaches for regeneration of infected osseous tissue. *Mater. Today* **2021**, *46*, 136–154. [CrossRef]
12. Ren, L.; Memarzadeh, K.; Zhang, S.Y.; Sun, Z.Q.; Yang, C.G.; Ren, G.G.; Robert, P.A.; Yang, K. A novel coping metal material CoCrCu alloy fabricated by selective laser melting with antimicrobial and antibiofilm properties. *Mater. Sci. Eng. C* **2016**, *67*, 461–467. [CrossRef] [PubMed]
13. Alejandra, R.-C.; Diego, T.; Belal, R.; Monica, O.H.; Maria, P.G.; Jose, A.C.; Jose, M.M.; Elisa, R. Bioactivity and antibacterial properties of calcium and silver-doped coatings on 3D printed titanium scaffolds. *Surf. Coat. Technol.* **2021**, *421*, 127476. [CrossRef]
14. Feng, J.; Cheng, L.; Zhou, X.D.; Xu, H.H.K.; Michael, D.W.; Markus, M.; Hans, M.; Li, Q.; Hannig, M.; Rupf, S. In situ antibiofilm effect of glass-ionomer cement containing dimethyllaminododecyl methacrylate. *Dent. Mater.* **2015**, *31*, 992–1002. [CrossRef]
15. Liang, Y.; Song, Y.Y.; Wang, L.; Wei, C.; Zhou, X.; Feng, Y.H. Research progress on antibacterial activity of medical titanium alloy implant materials. *Odontology* **2023**, *111*, 813–829. [CrossRef] [PubMed]
16. Arciola, C.R.; Campoccia, D.; Montanaro, L. Implant infections: Adhesion, biofilm formation and immune evasion. *Nat. Rev. Microbiol.* **2018**, *16*, 397–409. [CrossRef] [PubMed]
17. Chen, C.H.; Lu, T.K. Development and challenges of antimicrobial peptides for therapeutic applications. *Antibiotics* **2020**, *9*, 24. [CrossRef] [PubMed]
18. Jeannine, H. Rational approaches towards inorganic and organometallic antibacterials. *Biol. Chem.* **2022**, *403*, 363–375. [CrossRef]
19. Dizaj, S.M.; Lotfipour, F.; Mohammad, B.J.; Mohammad, H.Z.; Khosro, A. Antimicrobial activity of the metals and metal oxide nanoparticles. *Mater. Sci. Eng. C* **2014**, *44*, 278–284. [CrossRef] [PubMed]
20. Kim, J.D.; Yun, H.; Kim, G.C.; Lee, C.W.; Choi, H.C. Antibacterial activity and reusability of CNT-Ag and GO-Ag nanocomposites. *Appl. Surf. Sci.* **2013**, *283*, 227–233. [CrossRef]
21. Saidin, S.; Jumat, A.M.; Amin, M.A.A.N.; Abdullah, S.S.A.H. Organic and inorganic antibacterial approaches in combating bacterial infection for biomedical application. *Mater. Sci. Eng. C* **2021**, *118*, 111382. [CrossRef]
22. Luo, J.H.; Mamat, B.; Yue, Z.H.; Zhang, N.Y.; Xu, X.J.; Li, Y.L.; Su, Z.; Ma, C.; Zang, F.Z.; Wang, Y.B. Multi-metal ions doped hydroxyapatite coatings via electrochemical methods for antibacterial and osteogenesis. *Colloid Interface Sci. Commun.* **2021**, *43*, 100435. [CrossRef]
23. Chen, M.; Zhang, E.L.; Zhang, L. Microstructure, mechanical properties, bio-corrosion properties and antibacterial properties of Ti-Ag sintered alloys. *Biol. Adv.* **2016**, *62*, 350–360. [CrossRef] [PubMed]

24. Chouirfa, H.; Bouloussa, H.; Migonney, V.; Daudre, C.F. Review of titanium surface modification techniques and coatings for antibacterial applications. *Acta Biomater.* **2019**, *83*, 37–54. [CrossRef] [PubMed]
25. Hadidi, M.; Bigham, A.; Saebnoori, E.; Tabrizi, S.A.H.; Rahmati, S.; Alizadeh, Z.M.; Nasirian, V.; Rafienia, M. Electrophoretic-deposited hydroxyapatite-copper nanocomposite as an antibacterial coating for biomedical applications. *Surf. Coat. Technol.* **2017**, *321*, 171–179. [CrossRef]
26. Hou, X.G.; Mao, D.; Ma, H.Y.; Ai, Y.K.; Zhao, X.L.; Deng, J.H.; Li, D.J.; Liao, B. Antibacterial ability of Ag-TiO_2 nanotubes prepared by ion implantation and anodic oxidation. *Mater. Lett.* **2015**, *161*, 309–312. [CrossRef]
27. Kaczmarek, M.; Jurczyk, K.; Purwin, D.; Koper, J.K.; Romaniuk, A.; Lipinska, N.; Jakubowicz, J.; Jurczyk, M.U. Molecular analysis of biocompatibility of anodized titanium with deposited silver nanodendrites. *Biol. Adv.* **2018**, *93*, 437–444. [CrossRef]
28. Xu, J.Y.; Zhang, J.W.; Shi, Y.F.; Tang, J.C.; Huang, D.N.; Yan, M.; Dargusch, M.S. Surface modification of biomedical Ti and Ti alloys: A review on current advances journal of metals. *Materials* **2022**, *15*, 1749. [CrossRef] [PubMed]
29. Zhang, E.L.; Zhao, X.T.; Hu, J.L.; Wang, R.X.; Fu, S.; Qin, G.W. Antibacterial metals and alloys for potential biomedical implants. *Bioact. Mater.* **2021**, *6*, 2569–2612. [CrossRef]
30. Zhang, L.C.; Chen, L.Y. A Review on Biomedical Titanium Alloys: Recent Progress and Prospect. *Adv. Eng. Mater.* **2019**, *21*, 18012115. [CrossRef]
31. Zhang, E.L.; Ren, J.; Li, S.Y.; Yang, L.; Qin, G.W. Optimization of mechanical properties, biocorrosion properties and antibacterial properties of as-cast Ti-Cu alloys. *Biomed. Mater.* **2016**, *11*, 065001. [CrossRef] [PubMed]
32. Andrii, L.; Cecilia, F.; Oleksii, S.; Fabrizia, G.; Dario, B.; Raymond, J.T. Comparison of Antimicrobial and Antibiofilm Activity of Proflavine Co-crystallized with Silver, Copper, Zinc, and Gallium Salts. *ACS Appl. Bio Mater.* **2022**, *5*, 4203–4212. [CrossRef]
33. Wang, H.; Gao, W.W.; Zhang, X.Y.; LI, Y.; Zhang, S.Y.; Ren, L.; Yang, K. Enhanced Corrosion Resistance and Biological Properties of Ultrafine-Grained Ti15Zr5Cu Alloy. *Metals* **2022**, *12*, 1144. [CrossRef]
34. Hu, J.L.; Li, H.X.; Wang, X.Y.; Yang, L.; Chen, M.; Wang, R.X.; Qin, G.W.; Chen, D.F.; Zhang, E.L. Effect of ultrasonic micro-arc oxidation on the antibacterial properties and cell biocompatibility of Ti-Cu alloy for biomedical application. *Biomater. Adv.* **2020**, *115*, 110921. [CrossRef]
35. Schwartz, A.; Kossenko, A.; Zinigrad, M.; Danchuk, V.; Sobolev, A. Cleaning Strategies of Synthesized Bioactive Coatings by PEO on Ti-6Al-4V Alloys of Organic Contaminations. *Materials* **2023**, *16*, 4624. [CrossRef]
36. Wang, N.; Li, H.Y.; Lv, W.L.; Li, J.H.; Wang, J.S.; Zhang, Z.T.; Liu, Y.R. Effects of TiO_2 nanotubes with different diameters on gene expression and osseointegration of implants in minipigs. *Biomaterials* **2011**, *32*, 6900–6911. [CrossRef]
37. Fu, S.; Zhang, Y.; Qin, G.W.; Zhang, E.L. Antibacterial effect of Ti-Ag alloy motivated by Ag-containing phases. *Biomater. Adv.* **2021**, *128*, 112266. [CrossRef]
38. Biesinger, M.C.; Lau, L.W.; Gerson, A.R.; Smart, R.S.C. Resolving surface chemical states in XPS analysis of first row transition metals oxides and hydroxides: Sc, Ti, V, Cu and Zn. *Appl. Surf. Sci.* **2010**, *257*, 887–898. [CrossRef]
39. Jarosz, M.; Zaraska, L.; Koziel, M.; Simka, W.; Sulka, G. Electrochemical Oxidation of Ti15Mo Alloy-The Impact of Anodization Parameters on Surface Morphology of Nanostructured Oxide Layers. *Nanomaterials* **2020**, *11*, 68. [CrossRef] [PubMed]
40. Sul, Y.-T.; Johansson, C.B.; Jeong, Y.; Tomas, A. The electrochemical oxide growth behavior on titanium in acid and alkaline electrolytes. *Med. Eng. Phys.* **2001**, *23*, 329–346. [CrossRef]
41. Chan, C.W.; Carson, L.; Smith, G.C.; Alessio, M.; Lee, S. Enhancing the antibacterial performance of orthopaedic implant materials by fibre laser surface engineering. *Appl. Surf. Sci.* **2017**, *404*, 67–81. [CrossRef]
42. Han, M.-K.; Im, J.-B.; Hwang, M.-J.; Kim, B.J.; Kim, H.Y.; Park, Y.J. Effect of indium content on the microstructure, mechanical properties and corrosion behavior of titanium alloys. *Metals* **2015**, *5*, 850–862. [CrossRef]
43. Mato, S.; Thompson, G.E.; Skeldon, P.; Shimizu, K.; Habazaki, H.; Masheder, D. Enrichment of alloying elements beneath anodic oxides: Investigation of Ta-1.5 at.% Cu alloy. *Corros. Sci.* **2001**, *43*, 993–1002. [CrossRef]
44. Yahalom, J.; Zahavi, J. Electrolytic breakdown crystallization of anodic oxide films on A1, Ta and Ti. *Electrochim. Acta* **1970**, *15*, 1429–1435. [CrossRef]
45. Zheng, Q.C.; Mao, L.L.; Shi, Y.T.; Fu, W.H.; Hu, Y.H. Biocompatibility of Ti-6Al-4V titanium alloy implants with laser microgrooved surfaces. *Mater. Technol.* **2022**, *37*, 2039–2048. [CrossRef]
46. Deligianni, D.D.; Katsala, N.D.; Koutsoukos, P.G.; Missirlis, Y.F. Effect of surface roughness of hydroxyapatite on human bone marrow cell adhesion, proliferation, differentiation and detachment strength. *Biomaterials* **2001**, *22*, 87–96. [CrossRef]
47. Bernard, S.A.; Balla, V.K.; Davies, N.M.; Bose, S.; Bandyopadhyay, A. Bone cell-materials interactions and Ni ion release of anodized equiatomic NiTi alloy. *Acta Biomater.* **2011**, *7*, 1902–1912. [CrossRef] [PubMed]
48. Prando, D.; Brenna, A.; Bolzoni, F.M.; Diamanti, M.V.; Pedeferri, M.; Ormellesw, P.M. Electrochemical anodizing treatment to enhance localized corrosion resistance of pure titanium. *J. Appl. Biomater. Func.* **2017**, *15*, 19–24. [CrossRef]
49. Benea, L.; Ravoiu, A.; Celis, J.-P. Anticorrosion Performance of the Electrochemically Grown Mixed Porous Oxide Films on Titanium Alloy in Biological Solution. *ACS Biomater. Sci. Eng.* **2019**, *5*, 5925–5934. [CrossRef]
50. Kumar, S.; Chattopadhyay, K.; Singh, V. Effect of surface nanostructuring on corrosion behavior of Ti-6Al-4V alloy. *Mater. Charact.* **2016**, *121*, 23–30. [CrossRef]
51. Wan, Y.Q.; Wang, Y.; Liu, Z.M.; Qu, X.; Han, B.X.; Bei, J.Z.; Wang, S.G. Adhesion and proliferation of OCT-1 osteoblast-like cells on micro- and nano-scale topography structured poly(L-lactide). *Biomaterials* **2005**, *26*, 4453–4459. [CrossRef] [PubMed]

52. Shi, A.Q.; Zhu, C.S.; Fu, S.; Wang, R.X.; Qin, G.W.; Chen, D.W.; Zhang, E.L. What controls the antibacterial activity of Ti-Ag alloy, Ag ion or Ti_2Ag particles? *Biomater. Adv.* **2020**, *109*, 110548. [CrossRef] [PubMed]
53. Wang, G.K.; Hou, H.M.; Wang, S.L.; Yan, C.L.; Liu, Y.F. Exploring the interaction of silver nanoparticles with lysozyme: Binding behaviors and kinetics. *Colloids Surf. B* **2017**, *157*, 138–145. [CrossRef]
54. Shang, L.; Nienhaus, K.; Nienhaus, G.U. Engineered nanoparticles interacting with cells: Size matters. *J. Nanobiotechnol.* **2014**, *12*, 5. [CrossRef] [PubMed]
55. Avalos, A.; Haza, A.I.; Mateo, D.; Paloma, M. Cytotoxicity and ROS production of manufactured silver nanoparticles of different sizes in hepatoma and leukemia cells. *J. Appl. Toxicol.* **2014**, *34*, 413–423. [CrossRef] [PubMed]
56. Prasher, P.; Singh, M.; Mudila, H. Oligodynamic Effect of Silver Nanoparticles: A Review. *BioNanoScience* **2018**, *8*, 951–962. [CrossRef]
57. Loo, C.-Y.; MRohanizadeh, R.; Young, P.M.; Traini, D.; Cavaliere, R.; Cynthia, B.W.; Lee, W.H. Combination of Silver Nanoparticles and Curcumin Nanoparticles for Enhanced Anti-biofilm Activities. *J. Agric. Food Chem.* **2016**, *64*, 2513–2522. [CrossRef]
58. Tripathi, D.K.; Tripathi, A.; Shweta; Singh, S.; Singh, Y.; Vishwakarma, K.; Yadav, G.; Sharma, S.; Singh, V.K.; Mishra, R.K.; et al. Uptake, Accumulation and Toxicity of Silver Nanoparticle in Autotrophic Plants, and Heterotrophic Microbes: A Concentric Review. *Front. Microbiol.* **2017**, *8*, 243339. [CrossRef]
59. Rahisuddin; Thabaiti, S.A.; Khan, Z.; Manzoor, N. Biosynthesis of silver nanoparticles and its antibacterial and antifungal activities towards Gram-positive, Gram-negative bacterial strains and different species of Candida fungus. *Bioproc. Biosyst. Eng.* **2015**, *38*, 1773–1781. [CrossRef] [PubMed]
60. Phanjom, P.; Ahmed, G. Effect of different physico-chemical conditions on the synthesis of silver nanoparticles using fungal cell filtrate of Aspergillus oryzae (MTCC No. 1846) and their antibacterial effect. *Adv. Nat. Sci. Nanosci. Nanotechnol.* **2017**, *8*, 045016. [CrossRef]
61. Hao, Y.; Winans, S.C.; Glick, B.R.; Charles, T.C. Identification and characterization of new LuxR/LuxI-type quorum sensing systems from metagenomic libraries. *Environ. Microbiol.* **2010**, *12*, 105–117. [CrossRef] [PubMed]
62. Lara, H.H.; Trevino, E.N.G.; Turrent, L.L.; Singh, D.K. Silver nanoparticles are broad-spectrum bactericidal and virucidal compounds. *J. Nanobiotechnol.* **2011**, *9*, 30. [CrossRef] [PubMed]
63. Wang, J.; Shu, K.H.; Zhang, L.; Si, Y.B. Effects of Silver Nanoparticles on Soil Microbial Communities and Bacterial Nitrification in Suburban Vegetable Soils. *Pedosphere* **2017**, *27*, 482–490. [CrossRef]
64. Pramanik, S.; Chatterjee, S.; Saha, A.; Devi, P.S.; Kumar, G.S. Unraveling the Interaction of Silver Nanoparticles with Mammalian and Bacterial DNA. *J. Phys. Chem. B* **2016**, *120*, 5313–5324. [CrossRef] [PubMed]
65. Cui, S.S.; Liu, S.; Nie, J.J.; Chen, D.F.; Wu, X.B.; Qin, G.W.; Zhang, E.L. Design and preparation of a biomedical titanium alloy with low elastic modulus and high antibacterial property based on Ti-Mo-Ag system. *J. Alloys Compd.* **2022**, *908*, 164639. [CrossRef]
66. Patel, H.; Joshi, J. Green and chemical approach for synthesis of Ag_2O nanoparticles and their antimicrobial activity. *J. Sol-Gel Sci. Technol.* **2023**, *105*, 814–826. [CrossRef]
67. Fayyadh, A.A.; Alzubaidy, M.H.J. Green-synthesis of Ag_2O nanoparticles for antimicrobial assays. *J. Mech. Behav. Mater.* **2021**, *30*, 228–236. [CrossRef]
68. Heidenau, F.; Mittelmeier, W.; Detsch, R.; Haenle, M.; Stenzel, F.; Ziegler, G.; Gollwitzer, H. A novel antibacterial titania coating: Metal ion toxicity and in vitro surface colonization. *J. Mater. Sci. Mater. Med.* **2005**, *16*, 883–888. [CrossRef]
69. Hardes, J.; Streitburger, A.; Ahrens, H.; Nusselt, T.; Gebert, C.; Winkelmann, W.; Battmann, A.; Gosheger, G. The influence of elementary silver versus titanium on oste-oblasts behavior in vitro using human osteosarcoma cell lines. *Sarcoma* **2007**, *2007*, 26539. [CrossRef]
70. Qin, H.; Cao, H.L.; Zhao, Y.C.; Zhu, C.; Cheng, T.; Wang, Q.J.; Peng, X.C.; Cheng, M.Q.; Wang, J.X.; Jin, G.D.; et al. In vitro and in vivo anti-biofilm effects of silver nanoparticles immobilized on titanium. *Biomaterials* **2021**, *269*, 120315. [CrossRef]
71. Greulich, C.; Braun, D.; Peetsch, A.; Diendorf, J.; Siebers, B.; Epple, M.; Manfred, K. The toxic effect of silver ions and silver nanoparticles towards bacteria and human cells occurs in the same concentration range. *RSC Adv.* **2012**, *2*, 6981–6987. [CrossRef]
72. Akter, M.; Sikder, M.T.; Rahman, M.M.; Ullah, A.K.M.; Hossain, K.F.B.; Banik, S.; Hosokawa, T.; Saito, T.; Kurasaki, M. A systematic review on silver nanoparticles-induced cytotoxicity: Physicochemical properties and perspectives. *J. Adv. Res.* **2018**, *9*, 1–16. [CrossRef]
73. Hamad, A.; Khashan, K.S.; Hadi, A. Silver Nanoparticles and Silver Ions as Potential Antibacterial Agents. *J. Inorg. Organomet. Polym.* **2020**, *30*, 4811–4828. [CrossRef]
74. Liao, C.Z.; Li, Y.C.; Tiong, S. Bactericidal and Cytotoxic Properties of Silver Nanoparticles. *Int. J. Mol. Sci.* **2019**, *20*, 449. [CrossRef]
75. Xing, J.H.; Xia, Z.B.; Hu, J.F.; Zhang, Y.H.; Zhong, L. Time dependence of growth and crystallization of anodic titanium oxide films in potentiostatic mode. *Corros. Sci.* **2013**, *75*, 212–219. [CrossRef]

coatings

MDPI

Article

Hybrid Polymer–Inorganic Coatings Enriched with Carbon Nanotubes on Ti-6Al-4V Alloy for Biomedical Applications

Dominika Träger, Dagmara Słota *, Karina Niziołek, Wioletta Florkiewicz and Agnieszka Sobczak-Kupiec

Department of Materials Engineering, Faculty of Materials Engineering and Physics, Cracow University of Technology, 37 Jana Pawła II Av., 31-864 Krakow, Poland; dominika.trager@student.pk.edu.pl (D.T.); karina.niziolek@doktorant.pk.edu.pl (K.N.); agnieszka.sobczak-kupiec@pk.edu.pl (A.S.-K.)

* Correspondence: dagmara.slota@doktorant.pk.edu.pl

Abstract: Bone tissue degeneration, caused by disease as well as trauma, is a problem affecting many social groups in the 21st century. It involves pain and reduced patient comfort. Developments in materials engineering allow for the design of novel, innovative materials that can be used in therapies to promote bone regeneration. This work presents the preparation of a ceramic–polymer coating modified with carbon nanotubes on a titanium alloy for biomedical applications. The ceramic part is hydroxyapatite synthesized by the wet precipitation method using orthophosphate and calcium hydroxide. The polymer of choice was polyethylene glycol. A UV light synthesis method was successfully applied to obtain coatings characterized by continuity and full crosslinking. Extensive physicochemical analysis and incubation studies were carried out. Interactions between coatings and fluids mimicking artificial biological environments were analyzed for 9 days, i.e., in fluids such as SBF solution, artificial saliva, and distilled water. During the in vitro incubation, changes in pH values were measured by potentiometric tests, and ionic conductivity was measured by analyzing conductometry. After incubation, the surface morphology was studied by scanning electron microscopy (SEM) together with energy-dispersive (EDS) microanalysis, which made it possible to determine the presence of individual elements on the surface, as well as to observe the appearance of new apatite layers. Fourier-transform infrared (FT-IR) spectrometry was also performed before and at the end of the incubation period. On the basis of the presented studies, it was concluded that coatings that contain nanotubes are bioactive and do not negatively affect the properties of the coatings. Bioactivity was confirmed microscopically by observing new apatite layers after incubation in SBF, which were identified as phosphorus and calcium deposits. Degradation of the polymer phase was observed in the artificial saliva. These materials require further study, including safety analysis, but they demonstrate potential for further work.

Keywords: composites; coatings; polyethylene glycol; hydroxyapatite; carbon nanotubes

1. Introduction

The role of a supporting structure in the human body is played by bones, which are connective and supportive tissues. They consist of three types of cells: osteoblasts, osteoclasts, and osteocytes. In addition, they include collagen fibers and compounds of elements such as calcium, phosphorus, and magnesium. The cells that create bone are responsible for its remodeling, which is necessary to maintain its normal mechanical properties [1].

Unfortunately, bone tissue can be destroyed by injury or disease. Bone overload syndrome refers to bone tissue damage that occurs as a result of prolonged fatigue. It is characterized by the appearance of cracks or fractures, accompanied by pain. Stress fractures are associated with overtraining, so the groups most often affected are athletes [2].

In the aging population, more people are suffering due to osteoporosis. This disease causes a decrease in the strength of bone tissue, causing bone fragility. Lack of adequate micronutrient supplementation, insufficient exercise, and advanced age increase the risk of this disease [3].

In order to accelerate bone regeneration and to fill defects caused by degradation, a frequently used material is hydroxyapatite (HA), with the chemical formula $Ca_{10}(PO_4)_6(OH)_2$. Thanks to its calcium and phosphorus content, HA is a highly biocompatible material and is biomimetic to the inorganic part of natural tissue [4]. Its advantages also include its bioactivity. Hydroxyapatite bonds well to bone tissue and promotes cell adhesion [5–7]. In addition, HA combines well with other biomaterials, forming a combination with high bioactive properties [8].

Polyethylene glycol (PEG) is a synthetic polymer belonging to the polyether group, formed by ethylene oxide polymerization [9]. It is biocompatible, non-toxic, non-immunogenic, and amphipathic, which makes it useful for biomaterials research and as a drug carrier [10]. Its long hydrophilic chain helps biodegrade the composite [11].

To improve the mechanical properties of a ceramic–polymer interface, the recommended additive is a titanium (90%) aluminum (6%) vanadium (4%) alloy (Ti-6Al-4V). It has excellent biocompatibility and corrosion resistance in physiological environments, as well as a low modulus of plasticity. Thanks to these properties, Ti-6Al-4V alloy is often used as a load-bearing implant following the reduction of shielding stress [12–16]. Therefore, it has found applications in orthopedic surgery [17,18]. Furthermore, titanium improves cell adhesion and supports cell proliferation [19,20]. This is a very important aspect in the creation of a material to promote bone tissue regeneration. Positive effects of combining Ti-6AL-4V with HA have also been reported [8].

An interesting material that blends very well with HA and synthetic or natural polymers are carbon nanotubes (CNTs) [21–23]. Made of graphene, the thinnest material in existence, they are classified into single-walled nanotubes (SWCNTs) and multi-walled nanotubes (MWCNTs), and they find applications in many scientific fields [24,25]. They are characterized by very good mechanical properties, such as high strength and flexibility [22,26]. The use of carbon nanotubes in tissue engineering is made possible by their biocompatibility and lack of toxicity in their interactions with cells [27,28]. When CNTs are placed in contact with bone tissue, not only cell adhesion but also support of osteoblast growth has been reported [29–32]. Due to their conductive properties, CNTs help stimulate osteoblastic cells during bone regeneration [33].

The global demand for biomaterials is growing every year. According to Stradvierd Research (the Market Research Report), the total size of the biomaterials market will reach $123 billion in 2021. By 2028, the market is expected to be worth $219 billion. According to the Market Research Report, the orthopedic biomaterials market will be worth USD 19.2 billion in 2022. The compound annual growth rate (CAGR) is expected to remain at 7.8% until 2030.

One of the reasons for the growing demand is the increasing proportion of the aging population. According to the World Health Organization, by 2022, there were more elderly people over the age of sixty than children under the age of five. There has also been an increase in the number of sports injuries. According to National Safety Statistics, there was a 20% increase in injuries in 2021 compared to the previous year.

Among the orthopedic material categories, ceramics was the most popular segment, along with bioglass. This segment accounted for 32.7% of revenue in 2022. The second-fastest growing and currently second-most popular group is polymers, which accounted for the most revenue in 2021. It is expected to dominate the biomaterials market in the coming years.

In addition to the traditional use of biomaterials in orthopedics, the concept of using biomaterials for the early detection of disease is being developed to prevent further progression of the disease. Furthermore, biomaterials used for diagnostics and theragnostics can incorporate scaffolds or nanoparticles, which expand the range of applications [34].

The aim of the presented paper was to determine the composition and conditions for the synthesis of hybrid ceramic–polymer coatings with the characteristics of biomaterials.

The polymeric phase was enriched with MWCNTs to determine the potential of their use as a modifier. Physicochemical analysis, as well as incubation studies, were carried out. These coatings are novel, and so far, no material described in the literature with the same composition and preparation method has been found.

2. Materials and Methods

2.1. Materials

Polyethylene glycol (PEG, Mn = 8000) (Acros Organic, Geel, Belgium) was used for the synthesis of the coatings, poly(ethylene glycol) diacrylate (PEGDA, Mn = 700) was used as a crosslinking agent, and 2-hydroxy-2-methylopropiophenone (Darocur 1173) was used as a photoinitiator (Sigma Aldrich, Darmstadt, Germany). MWCN > 98%, O.D. × L 6–13 nm × 2.5–20 μm (Sigma Aldrich, Darmstadt, Germany) were suspended in a polymer matrix. The components used to create HA were: calcium hydroxide ($Ca(OH)_2$), phosphoric acid V(H_3PO_4 85%), and ammonia water (NH_4OH 25%) (POCH S.A., Gliwice, Poland).

2.2. Methods

2.2.1. Preparation of Hydroxyapatite

The hydroxyapatite used to form composite coatings was obtained using the wet precipitation method. A detailed description of the formation of HA, together with a detailed physiochemical analysis of the powder, was given previously [35].

2.2.2. Preparation of Coatings

A PEG (20%) solution was prepared by mixing the polymer powder with distilled water. Then PEGDA was added to the resulting solution as a crosslinking agent, as well as HA in powder form. The exact addition rates of the substrates used are demonstrated in Table 1. The received solution was homogenized using a Polytron PT 2500 E homogenizer. A photoinitiator was added during the homogenization process using an automatic pipette. Then, 200 μL of the solution obtained was applied and distributed using an automatic pipette to rectangular titanium plates with dimensions of 2 cm × 3 cm. The coatings were exposed to the photocrosslinking process using a Medilux UV 436 HF (Medilux, Korntal-Münchingen, Germany) lamp for 3 min. After the crosslinking process, the coatings were prepared for further testing.

Table 1. Composition of the coatings.

Sample Symbol	PEG 20% [mL]	PEGDA [mL]	HA [g]	Photoinitiator [μL]	MWCNTs [g]
10	10	1.8	-	200	-
20			0.5		-
41					0.01
42					0.02
43					0.03

In total, the compositions of one polymer and four composite coatings with increasing MWCNTs were developed. Carbon nanotubes were added to the polymer solution in solid form and mixed thoroughly to distribute them evenly in the polymer using a homogenization technique. Homogenization ensured good mixing of all components as well as the uniformity of the samples. The MWCNTs were added before adding the hydroxyapatite. The limit value of MWCNTs was 0.03 g, because above this amount, it was not possible to obtain fully crosslinked materials.

Figure 1 presents a scheme for obtaining a composite coating on a Ti-6Al-4V alloy and a photo of an example of the resulting coating.

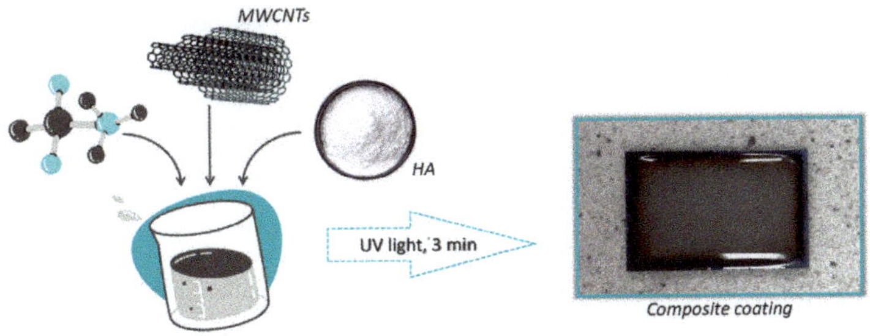

Figure 1. Scheme for obtaining a coating biomaterial.

2.2.3. In Vitro Incubation

The obtained coatings were incubated in vitro at the temperature of 36.6 °C in a POL-EKO incubator, model ST 5 B SMART (Wodzisław Śląski, Poland) for 9 days in artificial saliva, SBF solution, or distilled water. The artificial saliva and SBF compositions are presented in Table 2.

Table 2. Compositions of artificial biological fluids [36].

Artificial Saliva	
Component	**Amount [g/L]**
NaCl	0.400
KCl	0.400
$CaCl_2{\cdot}H_2O$	0.795
$Na_2S{\cdot}9H_2O$	0.005
$Na_2HPO_4{\cdot}H_2O$	0.780
CH_4N_2O	1.000
SBF	
Component	**Amount [g/L]**
NaCl	8.035
$NaHCO_3$	0.355
KCl	0.225
$K_2HPO_4{\cdot}3H_2O$	0.231
$MgCl_2{\cdot}6H_2O$	0.311
1 M HCl	39 mL
$CaCl_2$	0.292
Na_2SO_4	0.072
Tris	8.118

Potentiometric analysis to control the pH value of individual fluids was carried out to evaluate the bioactivity of the resulting composite coatings. Conductometric analysis, measuring the ionic conductivity of the fluids, was carried out simultaneously.

The purpose of these studies was to confirm the interactions occurring between the sample and the incubation medium. An increase in conductivity values suggests ion exchange between the biomaterial and the fluid. This can cause changes in the pH. However, pH changes can also be caused by leaching of individual components from the coatings.

Coatings applied to Ti6Al4V plates were placed in sterile containers filled with fluids. The pH and conductivity values were measured using an Elmetron CX-701 multifunctional device (Zabrze, Poland). Five replicates were performed for each composition.

2.2.4. Morphology Analysis

Morphological studies were carried out on the resulting coatings before and after the incubation period in SBF. A Hitachi TM3000 (Tokyo, Japan) tabletop scanning electron microscope (SEM) equipped with a Quantax 400 V EDS system was used to perform the morphological analysis. In order to visualize the morphology of the coatings and to capture the potential deposits formed on top of the samples, an EDS microanalysis was also performed. Before SEM measurement, a layer of gold was sputtered onto the surface of the dried sample. However, during the EDS measurement, the presence of Au was not considered in the analyses. Spot EDS analysis was performed to determine the percentage of each element as well as to perform surface mapping.

2.2.5. FT-IR Analysis

Fourier-transform infrared spectroscopy (FT-IR) was used to determine the functional groups and analyze the composition before and after incubation. A Thermo Scientific Nicolet iS5 FTIR (Loughborough, UK) spectrophotometer equipped with an iD7 was used. Due to the monolithic diamond crystal, there is a high optical contact between the sample and the diamond, leading to a clear spectrum. The wavelength range recorded was 4000–400 cm^{-1}. The study was performed at room temperature with conditions of 32 scans at 4.0 cm^{-1} resolution. The FT-IR spectra were collected for three random locations on the surface of each sample.

3. Results

3.1. In Vitro Incubation

3.1.1. pH Metric Analysis

During in vitro incubation, pH changes were measured in the solutions to determine the stability of the coatings under conditions similar to the human body. The exact results of the test are demonstrated in Figure 2. It was observed that coatings with HA had a more alkaline character than unsupplemented ones. This is related to the washing-out process of hydroxyapatite, which results in a hardly soluble precipitate [35].

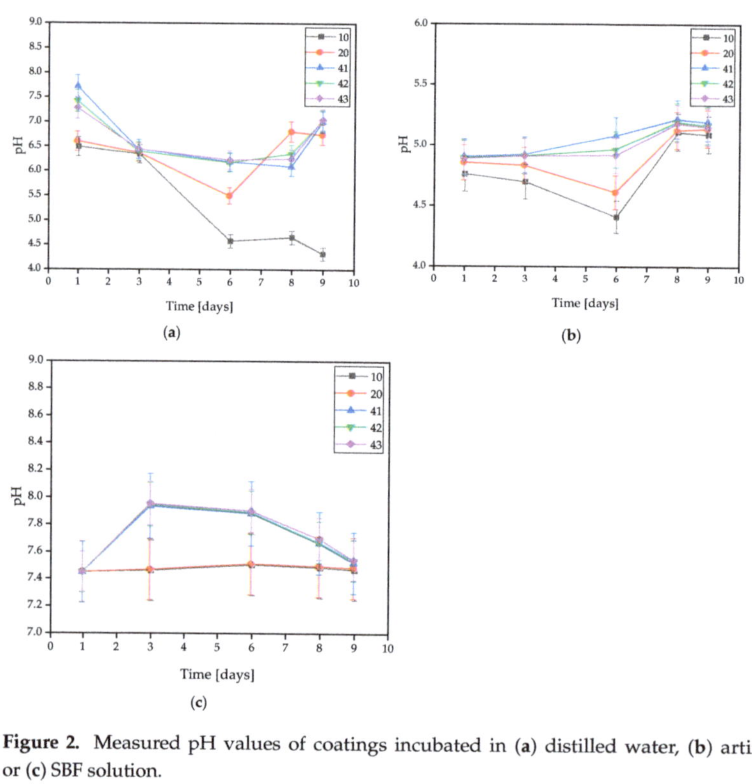

Figure 2. Measured pH values of coatings incubated in (**a**) distilled water, (**b**) artificial saliva, or (**c**) SBF solution.

On the basis of the MWCNTs-containing samples tested, it was found that the presence of nanotubes in the coating caused changes in the pH value of the solutions. The only rapid change in pH level observed was in the SBF solution during the 3 days of incubation, and its value was the highest at that point. For all three samples, the pH level remained between 7.939 and 7.955. However, a stabilization of the process was observed at later stages.

In the artificial saliva solution, the pH value increased gradually throughout the incubation. Samples immersed in distilled water exhibited a slight change in pH throughout the entire incubation. This is due to the high purity of the water, which results in a low concentration of free ions reacting with the coating and causing significant pH changes. Spikes in pH values in water are likely due to leaching of residual uncrosslinked polymer from the interior of the material. Smaller pH shifts in SBF may be due to its buffering nature. On this basis, it was concluded that the addition of MWCNTs caused a slight increase in pH value.

The stability of the resulting coatings in all solutions was also observed due to a lack of large pH changes throughout the entire incubation process. An interesting behavior of the materials was observed for SBF, as the pH value curves for coatings without MWCNTs, i.e., 10 and 20, almost completely overlap, while the pH values for coatings containing MWCNTs, i.e., 41, 42, and 43, also overlap, reach slightly higher values. This suggests that despite the buffering properties that SBF possesses, even a small presence of carbon nanotubes affects the pH change. However, the values are still within safe limits for the body.

3.1.2. Conductivity Analysis

The incubated samples were subjected to an electrolyte conductivity test. The exact results are presented in Figure 3. This test is based on the changes in the ion concentration in the solution.

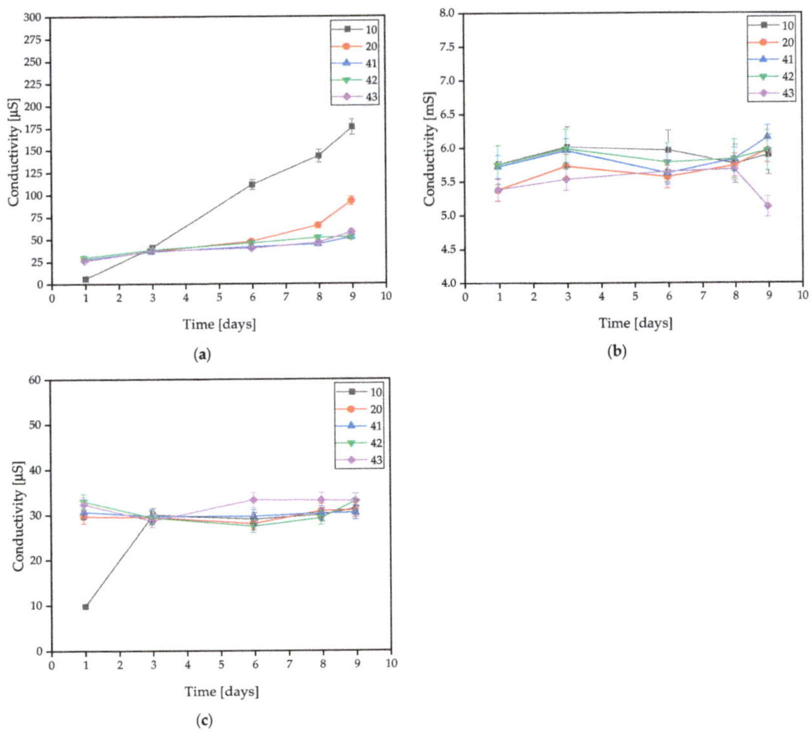

Figure 3. Measured conductivity values of coatings incubated in (**a**) distilled water, (**b**) artificial saliva, or (**c**) SBF solution.

Measurements for samples 10 and 20 demonstrated that the addition of HA resulted in a minimal improvement in stability and, in the case of distilled water, a slight decrease in conductivity. For coatings containing MWCNTs, in distilled water, the samples revealed very strong stability throughout the incubation process. This is due to the low concentration of free ions in the distilled water. Minimal conductivity changes were observed in the SBF solution, but these were low enough to conclude that the materials were stable in this solution. The same conclusion was reached with the artificial saliva solution. The conductivity of the sample increased minimally without sudden changes. This is related to the degradation process, which is more intense than in the other solutions.

3.2. Morphology Analysis

3.2.1. Coatings Morphology before Incubation

Figure 3 presents images of the surface morphology before incubation in artificial saliva and SBF solution. Figure 4b demonstrates a change in the surface morphology due to HA crystals suspended in the polymer area, while surface Figure 4a is completely smooth. On the surfaces Figure 4c–e, these crystals are also visible. The surface areas for Figure 4a,b were also compared to an earlier publication with l-cysteine modified coatings [35].

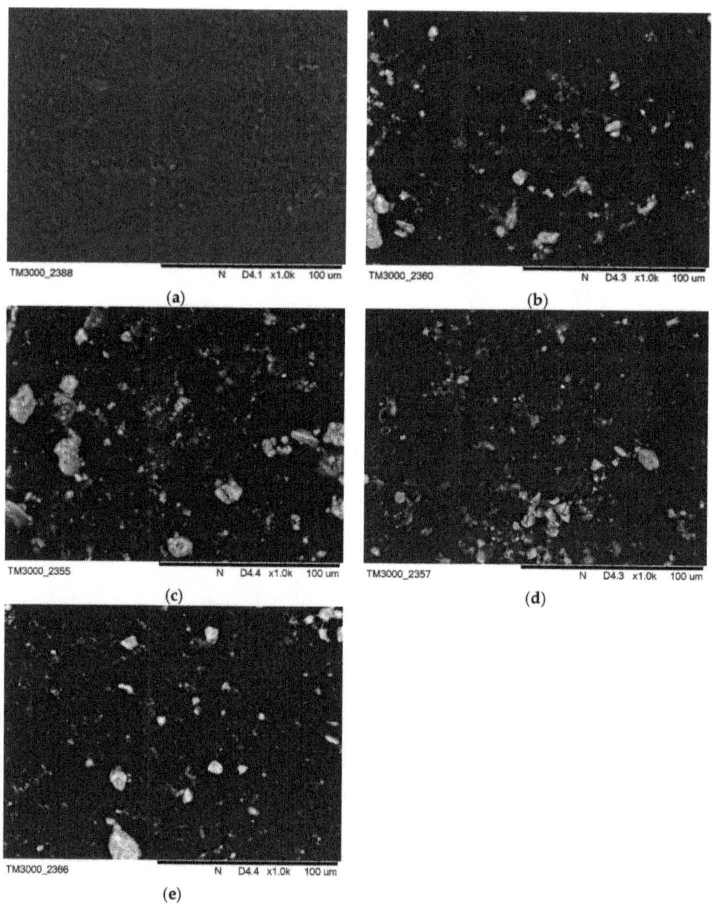

Figure 4. SEM images of (**a**) PEG-based coating 10, (**b**) PEG coating containing HA, (**c**) coating with 0.01 g CNTs, (**d**) coating with 0.02 g CNTs, and (**e**) coating with 0.03 g CNTs.

3.2.2. Coatings Morphology after Incubation

After the incubation process, changes were observed in the coatings of the samples containing MWCNTs. Images after incubation of the samples are demonstrated in Figures 5 and 6.

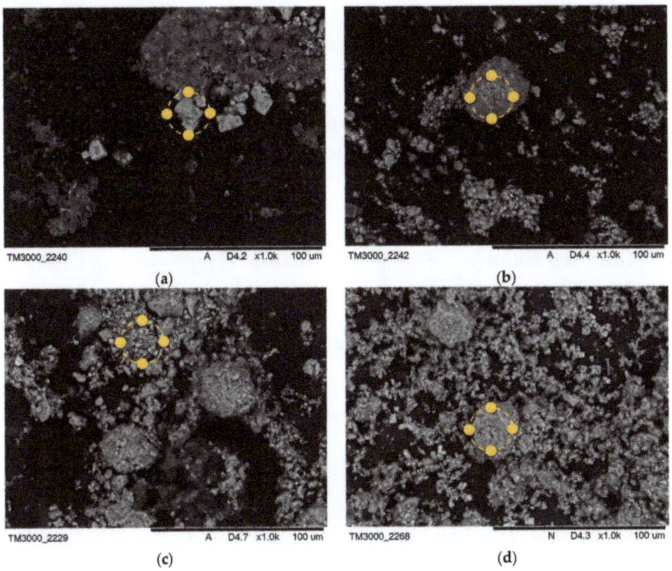

Figure 5. SEM images of samples after incubation in SBF solution (**a**) PEG-based coating 10, (**b**) PEG coating containing HA, (**c**) coating with 0.02 g CNTs, and (**d**) coating with 0.03 g CNTs. Yellow markers indicate the location of the EDS elemental analysis.

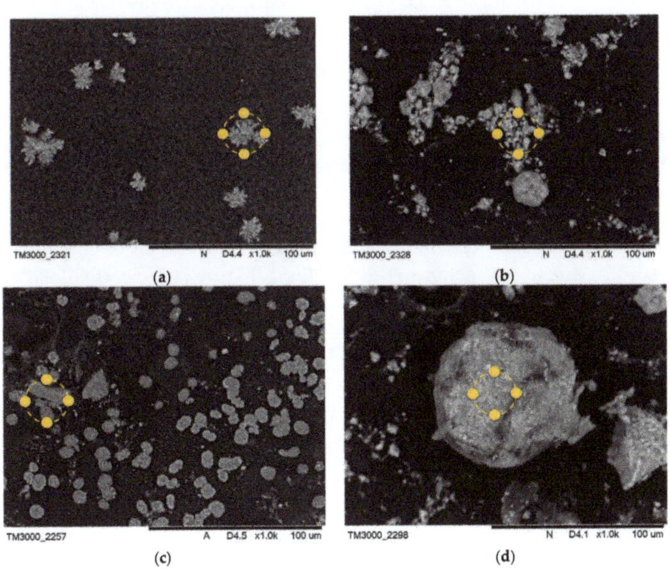

Figure 6. SEM images of samples after incubation in artificial saliva (**a**) PEG-based coating 10, (**b**) PEG coating containing HA, (**c**) coating with 0.02 g CNTs, (**d**) coating with 0.03 g CNTs. Yellow markers indicate the location of the EDS elemental analysis.

The formation of pits and grooves was observed on the surfaces, indicating the recrystallization process of the apatite. In all samples containing CNTs incubated in SBF solution, the appearance of large agglomerations of apatite crystals was noted. This is related to the specificity of the SBF solution, which simulates a human plasma environment with a high ion concentration. Calcium ions react with negative HA, later involving a phosphorus ion. This results in the compound being converted into apatite [33]. In biomaterials incubated in artificial saliva, the formation of a visible accumulation of apatite crystals was observed, as in the SBF solution.

Along with the morphological examination, an EDS microanalysis was performed. Its purpose was to identify the elements present and to determine their abundance in the coating. The exact results of the microanalysis are presented in Tables 3 and 4. The most frequent elements present were C and O. This indicates the chemical composition of the polymer that was used. Na and Cl ions were also observed. This appears to indicate that the biomaterial reacts with the solution in which it is incubated. The presence of Ca and P in the biomaterial results in the coatings having a bioactive property. The presence of Ca and P in material 10 after the incubation period is due to the composition of the incubation fluids, as the fluids contained trace amounts of these elements.

Table 3. Elemental composition of the tested coatings after incubation in SBF solution.

Sample	Mass Percentage [w%]
10	C: 45.88, O: 19.31, Cl: 19.06, Na: 14.00, Ca: 1.33, P: 0.42
20	O: 45.33, Ca: 24.87, C: 16.55, P: 9.45, Cl: 2.53, Na: 0.8
42	O: 41.14, Ca: 29.88, C: 13.87, P: 9.89, Cl: 4.39, Na: 0.74
43	C: 21.79, O: 42.48, Ca: 18.77, P: 8.58, Cl: 4.44, Na: 3.93

Table 4. Elemental composition of the tested coatings after incubation in artificial saliva.

Sample	Mass Percentage [w%]
10	C: 40.29, O: 39.57, Ca: 11.03, P: 7.03, Cl: 1.68, Na: 0.39
20	O: 46.15, C: 28.64, Ca: 15.91, P: 7.12, Cl: 1.52, Na: 0.65
42	C: 43.69, O: 35.86, Ca: 13.14, P: 3.95, Cl: 2.31, K: 1.04, Na: 0.01
43	O: 47.94, C: 23.94, Ca: 17.56, P: 8.77, Cl: 0.96, Na: 0.82

Considering that in the context of orthopedic applications, behavior in the SBF fluid is a more important indicator that provides information regarding the application potential, EDS microanalysis with mapping of individual elements was performed, as can be observed in Figure 6. Elements such as Ca and P were marked on the mapping, confirming the presence of calcium phosphate, such as HA, in the material, as well as the formation of new apatite layers on the surface due to interactions with SBF fluid. Figure 7a,b clearly illustrate the overlap on the surface.

The presence of Na, Mg, or Cl is a direct result of the composition of the SBF fluid, as these ions are present in the salts used to prepare it. The overlap of the spectrum of Mg with Cl or Na with Cl may indicate the formation of chlorides. This is also evidenced by the characteristic geometric shape of the crystals formed.

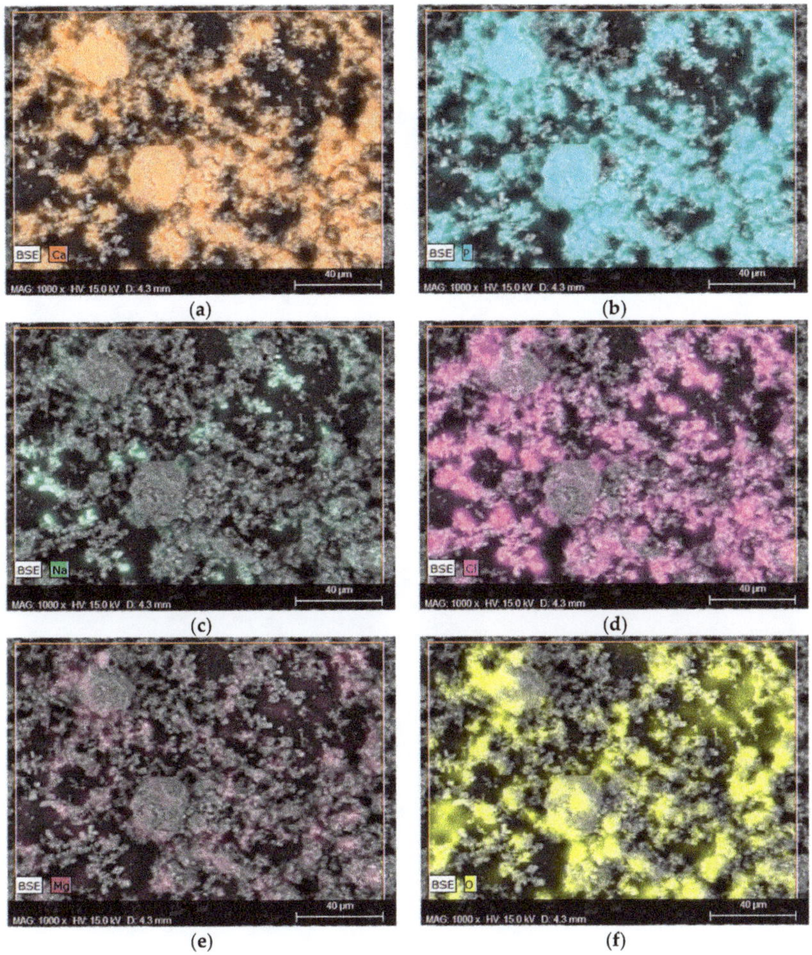

Figure 7. Elemental EDS microanalysis and determination of elements on the surface of material 43 after incubation in SBF (**a**) calcium, (**b**) phosphorus, (**c**) sodium, (**d**) chlorine, (**e**) magnesium, and (**f**) oxygen.

3.3. *FT-IR Analysis*

3.3.1. Spectrum of Coatings before Incubation

Using FT-IR spectrometric analysis to determine the chemical composition of the resulting coatings, the biomaterial spectra are demonstrated in Figure 8. The spectra of the coatings presented are PEG, PEG with hydroxyapatite, and also coatings containing 0.01 g, 0.02 g, or 0.03 g of nanotubes. The spectra of pure hydroxyapatite and pure PEG are presented in publications related to the studies performed and are described in this article [36]. For all samples that are shown, a peak was observed at a wavelength of 1100 cm^{-1} and 3500 cm^{-1}, originating from the OH group associated with the presence of the polymer. In samples containing HA, a characteristic peak related to PO_4^{3-} oscillation was observed at a wavelength of 560 cm^{-1}. In coatings containing CNTs, the featured peaks were more pronounced. This behavior was observed at wavenumbers of 1030 cm^{-1} and 560 cm^{-1}, respectively.

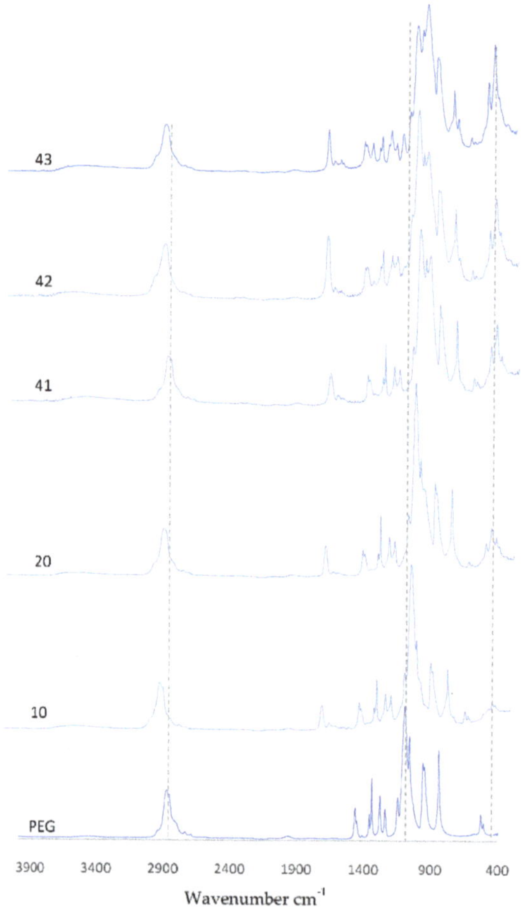

Figure 8. FT-IR spectra of coatings before incubation.

Finally, it is significant that FT-IR spectra were collected for three random locations on the surface of each sample. The spectra overlapped, which suggests that all of the coatings were homogeneous. Otherwise, the individual peaks would not have overlapped, or shifts would have been observed.

3.3.2. Spectrum of Coatings after Incubation

After a 9-day incubation carried out in SBF solution, artificial saliva, or distilled water, FT-IR analysis was carried out. Figure 9 shows the obtained spectra of the incubated coatings. An increase in absorbance was observed for the peak at wavelength 1730 cm^{-1}. For samples 10 and 20 incubated in SBF solution, the vibration modulus increased in the range of 700 cm^{-1}–400 cm^{-1}. In the spectra of coatings containing nanotubes, there was a decrease in the number of visible peaks. It should be noted that the diagrams overlap with the spectra obtained for the coatings before incubation. This suggests that as a result of incubation, at the coating–incubation fluid interface, there are no potential degradation or destruction processes of the composition.

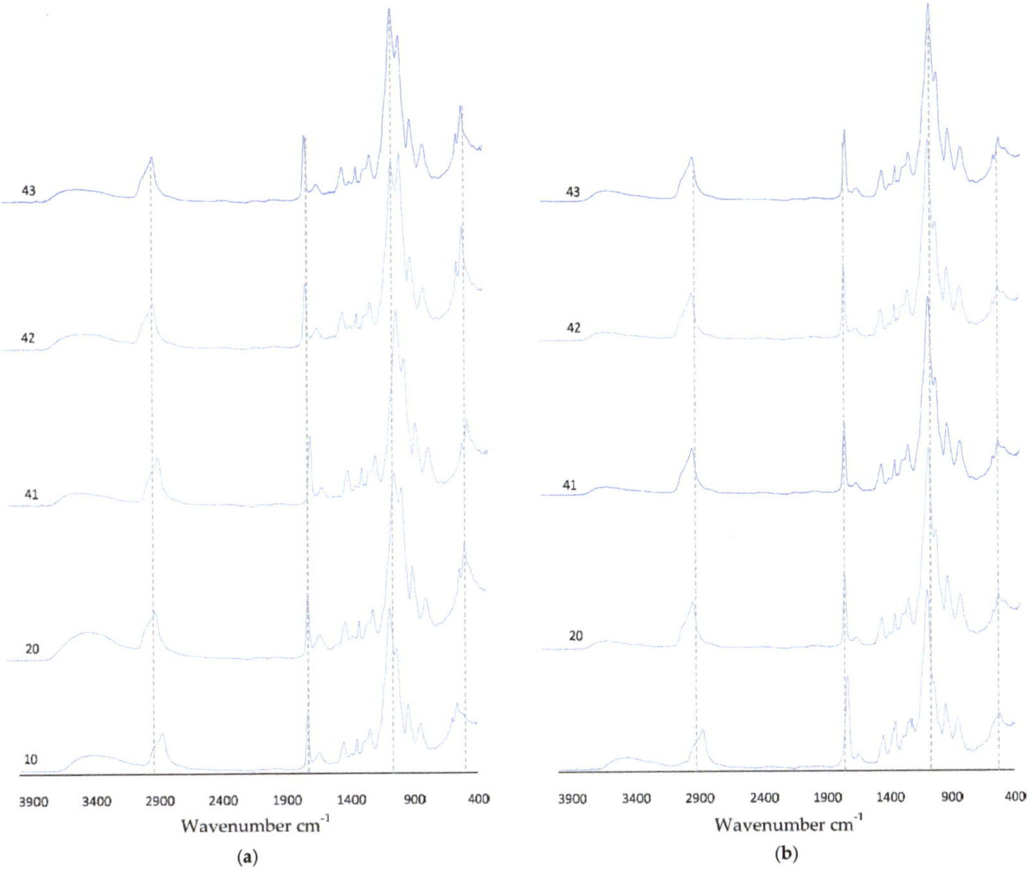

Figure 9. FT-IR spectra of coatings after incubation in (**a**) SBF solution or (**b**) artificial saliva.

4. Discussion

The results of this study confirm that the ceramic–polymer coating containing nanotubes can potentially be used as a biomaterial to support bone tissue regeneration. A 9-day in vitro study was performed to determine the effect of the coatings on the environment of artificial body fluids in terms of changing pH values and conductivity. The results of this study indicate proper interactions of the coatings with the solutions in which they were incubated.

In samples containing hydroxyapatite, the pH value increased minimally in the alkaline direction, which is related to the characteristics of the HA. A conductivity test verified the interaction of the coating ions with the ions of the prepared solutions, indicating a proper interaction without rapid changes throughout the incubation period. The changes in the pH and conductivity values confirm that the developed coatings are not inert and interact with the selected fluids. However, these interactions do not cause degradation or destruction of the coating surfaces, as confirmed by the overlapping FTIR spectra.

Morphological studies, together with EDS microanalysis, were carried out to determine the changes in the coatings caused by the incubation, and to identify and quantify the elements present. On the basis of this study, apatite agglomerations were found to appear on the surface of the samples, formed due to ion reactions between the HA and the solution. The microanalysis confirmed the occurrence of calcium and phosphorus. Elemental mappings in particular confirmed the formation of new calcium phosphate values. This effect was likely further supported by the presence of these ions in the composition of the

SBF fluid. The other elements appearing are related to the structure of the polymer and the normal reactions of the coatings with the solutions.

FT-IR spectrometric examination confirmed the occurrence of the functional groups associated with the substrates used through the appearance of characteristic peaks, revealing the chemical composition of the biomaterial.

The performed studies were compared with similar research in terms of the connection between carbon nanotubes and hydroxyapatite. The selected studies had differences in the method of coating synthesis. Despite the mentioned difference, similarities in the form of proper bonding of HA to the surface of the nanotubes was demonstrated [37–39].

5. Conclusions

In the presented study, the synthesis of coatings that have the potential to support bone tissue regeneration is reported. The technique developed was based on the use of UV light and enabled us to obtain composite coatings on titanium plates under UV light. The selected parameters, i.e., crosslinking time and the types of crosslinking agents and photoinitiators, made it possible to obtain completed materials with satisfactory organoleptic properties (without roughness or holes, fully continuous and crosslinked). This study confirmed that the addition of nanotubes to a ceramic–polymer base had no negative effect on the results during incubation in biological fluids. Coatings containing MWCNTs demonstrated bioactivity, which is an important feature of a biomaterial. Bioactivity was confirmed by observing new apatite layers that appeared following incubation. Further biological studies should focus on the reactions of cells in contact with the coatings in order to determine whether the biomaterial influences the adhesion and growth of bone tissue.

Author Contributions: Conceptualization, D.T. and D.S.; methodology, D.S. and W.F.; software, D.T. and K.N.; validation, D.S., W.F. and K.N.; formal analysis, D.S. and W.F.; investigation, D.S. and W.F.; resources, W.F.; data curation, D.S.; writing—original draft preparation, D.T.; writing—review and editing, A.S.-K., D.S. and K.N.; visualization, D.T. and D.S.; supervision, A.S.-K.; project administration, A.S.-K.; funding acquisition, A.S.-K. All authors have read and agreed to the published version of the manuscript.

Funding: This research received no external funding.

Institutional Review Board Statement: Not applicable.

Informed Consent Statement: Not applicable.

Data Availability Statement: Data that support the findings of this study are contained within the article.

Acknowledgments: The research work was carried out within the SMART-MAT Functional Materials Science Club (BioMat section) at the Faculty of Materials Engineering and Physics of the Cracow University of Technology.

Conflicts of Interest: The authors declare no conflict of interest.

Abbreviations

Abbreviations	Description
HA	Hydroxyapatite
PEG	Polyethylene Glycols
CNTs	Carbon Nanotubes
SWCNTs	Single-Walled Carbon Nanotubes
MWCNTs	Multi-Walled Carbon Nanotubes
PEGDA	Poly(ethylene glycol) Diacrylate
SBF	Simulated Body Fluid
XRD	X-ray Diffraction Analysis
EDS	Energy Dispersive X-ray Spectroscopy
FT-IR	Fourier Transform Infrared Spectroscopy

References

1. Hill, P.A. Bone Remodelling. *Br. J. Orthod.* **1998**, *25*, 101–107. [CrossRef] [PubMed]
2. Rachner, T.D.; Khosla, S.; Hofb, L.C. Osteoporosis: Now and the Future. *Lancet* **2011**, *377*, 1276–1287. [CrossRef]
3. Johnson, R.J. Overuse Injuries in Sports: A Review. *Sports Med.* **1985**, *2*, 316–333.
4. Gibson, I.R.; Ke, S.; Best, S.M.; Bonfield, W. Effect of Powder Characteristics on the Sinterability of Hydroxyapatite Powders. *J. Mater. Sci. Mater. Med.* **2001**, *12*, 163–171. [CrossRef] [PubMed]
5. Moroni, A.; Caja, V.L.; Egger, E.L.; Trinchese, L.; Chao, E.Y.S. Histomorphometry of Hydroxyapatite Coated and Uncoated Porous Titanium Bone Implants. *Biomaterials* **1994**, *15*, 926–930. [CrossRef] [PubMed]
6. Tracy, B.M.; Doremus, R.H. Direct Electron Microscopy Studies of the Bone-Hydroxylapatite Interface. *J. Biomed. Mater. Res.* **1984**, *18*, 719–726. [CrossRef] [PubMed]
7. Kato, K.; Sc, D.; Aoki, H.; Eng Tabata, D.T.; Do, D.S.; Ogiso, M. Biocompatibility of Apatite Ceramics in mandibles. *Biomater. Med. Devices Artif. Organs* **1979**, *7*, 291–297. [CrossRef]
8. Silva, P.L.; Santos, J.D.; Monteiro, F.J.; Knowles, J.C. Adhesion and Microstructural Characterization of Plasma-Sprayed Hydroxyapatite/Glass Ceramic Coatings onto Ti-6A1-4V Substrates. *Surf. Coat. Technol.* **1998**, *102*, 191–196. [CrossRef]
9. Qiu, C.; Xiao, X.; Liu, R. Biomimetic Synthesis of Spherical Nano-Hydroxyapatite in the Presence of Polyethylene Glycol. *Ceram. Int.* **2008**, *34*, 1747–1751. [CrossRef]
10. Venkatasubbu, G.D.; Ramasamy, S.; Avadhani, G.S.; Ramakrishnan, V.; Kumar, J. Surface Modification and Paclitaxel Drug Delivery of Folic Acid Modified Polyethylene Glycol Functionalized Hydroxyapatite Nanoparticles. *Powder Technol.* **2013**, *235*, 437–442. [CrossRef]
11. Photos, P.J.; Bacakova, L.; Discher, B.; Bates, F.S.; Discher, D.E. Polymer Vesicles in Vivo: Correlations with PEG Molecular Weight. *J. Control Release* **2003**, *90*, 323–334. [CrossRef]
12. Mironov, S.; Sato, Y.S.; Kokawa, H. Friction-Stir Welding and Processing of Ti-6Al-4V Titanium Alloy: A Review. *J. Mater. Sci. Technol.* **2018**, *34*, 58–72. [CrossRef]
13. Okazaki, Y.; Nishimura, E.; Nakada, H.; Kobayashi, K. Surface analysis of Ti–15Zr–4Nb–4Ta Alloy after Implantation in Rat Tibia. *Biomaterials* **2001**, *22*, 599–607. [CrossRef]
14. Heinl, P.; Müller, L.; Körner, C.; Singer, R.F.; Müller, F.A. Cellular Ti-6Al-4V Structures with Interconnected Macro Porosity for Bone Implants Fabricated by Selective Electron Beam Melting. *Acta. Biomater.* **2008**, *4*, 1536–1544. [CrossRef] [PubMed]
15. Okazaki, Y.; Rao, S.; Ito, Y.; Tateishi, T. Corrosion Resistance, Mechanical Properties, Corrosion Fatigue Strength and Cytocompatibility of New Ti Alloys without Al and V. *Biomaterials* **1998**, *19*, 1197–1215. [CrossRef] [PubMed]
16. Mohseni, E.; Zalnezhad, E.; Bushroa, A.R.; Hamouda, A.M.; Goh, B.T.; Yoon, G.H. Ti/TiN/HA Coating on Ti–6Al–4V for Biomedical Applications. *Ceram. Int.* **2015**, *41*, 14447–14457. [CrossRef]
17. Ponader, S.; Von Wilmowsky, C.; Widenmayer, M.; Lutz, R.; Heinl, P.; Körner, C.; Singer, R.F.; Nkenke, E.; Neukam, F.W.; Schlegel, K.A. In Vivo Performance of Selective Electron Beam-Melted Ti-6Al-4V Structures. *J. Biomed. Mater. Res. A* **2010**, *92*, 56–62. [CrossRef]
18. Long, M.; Rack, H.J. Titanium Alloys in Total Joint Replacement—A Materials Science Perspective. *Biomaterials* **1998**, *19*, 1621–1639. [CrossRef]
19. Li, J.P.; Li, S.H.; Van Blitterswijk, C.A.; De Groot, K. A Novel Porous Ti6A14V: Characterization and Cell Attachment. *J. Biomed. Mater. Res. A* **2005**, *73*, 223–233. [CrossRef]
20. Wilke, A.; Landgraff, M.; Orth, J.; Poenitz, H.; Kienapfel, H.; Boelte, K.; Franke, R.P. Human Bone Marrow Cell Culture: A Sensitive Method for Determination of the Biocompatibility of Implant Materials. *Altern. Lab. Anim.* **1999**, *27*, 137–151. [CrossRef]
21. Pei, B.; Wang, W.; Dunne, N.; Li, X. Applications of Carbon Nanotubes in Bone Tissue Regeneration and Engineering: Superiority, Concerns, Current Advancements, and Prospects. *Nanomaterials* **2019**, *9*, 1501. [CrossRef]
22. Mattioli-Belmonte, M.; Vozzi, G.; Whulanza, Y.; Seggiani, M.; Fantauzzi, V.; Orsini, G.; Ahluwalia, A. Tuning Polycaprolactone-Carbon Nanotube Composites for Bone Tissue Engineering Scaffolds. *Mater. Sci. Eng. C* **2012**, *32*, 152–159. [CrossRef]
23. Venkatesan, J.; Pallela, R.; Kim, S.K. Applications of Carbon Nanomaterials in Bone Tissue Engineering. *J. Biomed. Nanotechnol.* **2014**, *10*, 3105–3123. [CrossRef] [PubMed]
24. Mohan, V.B.; Lau, K.T.; Hui, D.; Bhattacharyya, D. Graphene-Based Materials and Their Composites: A Review on Production, Applications and Product Limitations. *Compos. B Eng.* **2018**, *142*, 200–220. [CrossRef]
25. Pantano, A.; Parks, D.M.; Boyce, M.C. Mechanics of Deformation of Single- and Multi-Wall Carbon Nanotubes. *J. Mech. Phys. Solids* **2004**, *52*, 789–821. [CrossRef]
26. Hu, H.; Ni, Y.; Montana, V.; Haddon, R.C.; Parpura, V. Chemically Functionalized Carbon Nanotubes as Substrates for Neuronal Growth. *Nano Lett.* **2004**, *4*, 507–511. [CrossRef] [PubMed]
27. Chłopek, J.; Czajkowska, B.; Szaraniec, B.; Frackowiak, E.; Szostak, K.; Béguin, F. In Vitro Studies of Carbon Nanotubes Biocompatibility. *Carbon N. Y.* **2006**, *44*, 1106–1111. [CrossRef]
28. Wang, W.; Yokoyama, A.; Liao, S.; Omori, M.; Zhu, Y.; Uo, M.; Akasaka, T.; Watari, F. Preparation and Characteristics of a Binderless Carbon Nanotube Monolith and Its Biocompatibility. *Mater. Sci. Eng. C* **2008**, *28*, 1082–1086. [CrossRef]
29. Akasaka, T.; Yokoyama, A.; Matsuoka, M.; Hashimoto, T.; Abe, S.; Uo, M.; Watari, F. Adhesion of Human Osteoblast-like Cells (Saos-2) to Carbon Nanotube Sheets. *Biomed. Mater. Eng.* **2009**, *19*, 147–153. [CrossRef]

30. Aoki, N.; Yokoyama, A.; Nodasaka, Y.; Akasaka, T.; Uo, M.; Sato, Y.; Tohji, K.; Watari, F. Cell Culture on a Carbon Nanotube Scaffold. *J. Biomed. Nanotechnol.* **2006**, *1*, 402–405. [CrossRef]
31. Terada, M.; Abe, S.; Akasaka, T.; Uo, M.; Kitagawa, Y.; Watari, F. Development of a Multiwalled Carbon Nanotube Coated Collagen Dish. *Dent. Mater. J.* **2009**, *28*, 82–88. [CrossRef] [PubMed]
32. Kroustalli, A.A.; Kourkouli, S.N.; Deligianni, D.D. Cellular Function and Adhesion Mechanisms of Human Bone Marrow Mesenchymal Stem Cells on Multi-Walled Carbon Nanotubes. *Ann. Biomed. Eng.* **2013**, *41*, 2655–2665. [CrossRef] [PubMed]
33. Saito, N.; Usui, Y.; Aoki, K.; Narita, N.; Shimizu, M.; Hara, K.; Ogiwara, N.; Nakamura, K.; Ishigaki, N.; Kato, H.; et al. Carbon Nanotubes: Biomaterial Applications. *Chem. Soc. Rev.* **2009**, *38*, 1897–1903. [CrossRef] [PubMed]
34. Ackun-Farmmer, M.A.; Overby, C.T.; Haws, B.E.; Choe, R.; Benoit, D.S.W. Biomaterials for Orthopedic Diagnostics and Theranostics. *Curr. Opin. Biomed. Eng.* **2021**, *19*, 100308. [CrossRef] [PubMed]
35. Słota, D.; Florkiewicz, W.; Sobczak-Kupiec, A. Ceramic-Polymer Coatings on Ti-6Al-4V Alloy Modified with L-Cysteine in Biomedical Applications. *Mater. Today Commun.* **2020**, *25*, 101301. [CrossRef]
36. Slota, D.; Głąb, M.; Tyliszczak, B.; Dogulas, T.E.L.; Rudnicka, K.; Miernik, K.; Urbaniak, M.M.; Rusek-Wala, P.; Sobczak-upiec, A. Composites Based on Hydroxyapatite and Whey Protein Isolate for Applications in Bone Regeneration. *Materials* **2021**, *14*, 2317. [CrossRef]
37. Malekahmadi, O.; Kalantar, M.; Nouri-Khezrabad, M. Effect of Carbon Nanotubes on the Thermal Conductivity Enhancement of Synthesized Hydroxyapatite Filled with Water for Dental Applications: Experimental Characterization and Numerical Study. *J. Therm. Anal. Calorim.* **2021**, *144*, 2109–2126. [CrossRef]
38. Park, J.E.; Jang, Y.S.; Bae, T.S.; Lee, M.H. Biocompatibility Characteristics of Titanium Coated with Multiwalled Carbon Nanotubes-Hydroxyapatite Nanocomposites. *Materials* **2019**, *12*, 224. [CrossRef]
39. Balani, K.; Anderson, R.; Laha, T.; Andara, M.; Tercero, J.; Crumpler, E.; Agarwal, A. Plasma-Sprayed Carbon Nanotube Reinforced Hydroxyapatite Coatings and Their Interaction with Human Osteoblasts in Vitro. *Biomaterials* **2007**, *28*, 618–624. [CrossRef]

coatings

MDPI

Article

In Vitro Characterization of Doped Bioglass 45S5/HAp Coatings Obtained by CoBlast^{TM} Deposition

Ana Sofia Pádua [1,2,†], Sílvia Rodrigues Gavinho [3,*,†], Tânia Vieira [2], Imen Hammami [3], Jorge Carvalho Silva [2,*], João Paulo Borges [1] and Manuel Pedro Fernandes Graça [3]

1 CENIMAT | i3N, Department of Materials Science, School of Science and Technology, NOVA University Lisbon, 2829-516 Caparica, Portugal; as.padua@campus.fct.unl.pt (A.S.P.); jpb@fct.unl.pt (J.P.B.)
2 CENIMAT | i3N, Department of Physics, School of Science and Technology, NOVA University Lisbon, 2829-516 Caparica, Portugal; ts.vieira@fct.unl.pt
3 I3N and Physics Department, Aveiro University, 3810-193 Aveiro, Portugal; imenhammami@ua.pt (I.H.); mpfg@ua.pt (M.P.F.G.)
* Correspondence: silviagavinho@ua.pt (S.R.G.); jcs@fct.unl.pt (J.C.S.)
† These authors contributed equally to this work.

Abstract: Bone replacement is one of the major medical procedures in the oral surgery field due to the progressive ageing population and to illness or trauma in younger age groups. The use of implants without biological activity and effective osseointegration increases the chances of implant failure. This work aims to improve the interaction between implants and bone by using Bioglass 45S5 (BG)/hydroxyapatite (HAp) mixtures, including copper-, zinc-, and cerium-doped BG, as well as co-doping by the mentioned metals, as coatings produced by the CoBlast^{TM} technique. All coatings present a uniform coverage of the Ti-6Al-4V substrate. Furthermore, in vitro testing using human osteosarcoma Saos-2 cells indicated that BG/HAp coatings have no cytotoxic effect, and the used of doping agents did not alter cell adhesion, proliferation, or alkaline phosphatase (ALP) expression when compared to undoped coating. These results demonstrate that BG/HAp by CoBlast^{TM} can be a solution to improve implants' osseointegration.

Keywords: CoBlast^{TM}; hydroxyapatite; Bioglass 45S5; copper; zinc; cerium; implant coating

1. Introduction

The field of medical implants has witnessed remarkable advancements in recent years, revolutionizing patient care by restoring functionality and improving quality of life. However, the successful integration of these implants within the complex biological system of the human body remains a challenge. Significant research efforts have been invested in implant surface modification to enhance biocompatibility, promote osseointegration, and mitigate potential complications such as biofilm-related infection [1–3]. In fact, implant surfaces, topography, and chemical composition have the greatest influence on the interaction between the biomaterial and bone tissue and, therefore, stability [4–8]. It is well-established that surface roughness, in particular, exhibits capacity in determining a biological response as roughness improves the processes and proliferation of osteoblasts and reduces the activity of osteoclasts, thus promoting the mineralization process [9–11]. In addition, implant roughness contributes to the process of the differentiation of mesenchymal cells towards the osteoblastic phenotype [10,12]. Several surface treatments/coatings have been studied and tested, such as particle blasting, laser processing, and plasma spray [13–16]. Although the coating processing methods are diverse, the CoBlast^{TM} technique from ENBIO, Ireland, has been proven to be highly successful in the biomedical field [17–19]. This technique allows the metal's natural oxide layer removal by abrasion while also depositing the coating at room temperature. This method avoids the problems associated with other deposition techniques, especially problems related to a lack of mechanical stability and adhesion to the substrate [1,17,20]. The Coblast™ process shows very promising characteristics and potential, allowing the achievement of a coating with significantly better mechanical properties, not only in terms of the coating/substrate adhesion, but also of the intra-coating cohesion, at a considerably lower cost.

It has been reported that the use of bioactive glass as a coating material can stimulate the good functioning of the implant due to its ability to enhance tissue integration and promote regeneration [21–24]. Based on the inorganic composition of natural bone, Hench et al. developed Bioglass®, which is capable of replacing damaged bone tissue without being rejected by the human body [23,25,26]. This reaction mechanism arises from the bioactive glass' ionic exchange capability within the surrounding physiological environment, enabling the migration of Ca^{2+} and PO_4^{3-} ions to the surface, forming an initial layer of amorphous calcium phosphate, which subsequently transforms into a crystalline hydroxyapatite layer. Several studies have reported that the insertion of inorganic ions such as copper, zinc, cerium, etc., into the glass network improves its biological response [27–29]. Trace elements play a crucial role in preserving overall health and preventing a variety of diseases. These essential elements are required in small amounts, but their effects on the body's physiological processes are profound. They contribute to enzymatic reactions, cellular functions, and structural integrity maintenance. Trace element deficiencies or imbalances can result in a variety of health problems such as weakened immune system function, impaired cognitive abilities, and bone diseases like osteoporosis [30,31]. One of these elements is copper (Cu), which plays a crucial role in promoting angiogenesis and facilitating the regeneration of both hard and soft tissues. Furthermore, Cu exhibits potential antibacterial properties due to its ability to generate reactive oxygen species (ROS), which can induce oxidative stress and damage bacterial cellular components [32,33]. Cerium (Ce) has garnered growing attention due to its ability to protect cells from damage caused by reactive oxygen species (ROS) and to reverse oxidative stress following implantation in bone, thereby promoting osteogenesis and expediting the bone healing process [34,35]. Zinc (Zn) has antimicrobial properties and significantly contributes to the proliferation of osteoblast cells and the stimulation of mineralization and bone formation by preventing the proliferation of bone-resorbing osteoclasts [36,37].

Although bioglass has achieved remarkable success, its high production cost has remained a significant issue. It has been reported that composite combining bioglass with calcium phosphate-based materials, such as hydroxyapatite (HAp) ($Ca_5(PO_4)_3(OH)$), demonstrates enhanced biological properties compared to their individual phases, as reported in various studies [38,39]. Hydroxyapatite (synthetic or natural), which is the major inorganic component of human bones and teeth, has been used widely in orthopedic and dental applications, due to its excellent biocompatibility, and osteoconductive and osteoinductive properties [40–42]. The synergistic combination of bioglass and hydroxyapatite in coating applications induces rapid bone regeneration and functional integration, ultimately leading to improved clinical outcomes [38,43].

In this work, we attempt to demonstrate the potential of Bioglass 45S5/hydroxyapatite composite coatings using the CoBlast™ technique for application in both orthopedics and dental implants. For this purpose, bioglass modified by the doping and the co-doping of 2% mol Zn, 2% mol Ce, and 0.5% mol Cu was synthesized and subsequently mixed with the HAp. The selection of ion concentrations inserted in the bioglass is justified in our previous work [44–47]. The cytocompatibility of the coatings produced was assessed with the Saos-2 cell line. Moreover, cell adhesion and proliferation and alkaline phosphatase (ALP) activity were evaluated.

2. Materials and Methods

2.1. Materials Synthesis

The bioglass samples, based on the 45S5 formulation proposed by L. Hench ($46.1SiO_2$-$2.6P_2O_5$-$24.35Na_2O$-$26.91CaO$, mol%), were synthesized using the melt-quenching method. In this study, 7 types of samples were developed by adding several concentrations of cerium, zinc, and copper oxides to the 45S5 base formulation [25]. All precursor reagents used were purchased from Sigma Aldrich, USA. The concentrations of each compound are described in Table 1.

Table 1. Molar percentages of each reagent in the different bioglass samples.

	Reagents (mol%)							
	Sample	**SiO_2**	**P_2O_5**	**Na_2O**	**CaO**	**ZnO**	**CeO_2**	**CuO**
	BG	46.10	2.60	24.35	26.91	-	-	-
	2Zn	45.18	2.55	23.86	26.37	2		
Doped	2Ce	45.18	2.55	23.86	26.37		2	
	0.5Cu	45.87	2.59	24.25	26.78			0.5
	2Zn2Ce	44.26	2.49	23.38	25.83	2	2	-
Co-doped	2Zn0.5Cu	44.95	2.54	23.74	26.24	2		0.5
	2Ce 0.5Cu	44.95	2.54	23.74	26.24		2	0.5

Firstly, the high-purity SiO_2, Na_2CO_3, $CaCO_3$, and P_2O_5 starting materials that compose 45S5 (BG) were mixed in a ball mill system with agate vessels and balls for 1 h at 300 rpm. The mixed powders were calcined for 8 h at 800 °C in alumina crucibles. The oxides were added to calcinated BG and mixed at 300 rpm for 1 h. The several compositions were melted in a Pt crucible at 1300 °C for 1 h. The bioglass was re-melted to promote the homogeneity of the glass network. Quenching was carried out in a metal mold and the glass was subsequently ground. The glass was first ground in an agate mortar and then in a ball mill at 300 rpm for 1 h.

For the composites synthesis, all the bioglass compositions were mixed with hydroxyapatite (HAp) in the ball mill system at 200 rpm for 1 h, with the mass ratio shown in Table 2. The commercial hydroxyapatite used in the composites was supplied by Biocerамed, S.A., Portugal.

Table 2. Weight percentage of the bioglass and hydroxyapatite in the composite's samples.

Composites	**Bioglass**	**HAp**
BG/HAp		
2Zn/HAp		
2Ce/HAp		
0.5Cu/HAp	60	40
2Zn2Ce/HAp		
2Zn0.5Cu/HAp		
2Ce0.5Cu/HAp		

2.2. Coatings Synthesis

CoBlast^{TM} was the method used to coat the metal substrates due to its room temperature conditions, the one-step surface preparation, and deposition process. Furthermore, this method is able to coat components with a variety of geometries [19,48].

Figure 1 shows the CoBlast^{TM} equipment with a single nozzle configuration and Sulzer Metco's Single-10C powder feed system.

The substrates used in this study were Ti6Al4V grade 5 alloy (Jacquet, Ovar, Portugal) and had dimensions of 10 mm × 10 mm × 1 mm. Before deposition, the substrates were washed in 1 M HCl and then in acetone in an ultrasonic bath.

Since abrasion and powder deposition were carried out simultaneously, it was also necessary to add the abrasive in a 50/50 ratio to the composites. The abrasive used was Al_2O_3, also supplied by Biocerамed, S.A., Loures, Portugal.

Regarding the deposition conditions, the jet pressure was 4 bar, the distance between the nozzle and the substrate was 19 mm, and the nozzle was positioned at 90° to the substrates.

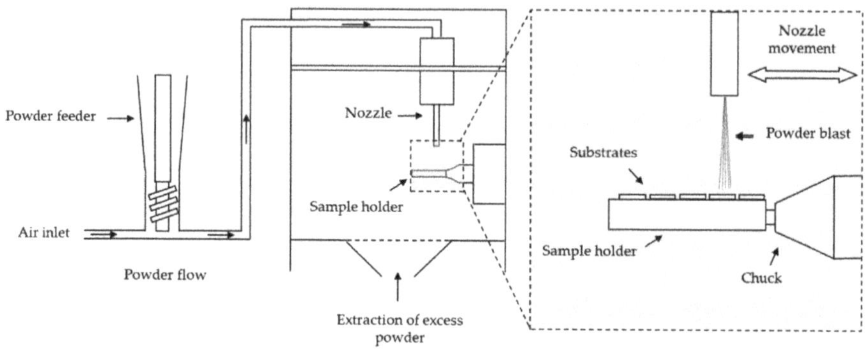

Figure 1. Schematic design of CoBlast^{TM} equipment.

2.3. *Morphological and Structural Characterizations*

In this work, coating surface morphology was assessed by SEM (Scanning Electron Microscopy) using a TESCAN model Vega 3 (TESCAN ORSAY HOLDING, Brno, Czech Republic) microscope. The elemental analysis of the coating surface was also analyzed by Energy Dispersive X-ray Spectroscopy (EDS) using a Bruker QUANTAX detector (model XFlash MIN SVE) at 5 kx and 15 kV.

The powders used as coatings were analyzed by X-ray diffraction (XRD) at room temperature with an Aeris-Panalytical diffractometer with CuKα radiation (λ = 1.54056 Å) at 40 kV and 15 mA. The diffractograms were obtained in a 2θ range of 10° up to 70° and a scan step of 0.002°.

The coatings structure was analyzed by Grazing Incidence X-ray Diffraction (GIXRD) with CuKα radiation—1.5406 Å. The data were collected by a Rigaku SmartLab diffractometer at 40 kV and 30 mA. The patterns were obtained in the range of $10° < 2\theta < 70°$, scan speed of 0.6° min^{-1}, and with an incident angle of 5°.

2.4. *Cell Culture*

2.4.1. Cytotoxicity Assay

A cytotoxicity assay was performed to compare cell viability of the different composite coatings using the extract method. Samples were sterilized at 120 °C for 2 h. For extract preparation, sterilized samples were immersed in McCoy 5A culture medium with an ex-posed area of 1 cm^2 for each ml of medium and then incubated at 37 °C for 48 h.

The human osteosarcoma Saos-2 cells were seeded at a concentration of 30 k cells/cm^2 and incubated for 24h. Afterward, the medium was replaced by the extracts and 4 dilutions (corresponding to equivalent concentrations of 0.5 cm^2/mL, 0.25 cm^2/mL, 0.125 cm^2/mL, and 0.063 cm^2/mL), each with 6 replicates. For the resazurin test, two controls were set. The negative control, where cells were cultured in standard medium, a non-cytotoxic environment, and the positive control, where cells were cultured in a cytotoxic environment, created through the addition of 10% DMSO, a cytotoxic agent, to McCoy culture medium.

The sample extracts and cell controls were incubated for 48 h and afterward the extracts were replaced by a 50% McCoy and 50% resazurin solution that was incubated for 3 h. Cell activity was evaluated by determining the medium absorbance at 570 nm and 600 nm in a microplate reader (Biotek ELx 800UV, Winooski, VT, USA).

2.4.2. Adhesion and Proliferation

The sample ability to support cell metabolism was assessed through cell adhesion and proliferation studies. Samples were sterilized as previously described for the cytotoxicity assay.

The human osteosarcoma Saos-2 cells were seeded at a concentration of 20 k cells/cm^2 over the coating's surface and directly on the wells for the cell controls. Seeded cells were incubated at 37 °C in a humidified 5% CO_2 atmosphere for 24 h in McCoy's medium.

Afterwards, cell adhesion rate was assessed using a resazurin absorbance. The resazurin assay was repeated at 3, 5, 7, and 9 days to follow cell proliferation behavior for all tested samples.

2.4.3. Alkaline Phosphatase Activity

ALP is an enzyme expressed by cells during osteogenesis, so it can be used as a differentiation marker. A colorimetric assay was used to evaluate ALP expression. This reaction used 1 mg/ml of 4-nitrophenyl phosphate disodium salt (Sigma-Aldrich, St. Louis, MO, USA) dissolved in tris-hydrochloric acid solution (pH 8.7). The first step in this essay consists of filtering the medium that was in contact with the samples, to remove any cell debris or dead cells. Then the absorbance at 405 nm was read to obtain the baseline. Next, the ALP solution was added in a 1:1 ratio to the medium and incubated for 30 min. Finally, the absorbance was measured at 405 nm.

The results were normalized to the populations determined on the previous day.

2.4.4. Immunofluorescence Study

For the immunofluorescence staining, the samples were fixed with paraformaldehyde for 20 min, washed with PBS, permeabilized using 0.5% Triton X-100 in PBS for 15 min, and washed again in PBS. Actin staining was carried out by incubating the cells in phalloidin conjugate (Phalloidin CruzFluor™ 488, Santa Cruz Biotechnology, Dallas, TX, USA) diluted in PBS in a 1:1000 ratio for 30 min in the dark and then the DNA was counterstained with 10 μg/mL DAPI for 5 min in the dark.

Actin and cell nuclei were observed using a Nikon Eclipse Ti-S fluorescence microscope equipped with a Nikon D610 digital camera. Images were obtained with a 40x objective.

2.4.5. Statistical Analysis

All data related to in vitro evaluation were statistically analyzed using the software OriginPro 2018 and presented using mean ± SD. Furthermore, the statistical analysis was performed by one-way analysis of variance (One-way ANOVA) using Tukey's multiple comparison test. If the results presented $p < 0.05$, samples were accepted as significantly different.

3. Results

3.1. Morphological and Structural Characterizations

SEM micrographs were taken to assess the surface morphology. Figure 2 shows the surfaces of all the coatings obtained by CoBlast^{TM} deposition.

All the samples have an identical morphology and show evidence of considerable roughness. The roughness caused by the abrasive allows for the tribochemical bonding and mechanical entrapment of the bioglass. The level of roughness promoted by this technique is also extremely useful for integrating implants into the bone [1,49].

Figure 3 shows the elemental analysis in mapping mode by EDS of the BG/HAp (a)–(f), 2Zn/HAp (g), 2Ce/HAp (h), and 0.5Cu/HAp (i) coatings. Figure 3a shows the presence of the alumina associated with the abrasive. The common elements of Bioglass 45S5 and hydroxyapatite (Ca and P) are presented in Figure 3c,d. The Si and Na presented in Figure 3e,f can distinguish Bioglass 45S5 from hydroxyapatite. The surface maps of the 2Zn/HAp, 2Ce/Hap, and 0.5Cu/HAp coatings were also produced, showing a homogeneous distribution of the additional elements in the Bioglass 45S5 network.

Figure 4 shows the diffractogram of the BG/Hap-coated sample using GIXRD. A significant contribution of Ti from the substrate and also the presence of alumina used as the abrasive is visible. Furthermore, it is possible to visualize the presence of hydroxyapatite. Bioglass, being an amorphous material, is not detected in the BG/HAp-coated sample.

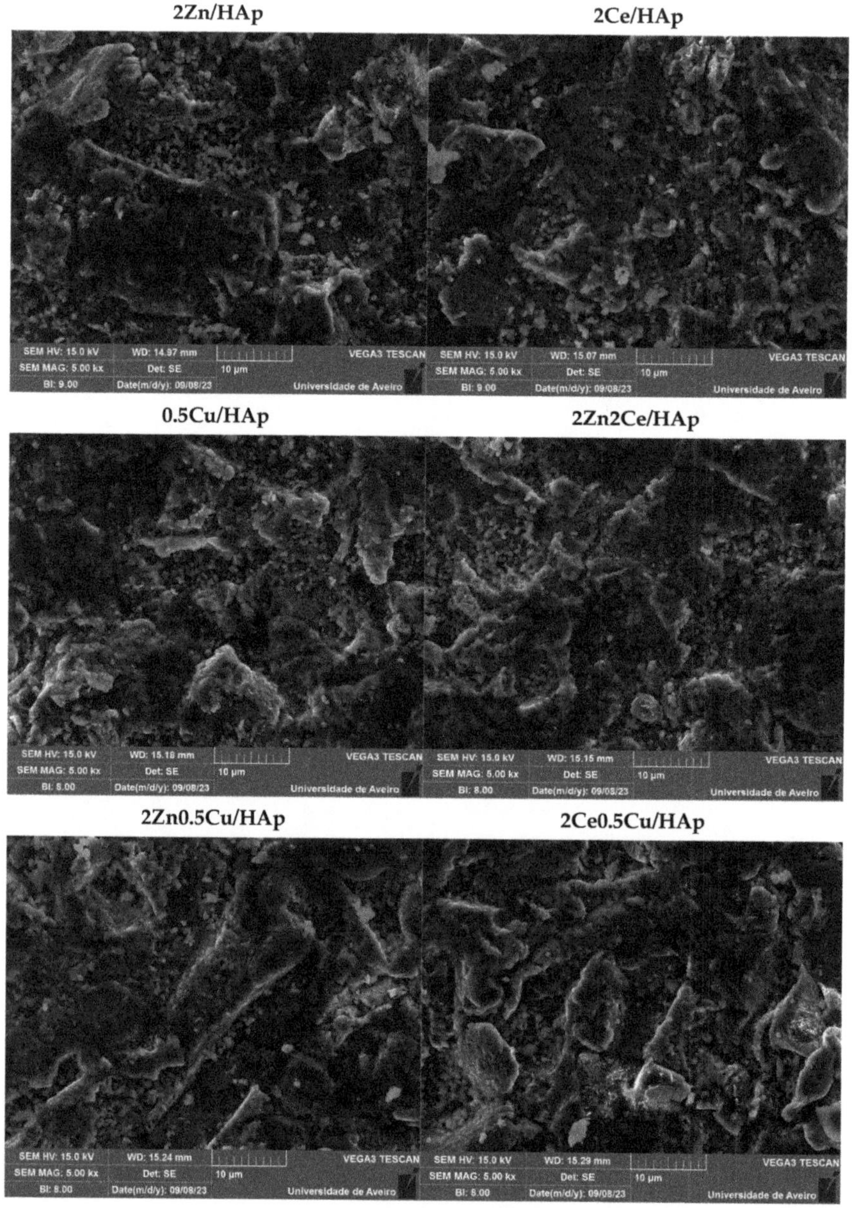

Figure 2. SEM micrographs of the coating surface (SEM magnification: 5kx).

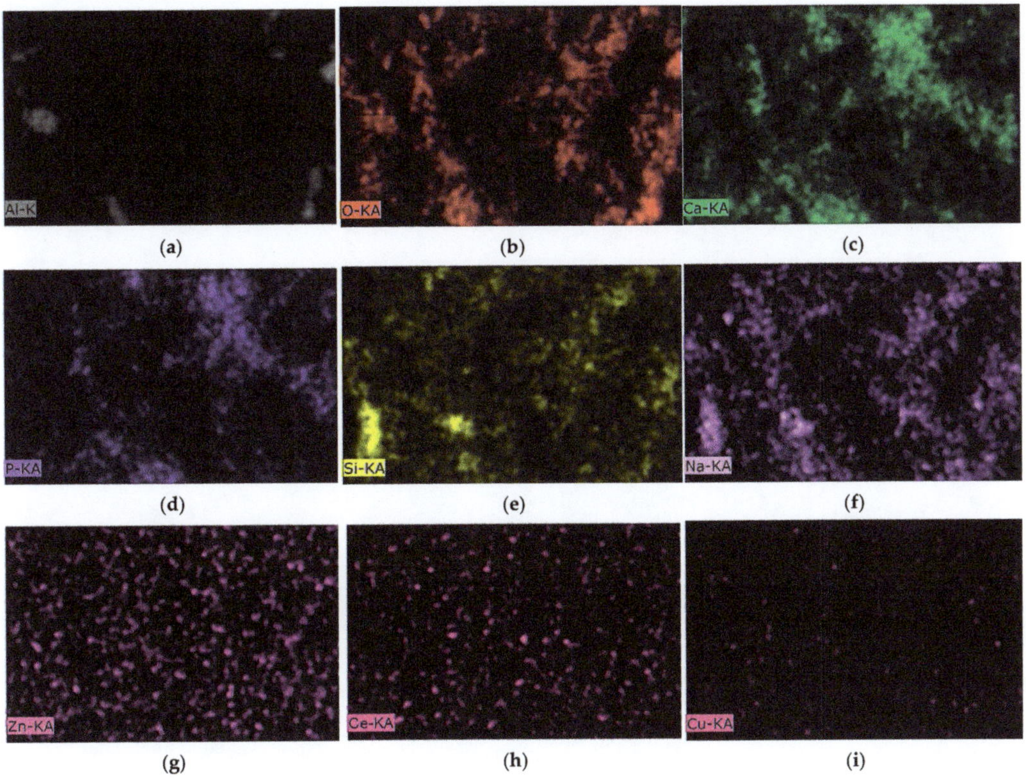

Figure 3. EDS mapping of BG/HAp samples (**a–f**), and sample and doping distribution of 2Zn/HAp (**g**), 2Ce/HAp (**h**), and 0.5Cu/HAp (**i**) (magnification: 5 kx at 15 kV).

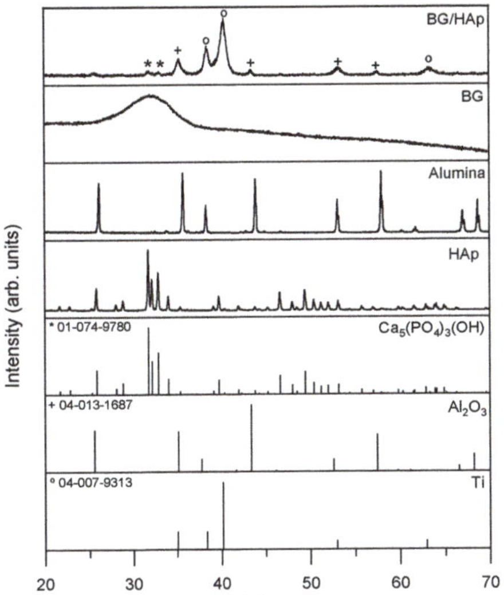

Figure 4. GIXRD of BG/HAp coating, XRD of powders (BG, Hap, and Alumina) used in CoBlast^{TM} deposition, and peak list of $Ca_5(PO_4)_3(OH)$ (ICDD: 01-074-9780), Al_2O_3 (ICDD: 04-013-1687), and Ti (IDCC: 04-007-9313).

3.2. Cell Culture

3.2.1. Cytotoxicity Assay

Cytotoxicity assays are one of the first steps in a biocompatibility assessment of a biomaterial. In this assay, the extracts obtained from the different BG/HAp coatings were placed in contact with human osteosarcoma Saos-2 cells. The cell populations were measured using the resazurin assay. The goal of this assay is to confirm that the composite coatings containing doped BG can be safely placed in contact with the organism without any harmful effects. The results obtained are presented in Figure 5.

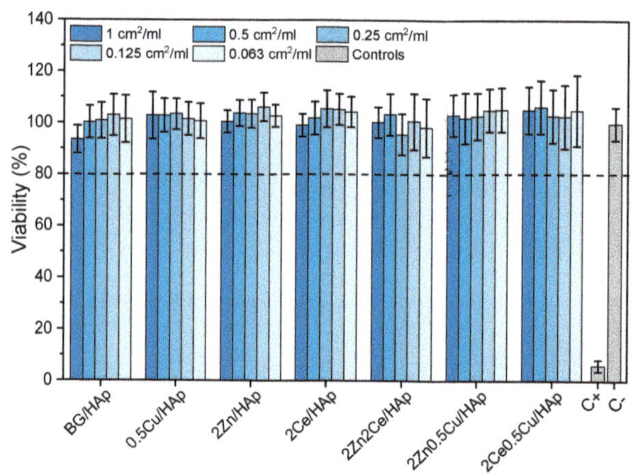

Figure 5. Relative cell viability after 48 h incubation with undoped, doped, and co-doped BG-coated sample extracts. C− is the negative and C+ is the positive control, both represented in gray. The dashed line is the minimum percentage of cell viability for which extracts are not deemed cytotoxic.

The results evidence that all tested coatings produced by CoBlast™ have relative cell viabilities above 80% in comparison with the negative control, which suggests that none of the materials tested provoked cell death and, therefore, are safe to be used as coatings for implants. The viability was as expected, since all BG powders used had already proved to be viable at concentrations up to 12.5 mg/mL for the BG, 2Ce, and 0.5Cu samples, and 50 mg/mL for the 2Zn sample [44,45,47].

The results obtained are corroborated by the findings previously described by Dias et al. where HA coatings by CoBlast™ were viable at concentrations equal or lower than 1 cm^2/mL [48].

3.2.2. Adhesion and Proliferation

The adhesion and proliferation of Saos-2 cells were determined to better understand the response of bone cells to the coated implants as well as foresee their ability to assist osteointegration in vivo. The results are shown in Figure 6.

The cell adhesion assessment shows that all samples attained similar adhesion rates to the BG/HAp coating, meaning that none of the doping agents hindered the cell's ability to connect with the coating's surface. Furthermore, the 2Ce0.5Cu/HAp coating presented a significant improvement in cell adhesion.

The proliferation analysis shows a progressive increase in cell population for all samples tested. After 14 days, cells already reached the confluency stage, where there is no more available space to further proliferate. At this stage, the only factor to determinate cell populations is sample surface area. Since all coatings were produced by CoBlast™ and powders used have the same dimensions, the resulting topographies of all coated samples are similar, as was previously established in the SEM analysis, and therefore, all samples have similar final proliferation results.

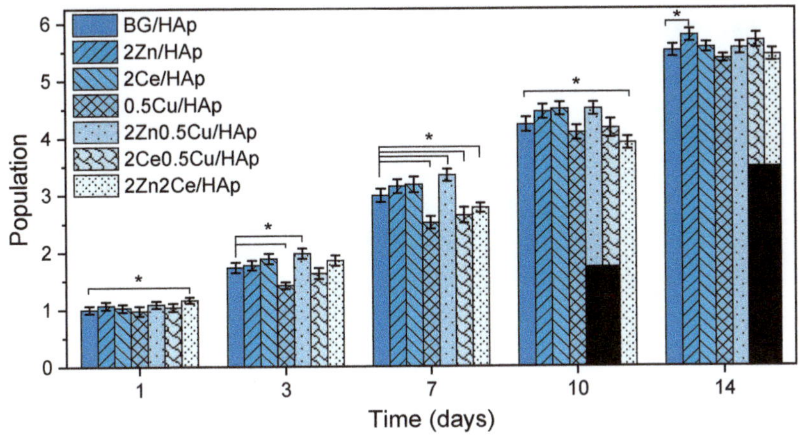

Figure 6. Comparison of populations of human osteosarcoma Saos-2 cells cultured for 14 days on all coatings. Absorbance values are normalized to the average absorbance of the BG/HAp sample on day 1. The vertical lines (whiskers) represent the standard deviation of the mean. The results of the statistical significance tests are represented with *, where $p < 0.05$.

O'Sullivan et al. [50] and Mesquita-Guimarães et al. [51] already demonstrated that cell proliferation was enhanced with coatings implementation: fibroblasts cell proliferation was studied over 7 days on Ti substrates, and Hap or BG coatings, showing a favorable response toward coated samples especially in BG samples, when compared with Ti substrates.

Cell population analysis in the first week indicated that the presence of Cu in the coatings tends to lead to worse results due to Cu's cytotoxic behavior [45,52], which can be seen in the 0.5Cu/HAp. However, with the implementation of co-doping this effect can be mitigated with the introduction of ions with osteogenic or anti-inflammatory properties [53–55]. In this test, 2Zn0.5Cu/HAp samples presented an increase in cell population compared to the 0.5Cu/HAp coating.

Overall, all samples evidenced a favorable cell response with high human osteosarcoma Saos-2 cells, and doping agents did not jeopardize proliferation and can also improve cells responses compared to uncoated metallic implants.

3.2.3. Alkaline Phosphatase Activity

ALP is an enzyme that reflects osteoblastic activity at different stages of osteoblast differentiation. During bone regeneration there is an upregulation of ALP expression, which acts as a local concentration regulator of inorganic ions, as well as a catalyst of apatite crystals growth [56,57]. These properties lead to biomineralization and bone matrix production [58]. The relative ALP expressed by the human osteosarcoma Saos-2 cells cultured on the seven different coating throughout the 15 days of culture time is displayed in Figure 7.

All cells grown on coatings revealed an increase in relative ALP expression throughout time. However, there are no statistically significant differences between samples, not even when comparing doped sample with the BG/HAp coating. This behavior was not anticipated since Zn- [59,60] and Cu-doped [61,62] BG are supposed to increase ALP expression. The lack of statistically significant differences between samples may be due to the need to increase doping concentrations since the amount of material present in each coating is low (2 mg/cm^2), which may lead to the release of minimal amounts of doping ions with very little impact on ALP expression.

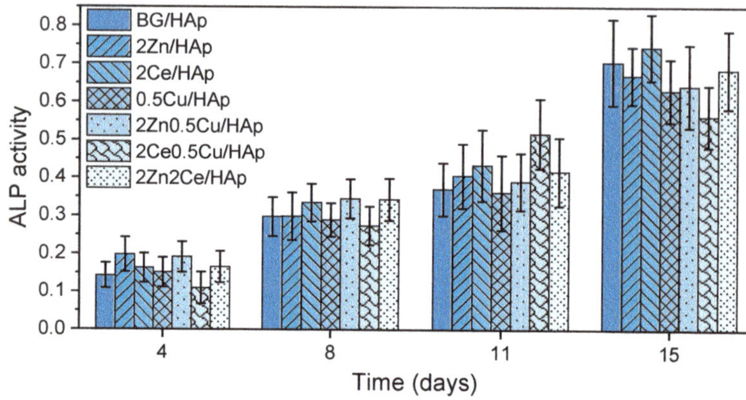

Figure 7. Relative ALP activity of human osteosarcoma Saos-2 cells on the different coatings on days 4, 8, 11, and 15. The data are normalized to the populations determined on the previous day. There are no statistically significant differences when comparing samples on each day.

3.2.4. Immunofluorescence Study

Immunofluorescence analysis is one of the most commonly used methods to study the shape, structure, size, and cell distribution on sample surfaces. For this purpose, human osteosarcoma Saos-2 cells were stained with phalloidin and DAPI to label the cytoskeleton and nuclear morphology, respectively. The cell cultures' immunofluorescence results are shown in Figure 8.

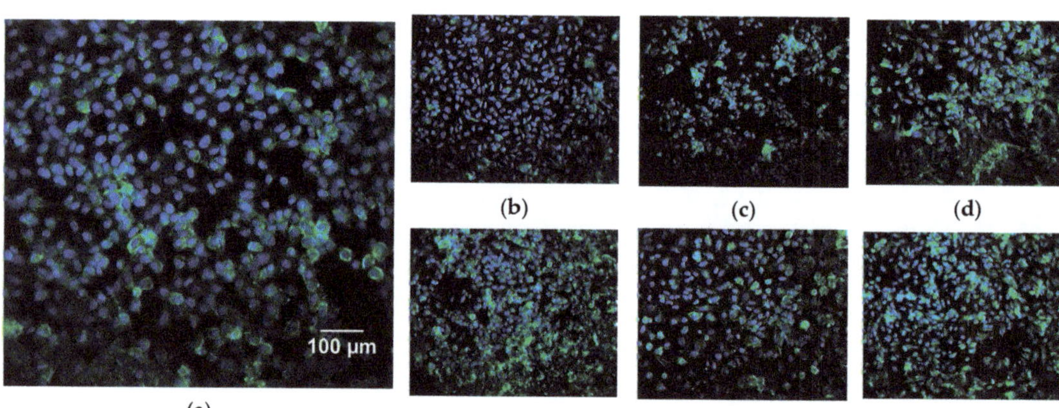

Figure 8. Immunofluorescence images after 15 days of culture of human osteosarcoma Saos-2 cells for different sample surfaces: BG/HAp (**a**), 2Zn/HAp (**b**), 2Ce/HAp (**c**), 0.5Cu/HAp (**d**), 2Zn2Ce/HAp (**e**), 2Zn0.5Cu/HAp (**f**), and 2Ce0.5Cu/HAp (**g**) by CoBlast^{TM} deposition. The cytoskeleton is stained with phalloidin (green) and the nucleus is stained with DAPI (blue).

All coated samples present a high cell confluency, observable by the number of nuclei (blue dots). This fact was already expected, as it is in accordance with cell populations calculated by resazurin readings. Regarding cell morphology, the confluency stage and coating topography alters cell cytoskeleton morphology, due to spatial constraints. These spatial constraints, such as the space available for each cell, sharp edges, sharp angles, and fine textural patterns, leads to the cytoskeleton occupying a smaller area with rounded geometry that is only slightly larger than the nucleus size, instead of the polygonal or polarized fibroblastic morphology typical of Saos-2 cell line morphology [50,63].

Samples (c), (e), and (g) in Figure 8 present cells that occupy a smaller area than on other samples. These three samples have in common the presence of cerium-doped, or co-doped, bioglass. The smaller area shows that cells on these samples have a higher difficulty interacting with the surface. The cause for this is unknown and does not seem to negatively affect cell adhesion, proliferation, nor ALP production.

4. Conclusions

Several BG/Hap- and BG-doped and co-doped/HAp coatings were successfully produced by CoBlast^{TM}. All coatings presented a similar surface morphology. The in vitro response showed no cytotoxic effects, excellent proliferation, and ALP expression. Cell proliferation and immunofluorescence assays demonstrated that none of the doping agents had a negative effect on cell populations and, within 14 days, all samples reached confluency. The effect of cerium doping of cell size and morphology should be further investigated. Overall, the results reveal that BG/HAp composite coatings produced by the Coblast^{TM} technique can improve the interaction between implants and bone for orthopedic and dental applications.

Author Contributions: Conceptualization, A.S.P. and S.R.G.; methodology, A.S.P., S.R.G., T.V. and I.H.; software, A.S.P. and S.R.G.; validation, M.P.F.G. and J.C.S.; formal analysis, A.S.P. and S.R.G.; investigation, A.S.P., S.R.G. and T.V.; data curation, A.S.P., S.R.G. and T.V.; writing—original draft preparation, A.S.P., S.R.G. and I.H.; writing—review and editing, M.P.F.G., J.C.S. and J.P.B.; supervision, J.C.S. and M.P.F.G. All authors have read and agreed to the published version of the manuscript.

Funding: FEDER funds through the COMPETE 2020 Program and National Funds through FCT—Portuguese Foundation for Science and Technology under the project LIS-BOA-01-0247-FEDER-039985/POCI-01-0247-FEDER-039985, LA/P/0037/2020, UIDP/50025/2020, and UIDB/50025/2020 of the Associate Laboratory Institute of Nanostructures, Nanomodelling, and Nanofabrication—i3N. S.R. Gavinho and A. Sofia Pádua acknowledge FCT—the Portuguese Foundation for Science and Technology for the PhD grant (SFRH/BD/148233/2019 and UI/BD/151287/2021, respectively).

Institutional Review Board Statement: Not applicable.

Informed Consent Statement: Not applicable.

Data Availability Statement: Not applicable.

Conflicts of Interest: The authors declare no conflict of interest.

References

1. Barry, J.N.; Twomey, B.; Cowley, A.; O'Neill, L.; McNally, P.J.; Dowling, D.P. Evaluation and Comparison of Hydroxyapatite Coatings Deposited Using Both Thermal and Non-Thermal Techniques. *Surf. Coatings Technol.* **2013**, *226*, 82–91. [CrossRef]
2. Dong, H.; Liu, H.; Zhou, N.; Li, Q.; Yang, G.; Chen, L.; Mou, Y. Surface Modified Techniques and Emerging Functional Coating of Dental Implants. *Coatings* **2020**, *10*, 1012. [CrossRef]
3. Joy-anne, N.O.; Su, Y.; Lu, X.; Kuo, P.H.; Du, J.; Zhu, D. Bioactive Glass Coatings on Metallic Implants for Biomedical Applications. *Bioact. Mater.* **2019**, *4*, 261–270. [CrossRef]
4. Guglielmotti, M.B.; Olmedo, D.G.; Cabrini, R.L. Research on Implants and Osseointegration. *Periodontology 2000* **2019**, *79*, 178–189. [CrossRef]
5. Liu, W.; Liu, S.; Wang, L. Surface Modification of Biomedical Titanium Alloy: Micromorphology, Microstructure Evolution and Biomedical Applications. *Coatings* **2019**, *9*, 249. [CrossRef]
6. Wang, Q.; Zhou, P.; Liu, S.; Attarilar, S.; Ma, R.L.W.; Zhong, Y.; Wang, L. Multi-Scale Surface Treatments of Titanium Implants for Rapid Osseointegration: A Review. *Nanomaterials* **2020**, *10*, 1244. [CrossRef]
7. Le Guéhennec, L.; Soueidan, A.; Layrolle, P.; Amouriq, Y. Surface Treatments of Titanium Dental Implants for Rapid Osseointegration. *Dent. Mater.* **2007**, *23*, 844–854. [CrossRef]
8. Gruber, R.; Bosshardt, D.D. *Dental Implantology and Implants—Tissue Interface*; Elsevier Inc.: Amsterdam, The Netherlands, 2015; ISBN 9780123977786.
9. Matos, G.R.M. Surface Roughness of Dental Implant and Osseointegration. *J. Maxillofac. Oral Surg.* **2021**, *20*, 1–4. [CrossRef]
10. Zhang, Y.; Chen, S.E.; Shao, J.; Van Den Beucken, J.J.J.P. Combinatorial Surface Roughness Effects on Osteoclastogenesis and Osteogenesis. *ACS Appl. Mater. Interfaces* **2018**, *10*, 36652–36663. [CrossRef]
11. Boyan, B.D.; Lotz, E.M.; Schwartz, Z. Roughness and Hydrophilicity as Osteogenic Biomimetic Surface Properties. *Tissue Eng. Part A* **2017**, *23*, 1479–1489. [CrossRef]

12. Kearns, V.R.; Williams, R.L.; Mirvakily, F.; Doherty, P.J.; Martin, N. Guided Gingival Fibroblast Attachment to Titanium Surfaces: An in Vitro Study. *J. Clin. Periodontol.* **2013**, *40*, 99–108. [CrossRef]
13. Dehghanghadikolaei, A.; Fotovvati, B. Coating Techniques for Functional Enhancement of Metal Implants for Bone Replacement: A Review. *Materials* **2019**, *12*, 1795. [CrossRef]
14. Prezas, P.R.; Soares, M.J.; Borges, J.P.; Silva, J.C.; Oliveira, F.J.; Graça, M.P.F. Bioactivity Enhancement of Plasma-Sprayed Hydroxyapatite Coatings through Non-Contact Corona Electrical Charging. *Nanomaterials* **2023**, *13*, 1058. [CrossRef]
15. Khan, A.A.; Al Kheraif, A.A.; Alhijji, S.M.; Matinlinna, J.P. Effect of Grit-Blasting Air Pressure on Adhesion Strength of Resin to Titanium. *Int. J. Adhes. Adhes.* **2016**, *65*, 41–46. [CrossRef]
16. Roy, M.; Vamsi Krishna, B.; Bandyopadhyay, A.; Bose, S. Laser Processing of Bioactive Tricalcium Phosphate Coating on Titanium for Load-Bearing Implants. *Acta Biomater.* **2008**, *4*, 324–333. [CrossRef]
17. Dunne, C.F.; Twomey, B.; O'Neill, L.; Stanton, K.T. Co-Blasting of Titanium Surfaces with an Abrasive and Hydroxyapatite to Produce Bioactive Coatings: Substrate and Coating Characterisation. *J. Biomater. Appl.* **2014**, *28*, 767–778. [CrossRef]
18. Tan, F.; Naciri, M.; Dowling, D.; Al-Rubeai, M. In Vitro and in Vivo Bioactivity of CoBlast Hydroxyapatite Coating and the Effect of Impaction on Its Osteoconductivity. *Biotechnol. Adv.* **2012**, *30*, 352–362. [CrossRef]
19. CoBlast-ENBIO. Available online: https://www.enbio.eu/coblast/ (accessed on 6 September 2023).
20. Dunne, C.F.; Twomey, B.; Stanton, K.T. Effect of a Blast Coating Process on the Macro- and Microstructure of Grade 5 Titanium Foam. *Mater. Lett.* **2015**, *147*, 75–78. [CrossRef]
21. Bano, S.; Romero, A.R.; Grant, D.M.; Nommeots-Nomm, A.; Scotchford, C.; Ahmed, I.; Hussain, T. In-Vitro Cell Interaction and Apatite Forming Ability in Simulated Body Fluid of ICIE16 and 13–93 Bioactive Glass Coatings Deposited by an Emerging Suspension High Velocity Oxy Fuel (SHVOF) Thermal Spray. *Surf. Coatings Technol.* **2021**, *407*, 126764. [CrossRef]
22. Jones, J.R.; Brauer, D.S.; Hupa, L.; Greenspan, D.C. Bioglass and Bioactive Glasses and Their Impact on Healthcare. *Int. J. Appl. Glas. Sci.* **2016**, *7*, 423–434. [CrossRef]
23. Jones, J.R. Reprint of: Review of Bioactive Glass: From Hench to Hybrids. *Acta Biomater.* **2015**, *23*, 53–82. [CrossRef] [PubMed]
24. Zafar, M.S.; Farooq, I.; Awais, M.; Najeeb, S.; Khurshid, Z.; Zohaib, S. *Bioactive Surface Coatings for Enhancing Osseointegration of Dental Implants*; Elsevier Ltd.: Amsterdam, The Netherlands, 2019; ISBN 9780081021965.
25. Hench, L.L. The Story of Bioglass®. *J. Mater. Sci. Mater. Med.* **2006**, *17*, 967–978. [CrossRef] [PubMed]
26. Gavinho, S.R.; Prezas, P.R.; Graça, M.P.F. Synthesis, Structural and Electrical Properties of the 45S5 Bioglass®. In *Electrical Measurements: Introduction, Concepts and Applications*; Nova Science Publisher: New York, NY, USA, 2017; ISBN 9781536129748.
27. Chitra, S.; Bargavi, P.; Balasubramaniam, M.; Chandran, R.R.; Balakumar, S. Impact of Copper on In-Vitro Biomineralization, Drug Release Efficacy and Antimicrobial Properties of Bioactive Glasses. *Mater. Sci. Eng. C* **2020**, *109*, 110598. [CrossRef] [PubMed]
28. Gavinho, S.R.; Soares, M.C.; Borges, J.P.; Silva, J.C.; Nogueira, I.S.; Graça, M.P.F. Preparation and Characterization of Zinc and Magnesium Doped Bioglasses. In *NATO Science for Peace and Security Series B: Physics and Biophysics*; Petkov, P., Achour, M.E., Popov, C., Eds.; Springer: Dordrecht, The Netherlands, 2020; pp. 465–475.
29. Zambon, A.; Malavasi, G.; Pallini, A.; Fraulini, F.; Lusvardi, G. Cerium Containing Bioactive Glasses: A Review. *ACS Biomater. Sci. Eng.* **2021**, *7*, 4388–4401. [CrossRef]
30. Shetty, S.R.; Babu, S.; Kumari, S.; Shetty, P.; Hegde, S.; Karikal, A. Status of Trace Elements in Saliva of Oral Precancer and Oral Cancer Patients. *J. Cancer Res. Ther.* **2015**, *11*, 146–149. [CrossRef]
31. Gumienna-Kontecka, E.; Rowińska-Żyrek, M.; Łuczkowski, M. The Role of Trace Elements in Living Organisms. In *Recent Advances in Trace Elements*; Chojnacka, K., Saeid, A., Eds.; Wiley: New York, NY, USA, 2018; pp. 177–206.
32. Salah, I.; Parkin, I.P.; Allan, E. Copper as an Antimicrobial Agent: Recent Advances. *RSC Adv.* **2021**, *11*, 18179–18186. [CrossRef]
33. Baino, F. Copper-Doped Ordered Mesoporous Bioactive Glass: A Promising Multifunctional Platform for Bone Tissue Engineering. *Bioengineering* **2020**, *7*, 45. [CrossRef]
34. Zheng, K.; Torre, E.; Bari, A.; Taccardi, N.; Cassinelli, C.; Morra, M.; Fiorilli, S.; Vitale-Brovarone, C.; Iviglia, G.; Boccaccini, A.R. Antioxidant Mesoporous Ce-Doped Bioactive Glass Nanoparticles with Anti-Inflammatory and pro-Osteogenic Activities. *Mater. Today Bio* **2020**, *5*, 100041. [CrossRef]
35. Malavasi, G.; Salvatori, R.; Zambon, A.; Lusvardi, G.; Rigamonti, L.; Chiarini, L.; Anesi, A. Cytocompatibility of Potential Bioactive Cerium-Doped Glasses Based on 45S5. *Materials* **2019**, *12*, 594. [CrossRef]
36. Chen, Y.H.; Tseng, S.P.; Wu, S.M.; Shih, C.J. Structure-Dependence of Anti-Methicillin-Resistant Staphylococcus Aureus (MRSA) Activity on ZnO-Containing Bioglass. *J. Alloys Compd.* **2020**, *848*, 156487. [CrossRef]
37. Neščáková, Z.; Zheng, K.; Liverani, L.; Nawaz, Q.; Galusková, D.; Kaňková, H.; Michálek, M.; Galusek, D.; Boccaccini, A.R. Multifunctional Zinc Ion Doped Sol—Gel Derived Mesoporous Bioactive Glass Nanoparticles for Biomedical Applications. *Bioact. Mater.* **2019**, *4*, 312–321. [CrossRef] [PubMed]
38. Palierse, E.; Roquart, M.; Norvez, S.; Corté, L. Coatings of Hydroxyapatite-Bioactive Glass Microparticles for Adhesion to Biological Tissues. *RSC Adv.* **2022**, *12*, 21079–21091. [CrossRef]
39. Bian, T.; Wang, L.; Xing, H. Preparation and Biological Assessment of a ZrO_2-Based Bone Scaffold Coated with Hydroxyapatite and Bioactive Glass Composite. *Mater. Chem. Phys.* **2021**, *267*, 124616. [CrossRef]

40. Ielo, I.; Calabrese, G.; De Luca, G.; Conoci, S. Recent Advances in Hydroxyapatite-Based Biocomposites for Bone Tissue Regeneration in Orthopedics. *Int. J. Mol. Sci.* **2022**, *23*, 9721. [CrossRef]
41. Gavinho, S.R.; Bozdag, M.; Kalkandelen, C.; Regadas, J.S.; Jakka, S.K.; Gunduz, O.; Oktar, F.N.; Graça, M.P.F. An Eco-Friendly Process to Extract Hydroxyapatite from Sheep Bones for Regenerative Medicine: Structural, Morphologic and Electrical Studies. *J. Funct. Biomater.* **2023**, *14*, 279. [CrossRef] [PubMed]
42. Cui, W.; Yang, L.; Ullah, I.; Yu, K.; Zhao, Z.; Gao, X.; Liu, T.; Liu, M.; Li, P.; Wang, J.; et al. Biomimetic Porous Scaffolds Containing Decellularized Small Intestinal Submucosa and Sr^{2+}/Fe^{3+}co-Doped Hydroxyapatite Accelerate Angiogenesis/Osteogenesis for Bone Regeneration. *Biomed. Mater.* **2022**, *17*, 025008. [CrossRef]
43. Bellucci, D.; Sola, A.; Anesi, A.; Salvatori, R.; Chiarini, L.; Cannillo, V. Bioactive Glass/Hydroxyapatite Composites: Mechanical Properties and Biological Evaluation. *Mater. Sci. Eng. C* **2015**, *51*, 196–205. [CrossRef]
44. Gavinho, S.R.; Pádua, A.S.; Sá-Nogueira, I.; Silva, J.C.; Borges, J.P.; Costa, L.C.; Graça, M.P.F. Fabrication, Structural and Biological Characterization of Zinc-Containing Bioactive Glasses and Their Use in Membranes for Guided Bone Regeneration. *Materials* **2023**, *16*, 956. [CrossRef]
45. Hammami, I.; Gavinho, S.R.; Jakka, S.K.; Valente, M.A.; Graça, M.P.F.; Pádua, A.S.; Silva, J.C.; Sá-Nogueira, I.; Borges, J.P. Antibacterial Biomaterial Based on Bioglass Modified with Copper for Implants Coating. *J. Funct. Biomater.* **2023**, *14*, 369. [CrossRef]
46. Gavinho, R.; Miguel, B.; Melo, G.; Silva, J.C.; Pedro, M.; Graça, F. Thermal, Structural, Morphological and Electrical Characterization of Cerium-Containing 45S5 for Metal Implant Coatings. *Coatings* **2023**, *13*, 294. [CrossRef]
47. Gavinho, S.R.; Pádua, A.S.; Sá-Nogueira, I.; Silva, J.C.; Borges, J.P.; Costa, L.C.; Graça, M.P.F. Biocompatibility, Bioactivity, and Antibacterial Behaviour of Cerium-Containing Bioglass®. *Nanomaterials* **2022**, *12*, 4479. [CrossRef] [PubMed]
48. Dias, I.J.G.; Pádua, A.S.; Pires, E.A.; Borges, J.P.M.R.; Silva, J.C.; Lança, M.C. Hydroxyapatite-Barium Titanate Biocoatings Using Room Temperature Coblasting. *Crystals* **2023**, *13*, 579. [CrossRef]
49. Flanagan, J.; Schütze, P.; Dunne, C.; Twomey, B.; Stanton, K.T. Use of a Blast Coating Process to Promote Adhesion between Aluminium Surfaces for the Automotive Industry. *J. Adhes.* **2020**, *96*, 580–601. [CrossRef]
50. O'Sullivan, C.; O'Hare, P.; Byrne, G.; O'Neill, L.; Ryan, K.B.; Crean, A.M.; O'Sullivan, C.; O'Hare, P.; Byrne, G.; O'Neill, L.; et al. A Modified Surface on Titanium Deposited by a Blasting Process. *Coatings* **2011**, *1*, 53–71. [CrossRef]
51. Mesquita-Guimarães, J.; Detsch, R.; Souza, A.C.; Henriques, B.; Silva, F.S.; Boccaccini, A.R.; Carvalho, O. Cell Adhesion Evaluation of Laser-Sintered HAp and 45S5 Bioactive Glass Coatings on Micro-Textured Zirconia Surfaces Using MC3T3-E1 Osteoblast-like Cells. *Mater. Sci. Eng. C* **2020**, *109*, 110492. [CrossRef]
52. Seo, J.J.; Mandakhbayar, N.; Kang, M.S.; Yoon, J.; Lee, N.-H.; Ahn, J.; Lee, H.-H.; Lee, J.; Kim, H.-W. Antibacterial, Proangiogenic, and Osteopromotive Nanoglass Paste Coordinates Regenerative Process Following Bacterial Infection in Hard Tissue. *Biomaterials* **2021**, *268*, 120593. [CrossRef]
53. Ali, A.; Ershad, M.; Vyas, V.K.; Hira, S.K.; Manna, P.P.; Singh, B.N.; Yadav, S.; Srivastava, P.; Singh, S.P.; Pyare, R. Studies on Effect of CuO Addition on Mechanical Properties and in Vitro Cytocompatibility in 1393 Bioactive Glass Scaffold. *Mater. Sci. Eng. C* **2018**, *93*, 341–355. [CrossRef]
54. Weng, L.; Boda, S.K.; Teusink, M.J.; Shuler, F.D.; Li, X.; Xie, J. Binary Doping of Strontium and Copper Enhancing Osteogenesis and Angiogenesis of Bioactive Glass Nanofibers While Suppressing Osteoclast Activity. *ACS Appl. Mater. Interfaces* **2017**, *9*, 24484–24496. [CrossRef]
55. Moghanian, A.; Ghorbanoghli, A.; Kazem-Rostami, M.; Pazhouheshgar, A.; Salari, E.; Saghafi Yazdi, M.; Alimardani, T.; Jahani, H.; Sharifian Jazi, F.; Tahriri, M. Novel Antibacterial Cu/Mg-substituted 58S-bioglass: Synthesis, Characterization and Investigation of in Vitro Bioactivity. *Int. J. Appl. Glas. Sci.* **2020**, *11*, 685–698. [CrossRef]
56. Tavares, F.J.T.M.; Soares, P.I.P.; Silva, J.C.; Borges, J.P. Preparation and In Vitro Characterization of Magnetic CS/PVA/HA/PSPIONs Scaffolds for Magnetic Hyperthermia and Bone Regeneration. *Int. J. Mol. Sci.* **2023**, *24*, 1128. [CrossRef]
57. Murray, E.; Provvedini, D.; Curran, D.; Catherwood, B.; Sussman, H.; Manolagas, S. Characterization of a Human Osteoblastic Osteosarcoma Cell Line (SAOS-2) with High Bone Alkaline Phosphatase Activity. *J. Bone Miner. Res.* **2009**, *2*, 231–238. [CrossRef] [PubMed]
58. Abe, Y.; Chiba, M.; Yaklai, S.; Pechayco, R.S.; Suzuki, H.; Takahashi, T. Increase in Bone Metabolic Markers and Circulating Osteoblast-Lineage Cells after Orthognathic Surgery. *Sci. Rep.* **2019**, *9*, 20106. [CrossRef] [PubMed]
59. Cerqueira, A.; Romero-Gavilán, F.; García-Arnáez, I.; Martinez-Ramos, C.; Ozturan, S.; Iloro, I.; Azkargorta, M.; Elortza, F.; Izquierdo, R.; Gurruchaga, M.; et al. Bioactive Zinc-Doped Sol-Gel Coating Modulates Protein Adsorption Patterns and in Vitro Cell Responses. *Mater. Sci. Eng. C* **2021**, *121*, 111839. [CrossRef] [PubMed]
60. Wang, X.; Li, X.; Ito, A.; Sogo, Y. Synthesis and Characterization of Hierarchically Macroporous and Mesoporous $CaO–MO–SiO_2–P_2O_5$ (M = Mg, Zn, Sr) Bioactive Glass Scaffolds. *Acta Biomater.* **2011**, *7*, 3638–3644. [CrossRef]
61. Wu, C.; Zhou, Y.; Xu, M.; Han, P.; Chen, L.; Chang, J.; Xiao, Y. Copper-Containing Mesoporous Bioactive Glass Scaffolds with Multifunctional Properties of Angiogenesis Capacity, Osteostimulation and Antibacterial Activity. *Biomaterials* **2013**, *34*, 422–433. [CrossRef]

62. Zhou, Y.; Han, S.; Xiao, L.; Han, P.; Wang, S.; He, J.; Chang, J.; Wu, C.; Xiao, Y. Accelerated Host Angiogenesis and Immune Responses by Ion Release from Mesoporous Bioactive Glass. *J. Mater. Chem. B* **2018**, *6*, 3274–3284. [CrossRef]
63. Jones, S.J.; Boyde, A. Colonization of Various Natural Substrates by Osteoblasts in Vitro. *Scan. Electron Microsc.* **1979**, *2*, 529–538.

coatings

MDPI

Article

Enhanced Multimodal Effect of Chemotherapy, Hyperthermia and Magneto-Mechanic Actuation of Silver-Coated Magnetite on Cancer Cells

Dumitru Daniel Herea [1], Camelia-Mihaela Zară-Dănceanu [1,*], Luminița Lăbușcă [1,2,*], Anca-Emanuela Minuti [1,3], Cristina Stavilă [1,3], Gabriel Ababei [1], Mihai Tibu [1], Marian Grigoraș [1], Mihaela Lostun [1], George Stoian [1], Oana-Georgiana Dragoș-Pînzaru [1], Gabriela Buema [1], Horia Chiriac [1] and Nicoleta Lupu [1]

1 Department of Magnetic Materials and Devices, National Institute of Research and Development for Technical Physics, 700050 Iasi, Romania
2 Orthopedics and Traumatology Clinic, County Emergency Hospital Saint Spiridon Iasi, 700111 Iasi, Romania
3 Faculty of Physics, Alexandru Ioan Cuza University, 700506 Iasi, Romania
* Correspondence: cdanceanu@phys-iasi.ro (C.-M.Z.-D.); llabusca@phys-iasi.ro (L.L.)

Abstract: Currently, various methods based on magnetic nanoparticles are being considered for the treatment of cancer. Among these, magnetic hyperthermia and magneto-mechanical actuation are the most tested physical methods that have shown promising results when applied both separately and in combination. However, combining them with specific drugs can further improve antitumor efficiency. In this study, we performed a systematic analysis to determine the best combination of hyperthermia, magneto-mechanical actuation of silver-coated magnetite nanoparticles (MNP@Ag) and chemotherapy (mitoxantrone) capable of destroying tumor cells in vitro while maintaining normal cells in their state of increased viability. The results showed that of the nine treatment configurations, the only one that satisfied the safety condition for normal cells (fibroblasts) and the highly cytotoxic condition for tumor cells (HeLa) was the combination of all three triggers. This combination led to the decrease in HeLa viability to about 32%, while the decrease in fibroblast viability reached 80%. It was observed that the cytotoxic effect was not a sum of the separate effects of each trigger involved, but the result of a nonlinear conjugation of the triggers in a dynamic regime imposed by the magneto-mechanical actuation of the nanoparticles. We conclude that by using such a treatment approach, the need for chemotherapeutic drugs can be substantially reduced while maintaining their therapeutic performance.

Keywords: silver-coated nanoparticles; hyperthermia; magneto-mechanic actuation; mitoxantrone; tumor cells

1. Introduction

Nanotechnology has imposed itself as a potent tool in a large variety of life science and medical applications. Nanomedicine, the application of nanotechnology in the medical field, is already enlarging the clinical field of diagnostic tools, drug delivery and medical sensors [1]. One particular field of development for nanomedical systems is cancer therapy [2]. Despite decades of research for the development of advanced therapies, cancer diagnostic and treatment remain an unsolved problem [3]. Even though cancer mortality has declined over the last three decades in high-income countries, there are persisting and substantial differences in mortality, incidence, and survival worldwide with more than 19.3 million new cases and almost 10.0 million cancer deaths occurring as per 2020 [4]. One of the most efficient approaches in treating malignancies is to combine different therapeutic principles. The so-called multimodal therapeutic approach has been recognized as a modality to increase efficiency, avoid side effects as well as the cancer relapse associated with cell heterogeneity within a tumor and/or metastatic dissemination [5]. Multimodal cancer therapy has benefited from noble metals such as gold nanoparticles which have shown a fine potential in backing the chemotherapy and radiotherapy treatments [6]. Magnetic nanoparticles (MNPs) are intensely scrutinized for their potential to serve as drug carriers and theranostic agents [7]. Tumor theranostics represent the combination of diagnosis and therapy to cure cancer, aiming to reduce delays in treatment and thus to improve patient care [8]. Due to their high versatility to be straightforwardly produced with tuned sizes, shapes and magnetic properties, MNPs have shown a significant role in improving a series of conventional or new biomedical methods and techniques. Ranging generally from clusters of several atoms to structures up to 100 nm—at least in one dimension—the MNPs are endowed with peculiar characteristics such as an enlarged surface-to-volume ratio, enhanced surface spin disorder, and various regular and irregular shapes, which allow, through subsequent surface-related modifications, the development of new core-shell magnetic materials for cutting-edge biomedical applications. MNPs, with their distinct class of iron oxides, were coated with amphiphilic polymers [9] polysaccharides [10], graphene [11], silica [12], gold [13], silver [14–19], etc. The latter displayed unique physical and chemical properties including optical, electrical, and thermal ones [20]. Consequently, silver nanoparticles are being intensively explored for various applications due to their specific physicochemical characteristics such as large surface-area-to-volume ratio, increased surface plasmon resonance, and ease of ligand functionalization [21]. Accordingly, they were tested for different biomedical applications due to their antibacterial, antiviral, antifungal, anti-inflammatory, anti-angiogenic, and antitumor effects [20,22]. Silver and silver nanoparticles have been recently shown to exert cytotoxic effects on cancer cells but not on normal cells, being proposed as a tumor theranostic agent [23].

By endowing silver nanoparticles with magnetic properties, their spectrum of biomedical applications is rapidly expanded to magnetic hyperthermia, magnetic drug delivery, magneto-mechanical actuation, magnetic separation, MRI imaging, and so on, virtually covering the entire spectrum of applications attributed to magnetic nanoparticles. Thus, these hybrid nanoparticles simultaneously provide all the specific magnetic benefits of the magnetic core together with all the advantages of the silver shell described above.

The remote magnetic actuation of MNPs incubated with tumor cells can trigger a mechanical hit and vibrations on tumor cells, leading to their irreversible damage. Basically, the alternating magnetic fields constantly change the position of the MNPs depending on the field gradients and frequency, to generate torques onto the cells [24]. Such torques can modify the level of calcium entry within cells, stimulate drug release, induce protein degradation, generate high-stemness tumor cells from different cancer cell lines, or destroy cancer cells [24,25].

We have previously reported on the efficiency of magnetically actuated MNPs in destroying cancer cells [26], on the ability of MNP loaded with an antitumoral agent (mitoxantrone) to target tumor-like tissue in vitro [27], and on the use of new alloy particles for self-controlled magnetic hyperthermia [28,29]. In this work, we sought to investigate systematically the in vitro action of a conventional antitumor drug, non-magnetic hyperthermia, and the magneto-mechanical actuation of silver-coated MNPs (MNP@Ag), applied separately or in combination, on HeLa-cell survival. Our previous experiments [30] have shown that the uncoated magnetite cannot be used to treat tumor cells simply because cancer cells grow and proliferate exceptionally well in the presence of an iron source. The final goal of the work was to find the best combination for the most efficient therapeutic effect seeking to decrease the antitumor drug dose to obtain similar tumor cell damage as in the case of a full dose.

2. Materials and Methods

All reagents used for the silver-coated MNP, i.e., ferrous chloride tetrahydrate, ($FeCl_2 \cdot 4H_2O$; 98%), ferric chloride hexahydrate, ($FeCl_3 \cdot 6H_2O$; 98%), sodium hydroxide ($NaOH$; 98%), mitoxantrone, fibroblast cells and HeLa tumor cells were purchased from

Sigma-Aldrich (Darmstadt, Germany) and used without any further purification. Silver nitrate was purchased from Alfa Aesar—Thermo Fisher (Kandel, Germany). For characterization, the particles were used either as powders or colloids. For dry state characterization, the magnetic nanoparticles were kept at 90 °C for 3 h. For colloid evaluation, the suspensions were sonicated in pulses.

2.1. Synthesis of Magnetic Nanoparticles

A measure of 1.1 g $FeCl_2 \cdot 4H_2O$ was dissolved in 10 mL of ultrapure water and filtered through a 220 nm filter. Then, 3 g of $FeCl_3 \cdot 6H_2O$ was mixed with the freshly prepared $FeCl_2$ solution and the resulting solution was filtered through a 220 nm filter. Next, 2 mL of ultrapure water was used to wash the filter and added to the iron solution. The solution was added over 400 mL of heated ultrapure water (about 90 °C), under magnetic stirring (800 rpm—MS7-H550-Pro, Onilab, City of Industry, CA, USA), followed by careful addition of solid NaOH (15 g). The solution immediately turned black. After 2 min, the heating was stopped, while the stirring continued for 60 min. Then, the suspension was sonicated for 1 h in an ultrasound bath. The magnetic nanoparticles were washed several times with ultrapure water until the pH was about 6. Afterwards, the obtained suspension (about 80 mL) was ultrasonicated for 40 min through an ultrasonic homogenizer with probe, at 100% amplitude.

2.2. Synthesis of Silver-Coated Magnetic Nanoparticles (MNP@Ag)

A measure of 40 mL of MNP suspension (14.5 mg/mL MNP) was mixed with 25 mL trisodium citrate (concentration 10 mg/mL, (Thermo Fisher, Kandel, Germany), ultrasonicated for 5 min, and added over 400 mL of heated ultrapure water (about 95 °C), under magnetic stirring (1000 rpm). After 0.5–1 min, 25 mL of filtered $AgNO_3$ (200 mg) (Thermo Fisher, Kandel, Germany) were quickly added to the MNP-citrate suspension. Then, the suspension was stirred for 2 h at 800 rpm and about 95 °C.

2.3. Cell Viability Evaluation

For cytotoxicity assays, the utilized suspensions of silver-coated MNPs were first submitted to autoclavation (121 °C, 1.5 bar). Dermal fibroblast cells and tumor cells, respectively, were plated in 96 well plates, at 2 × 104 cells/well and incubated for 48 h. Each well was filled with 190 µL complete cell culture media (CCM), i.e., Dulbecco Modified Essential Medium DMEM (Sigma Life Science, St. Louis, MO, USA) supplemented with 10% fetal bovine serum (FBS, Sigma Life Science, St. Louis, MO, USA) and 1% antibiotic-antimycotic (Sigma Aldrich, St. Louis, MO, USA). Then, 10 µL of silver-coated MNPs (concentration about 0.3 mg/mL) was added to each well, each sample in triplicate and further incubated for 24 h. Cell viability tests were performed using 5-Dimethylthiazol-2-yl-2, 5-diphenyltetrazolium bromide (MTT, Sigma Life Science, St. Louis, MO, USA) assay (Vybrant MTT cell proliferation assay Sigma Life Science, St. Louis, MO, USA) accordingly to supplier's instructions; dimethylsulfoxide (DMSO, Sigma Aldrich, St. Louis, MO, USA) was used as dilution agent. Absorbance was read at 570 nm. Cell viability (CV) as expressed by optical density (OD) was calculated using the formula $CV = 100 \times (ODs - ODb)/(ODc - ODb)$, where ODs = OD of particle treated cells; ODb = OD of blank (media only); ODc = OD of untreated cells. For particle-free viability evaluation, the protocol was identical, except that the wells were filled with 200 µL CCM.

2.4. Magneto-Mechanic Actuation of MNP@Ag

For magnetic field generation, a conventional Helmholtz system consisting of two coils was used. The proprietary software that controlled the coils allowed us to set the magnetic field intensity, its frequency and exposure time. The coil system was fed by waveforms generated using a code realized in LabView. The culture plates were placed in the center of the coil system, where the magnetic field is uniform. The intensity of the magnetic field was 12 G, and was shifted by 180° with a frequency of 2 Hz. The system can generate fields from 1 to 40 Oe. The coil was introduced inside a CO_2 cell culture incubator (Figure 1). The MMA was applied for 1h, five times, over three days (twice a day).

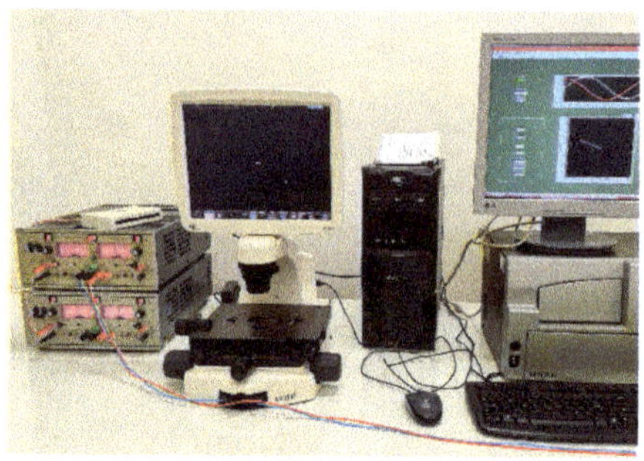

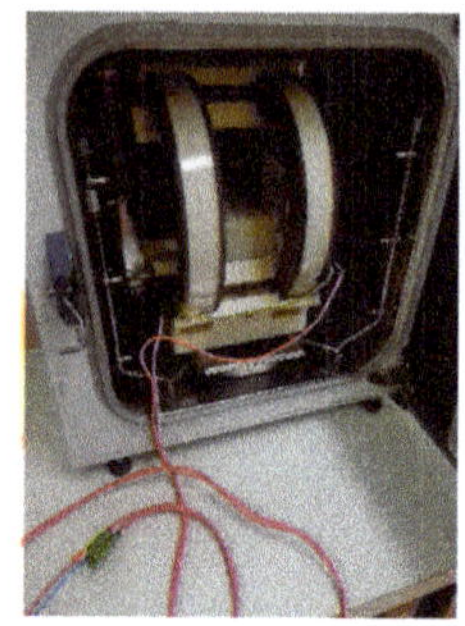

Figure 1. Experimental setup. Left—PC with the proprietary software to control the current passing through the coils; National Instruments acquisition board; and Kepco-Bop bipolar power amplifier. Right—the Helmholtz coil system inside the CO_2 cell culture incubator.

2.5. Hyperthermia Generation

In order to increase the temperature of the cells, we used a cell culture incubator. Compared to magnetic hyperthermia, this setup comes with several advantages: (a) the sterile conditions are continuously preserved for cell culture; (b) the temperature is well controlled and rigorously kept fixed throughout cell culture; (c) the 5% CO_2 atmosphere is continuously provided; (d) the humidity conditions are not affected. Practically, the conditions were kept similar for the control cells which were placed in another cell incubator at 37 °C. Therefore, the reliability of this approach affords the guarantee of a fine reproducibility of the experiments. The HT was applied for 1 h, five times, over three days (twice a day).

2.6. Characterization Equipment

X-ray diffraction patterns were recorded using a Brucker AXS D8-Advance powder X-ray diffractometer (CuKa radiation, k = 0.1541 nm, Brucker, Karlsruhe, Germany). Electron micrographs of the powders were taken by using a FIB/FE-SEM CrossBeam Carl Zeiss NEON 40 EsB equipped (Carl Zeiss SMT GmbH, Oberchoken, Germany) with an energy dispersive X-ray spectroscopy (EDS, Carl Zeiss SMT GmbH, Oberchoken, Germany) module for compositional studies. The magnetization data were acquired on a Lake Shore 7410 vibrating sample magnetometer (VSM, Lake Shore Cryotronics, Westerville, OH, USA). The morphology of the nanoparticles was determined through an ultra-high resolution transmission electron microscopy (UHR-TEM) (Libra 200 UHR-TEM, Carl Zeiss, Oberkochen, Germany). Mean sizes and size distribution of the as-prepared aqueous suspension of nanoparticles were evaluated through a dynamic light scattering (DLS) technique using a scattering spectrometer (Nanotrac NPA 252, Microtrac, Montgomeryville, PA, USA). UV–VIS absorption spectra were recorded with Synergy HTX multimode reader (BioTek Instruments, Santa Clara, CA, USA). Zeta potential was measured using a particle size analyzer (Beckman Coulter, Brea, CA, USA). Ultrasonication of the samples was performed with an ultrasound bath, and an ultrasonic homogenizer (UP50H, Hielscher Ultrasonics GmbH, Teltow, Germany), at 90% amplitude, respectively. The measurements and in vitro testing were performed from the stable, unsettled suspensions of nanoparticles.

Statistical analysis. All the experiments were performed in triplicate. Statistical analysis was performed using Microsoft Excel 2013 to calculate the mean ± SD (standard deviation), and Origin 2019 and Microsoft Excel 2013 to test for differences between means through analysis of variance (ANOVA) analysis with a post hoc Tukey's test, and Student *t*-test, respectively.

3. Results and Discussions

XRD analysis. The XRD patterns of the MNPs and MNP@Ag (Figure 2) showed specific diffraction peaks of magnetite at 30.2°, 35.5°, 43.2°, 53.6°, 57.2° and 62.9°, corresponding to (220), (311), (400), (422), (511) and (440) crystalline planes of magnetite phase, respectively [31], along with diffraction peaks of silver at 38.3°, 44.3°, 64.5°, corresponding to (111), (200) and (220) diffraction planes. The Bragg reflections pointed out a face-centered cubic structure of metallic silver.

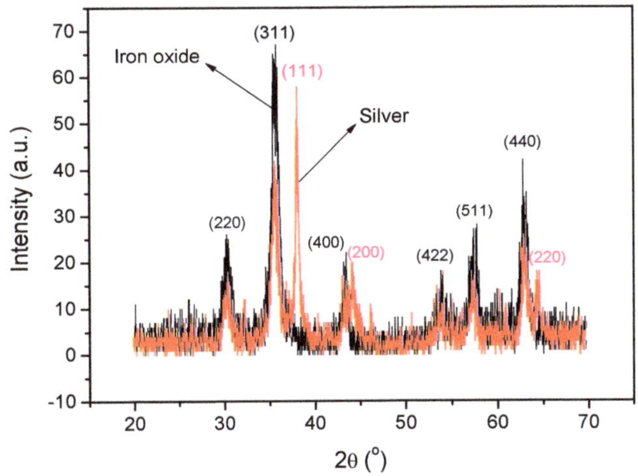

Figure 2. XRD patterns of iron oxide (MNP: black line) and MNP@Ag (red line).

UV–VIS spectroscopy. The spectra of MNP MNP@Ag solutions were recorded from 300 to 700 nm. Figure 3 shows an absorption peak at 402 nm for MNP@Ag. However, after subtracting the UV-VIS spectra of the MNP from MNP@Ag to obtain only the contribution of the silver shell, the absorption peak shifted to 412 nm. The obtained peaks were in the specific surface plasmon resonance range (400–500 nm) described for silver nanoparticles [31,32], demonstrating the presence of silver in the magnetic suspension. HRTEM mapping (Figure 4) also confirmed the presence of silver in the samples.

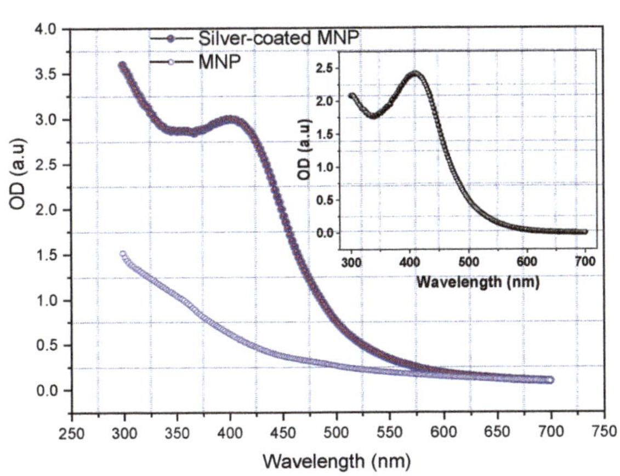

Figure 3. UV–Vis spectra of MNP and MNP@Ag, respectively. Inset: UV–Vis spectrum resulting from subtracting the UV-VIS spectra of the MNP from MNP@Ag.

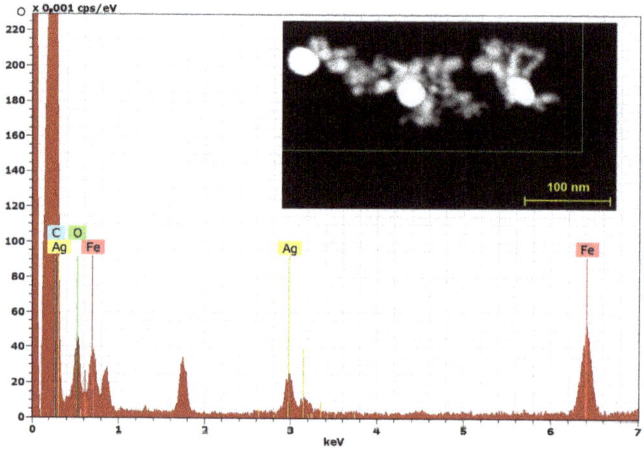

Figure 4. HRTEM element mapping of the silver-coated magnetite.

DLS analysis also used non-filtered (as-prepared) solutions of the nanoparticles, after short sonication. The obtained data showed a narrow size distribution for both samples with mean sizes of 39.2 nm for MNP and 40.5 nm for MNP@Ag (Figure 5). However, the mean sizes provided by the DLS method took into account the hydrodynamic diameter of the particles and, therefore, they seemed to be bigger. The size distribution data of MNP@Ag sample pointed out a low fraction (0.11%) of nanoparticles of about 102 nm—confirmed by TEM measurements—that were not present in the MNP sample. Most probably, this fraction contributed the most to the settlement formation. Otherwise, except for the tiny settlement, the suspension turned out to be very stable for several months.

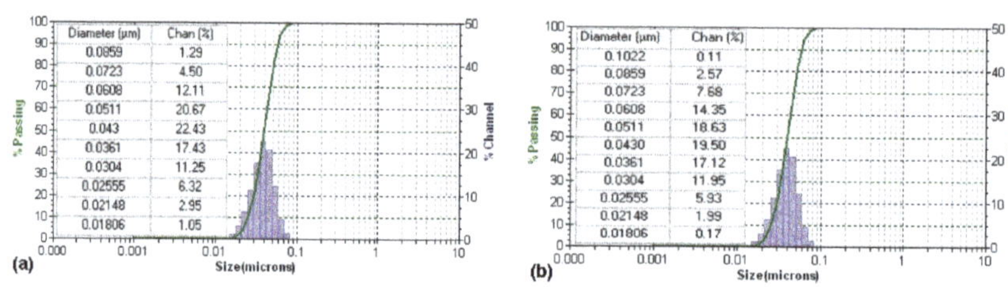

Figure 5. DLS histograms showing the size distribution of (**a**) MNP and (**b**) MNP@Ag.

The zeta potential of the MNP@Ag was found to be −40.26 mV, also indicating excellent stability in aqueous solutions. The strong electrostatic repulsion most probably overcame the low magnetic attraction between the nanoparticles. The stability extended to several months and was also empirically observed for the aqueous suspensions of MNP@Ag nanoparticles.

However, since it was not clear how stable the MNP@Ag nanoparticles were after incubation cells, we performed DLS measurements of MNP@Ag dispersed in cell culture medium. The size distribution of MNP@Ag was measured at different time intervals, including one after 24 h, without disturbing the suspension. The results showed that, after mixing with the culture medium, the hydrodynamic diameter of the particles increased from about 40 nm (measured in ultrapure water) to 80–90 nm (Figure 6).

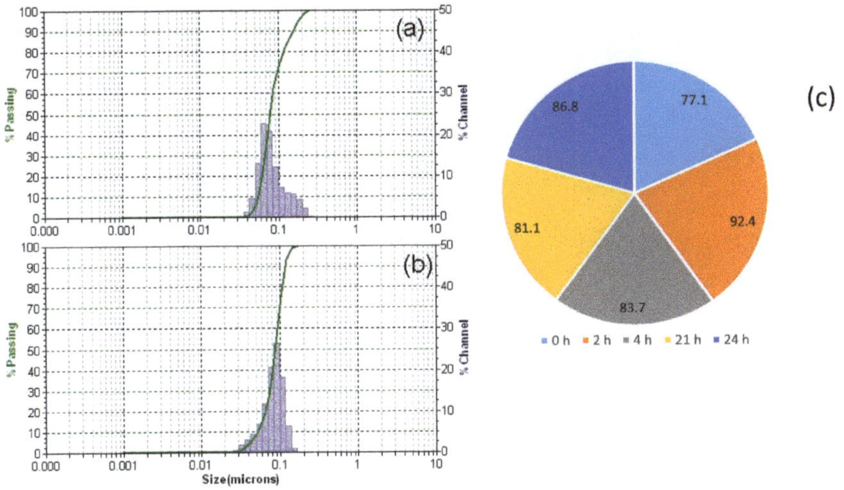

Figure 6. DLS histograms showing the size distribution of MNP@Ag nanoparticles immediately after mixing with cell culture medium (**a**), and after 24 h (**b**). The mean hydrodynamic diameters of the MNP@Ag nanoparticles at different times (**c**).

No sediment was observed in the cuvette of the Nanotrac/Microtrac sizer, containing the suspension, after 24 h. Therefore, the suspension was stable over 24 h in the cell culture medium, but the hydrodynamic diameter doubled. This can be attributed first to the formation of the protein corona on the surface of the nanoparticles. Second, clusters of several nanoparticles could not be excluded. However, the suspension remained stable over 24 h, while the size distribution remained almost unchanged for this period.

TEM and SEM analysis. Electron microscopy measurements showed magnetite nanoparticles with different prismatic shapes, and sizes below 30 nm (Figure 7a,b). Silver-coated magnetite nanoparticles displayed similar geometries as magnetite nanoparticles, with sizes slightly higher (Figure 7c,f). However, there was also observed a spot of bigger silver-coated nanoparticles (about 30–80 nm in diameter) with round shapes (Figure 7d). This fraction of nanoparticles was also detected and measured by the DLS technique (Figure 5). It was visible in a separate spot most probably because the provided aliquot sample for TEM measurements remained undisturbed for several days until carefully taken and dispersed on the grids. This time allowed the nanoparticles to discretely separate on the vertical axis of the suspension depending on their weight. After slowly spreading the sample on the grids, tiny volumes from the lower part of the suspension probably managed to remain separated from the upper part. It is also worth noting that after several days, a small amount of nanoparticles settled. Consequently, we analyzed the sediment as well (Figure 7e) and observed nanoparticles with bigger sizes (around 100 nm).

Magnetic measurements. The specific saturation magnetization (Figure 8) of the MNP (55.6 emu/g) was higher than that of MNP@Ag (48.7 emu/g). The saturation magnetization values were calculated by extrapolating the tangent to the plot of M versus 1/H at high fields, in the region where the dependence between M and 1/H is a straight line. The difference in saturation magnetizations represents an indirect proof of the presence of a non-magnetic silver shell on the MNPs. Furthermore, there were small differences in both coercive fields (Hc) and ratios between the residual and saturation magnetizations (squareness value, Mr/Ms) of the samples (Figure 8). The MNP@Ag sample showed lower values for Hc and Mr/Ms ratios, being even closer to a superparamagnetic behavior. The superparamagnetic behavior allowed a ferrofluid-like behavior of the MNP@Ag suspension (Figure 8a) after concentration by ultracentrifugation and ultra-sonication.

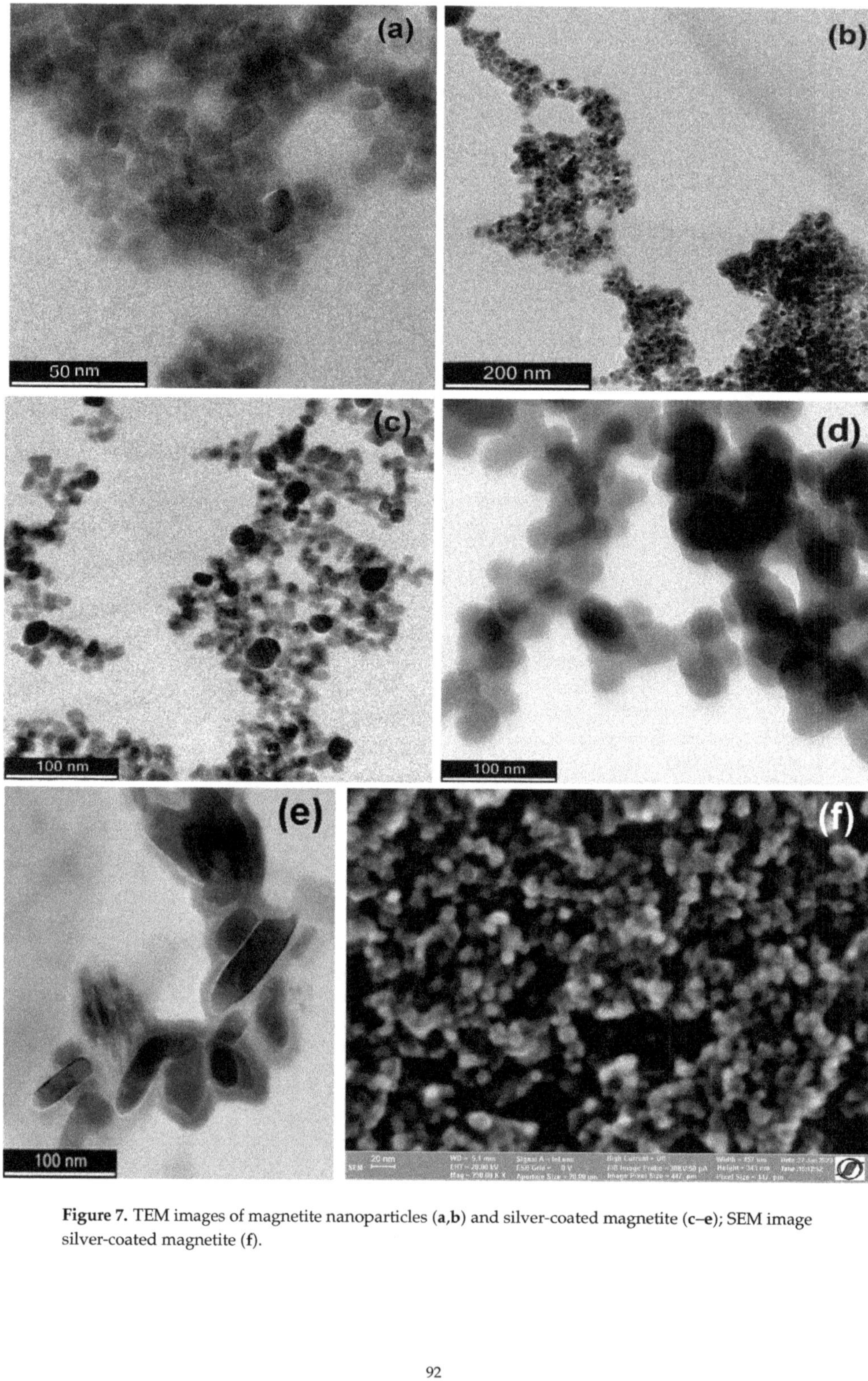

Figure 7. TEM images of magnetite nanoparticles (**a**,**b**) and silver-coated magnetite (**c**–**e**); SEM image silver-coated magnetite (**f**).

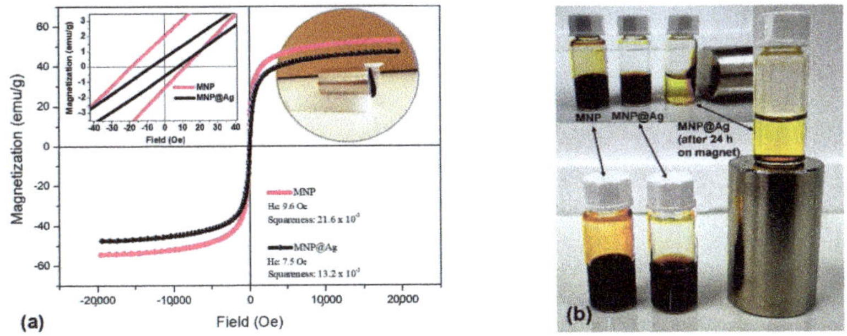

Figure 8. (**a**) Magnetization loops of the magnetic samples. Inset: details from the loops near zero magnetic field (top left) and ferrofluid−like behavior of concentrated MNP@Ag near strong NdFeB magnet; (**b**) images of the samples; the MNP@Ag sample was submitted to magnetic separation in its non-concentrated (i.e., as−prepared) condition.

The difference between specific saturation magnetizations of MNP and MNP@Ag was 12.4%. This means that a fraction of about 12.4% of the nanoparticles was formed by the non-magnetic silver. Since the theoretical percentage of the silver in the MNP@Ag nanoparticles was 17.9%, we can assume that the difference could be explained by a more abundant presence of the silver in the settled fraction, the latter not being taken into account by VSM measurement.

Evaluation of magnetic hyperthermia potential of MNP@Ag nanoparticles. The heating potential was investigated through an alternating magnetic field generator (HFG3 IGBT—Eldec, Germany) and an optical thermometer (Optocon, Germany). More details about the experimental setup can be found elsewhere [30]. Figure 9 shows the heating curve of the MNP@Ag nanoparticles. The alternating magnetic field (169 kA/m) was applied for 10 min until the temperature reached about 45 °C. The specific absorption rate (SAR) of the MNP@Ag nanoparticles (5 mg/mL) was 20.6 W/g. SAR is used to quantitatively measure the heating efficiency of a magnetic material and represents the mass-normalized rate of energy absorption by a (biological) object. Therefore, MNP@Ag nanoparticles showed good potential for magnetic hyperthermia applications.

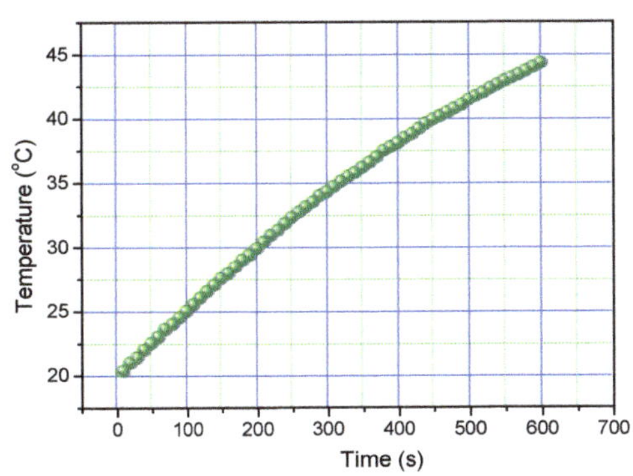

Figure 9. Time-dependent temperature of the heated MNP@Ag nanoparticles.

Cell Viability

Cell viability was assessed using MTT assay at 24 h after applying, separately or in combination, the respective trigger, i.e., MNP@Ag, magnetic field (MF), hyperthermia (HT), magneto-mechanic actuation (MMA), or mitoxantrone (MTX) on cell culture.

The first tests evaluated the influence of the MNP@Ag nanoparticles on cell viability. As Figure 10 shows, a progressive decrease in viability, inversely proportional to the nanoparticle concentration, was observed. However, the cell viability was preserved at high levels (higher than 95%) even for the highest concentration of the nanoparticles (300 µg/mL) for both types of cells. The statistical analysis (one-way ANOVA with Tukey's post hoc test) pointed out a statistically significant difference between the viabilities of the fibroblasts and the HeLa cells, indicating a better viability for the latter.

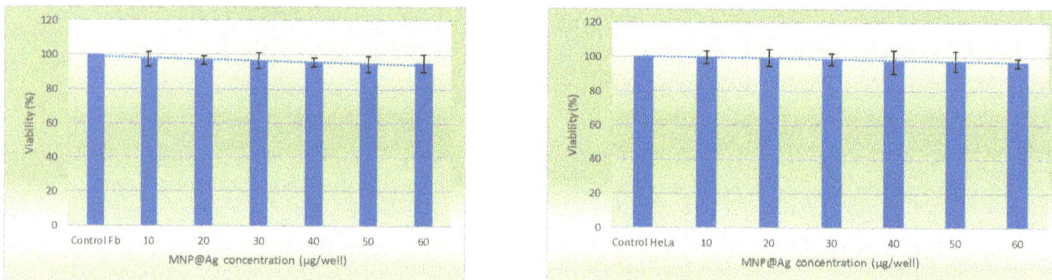

Figure 10. Cell viability of fibroblasts (**left**), and HeLa cells (**right**) after 24 h of incubation with MNP@Ag. Each well contained 0.2 mL cell culture medium. The difference between the two types of cells was statistically significant at the 0.05 level ($p < 0.05$).

For the following tests, the highest concentration of MNP@Ag, i.e., 0.3 mg/mL, was used. As shown in Figure 11, the magneto-mechanical actuation (MMA) of MNP@Ag remained almost without effect on cell viabilities when acting alone, compared to all other interventions, administered alone or in combination, which significantly decreased the viability of both cell types.

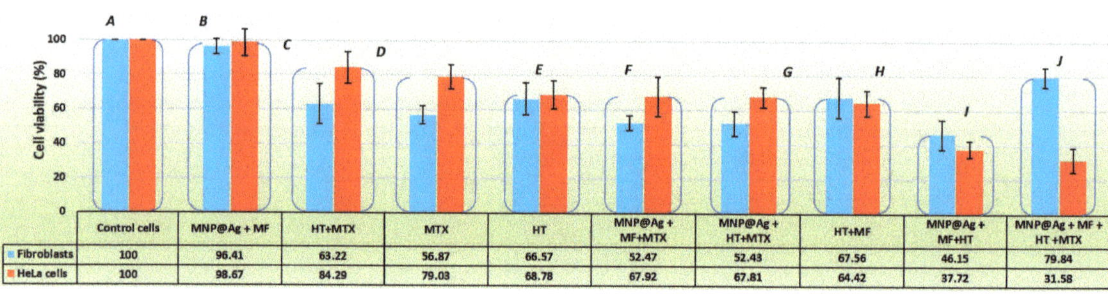

	Control cells	MNP@Ag + MF	HT+MTX	MTX	HT	MNP@Ag + MF+MTX	MNP@Ag + HT+MTX	HT+MF	MNP@Ag + MF+HT	MNP@Ag + MF + HT +MTX
Fibroblasts	100	96.41	63.22	56.87	66.57	52.47	52.43	67.56	46.15	79.84
HeLa cells	100	98.67	84.29	79.03	68.78	67.92	67.81	64.42	37.72	31.58

Figure 11. Cell viability of fibroblasts and HeLa cells after interventions with different trigger agents, i.e., magneto-mechanical actuation of MNP@Ag, hyperthermia, anti-tumor drug, and their combinations. MF stands for magnetic field while A–J letters represent the order number of each experiment.

It is interesting to note that HT alone was more effective in decreasing HeLa cell viability compared to both MTX and the combination of HT and MTX. The effect of HT was comparable to that generated by HT combined with MF, while fibroblast viability was over 60%. Furthermore, there was no difference between the magneto-mechanical actuation of nanoparticles combined with MTX (MNP@Ag + MF + MTX) on the one hand, and MNP@Ag combined with HT and MTX (MNP@Ag + HT + MTX) on the other hand.

The only intervention that appeared to be safe for fibroblasts, but highly cytotoxic for HeLa cells, was the combination of all the therapeutic triggers (MNP@Ag + MF + MT + MTX). The cytotoxic effect was not a sum of the separate effects of each involved trigger, but the result of a non-linear conjugation of the triggers in a dynamic regime imposed by the magneto-mechanic actuation of the nanoparticles.

When all the interventions (B–J) were analyzed, it could be observed that starting with H, there was a reversal in the decrease in cell viabilities. Thus, while as far as intervention

G, fibroblasts were more affected than HeLa cells, with a statistically significant difference between data sets (p, two-tailed, < 0.05), from H to J, it was the HeLa cells that started to suffer more. The key element common to the H, I, J interventions that induced this inversion was the combination of HT and magnetic field (MF). Furthermore, when the MMA of MNP@Ag (I) was included, the effect became more significant. Finally, when MTX (J) was added, the effect on HeLa cells was greatest, with a decrease in HeLa viability up to about 32%. Moreover, this last intervention came with an interesting result, namely the reduced decrease in fibroblast viability (only 80% decrease), with a statistically significant difference compared to control cells (chi test, $p < 0.001$, at $p < 0, 05$).

The last experiment, (J), was repeated with cells also in triplicate. The result was similar, with viabilities of about 24% for HeLa cells and about 85% for fibroblasts. This result shows that, compared to the intervention D, when only MTX was used, there was a significant improvement in therapeutic treatment against HeLa cells, suggesting a reduced need for MTX when combined with HT and MMA of MNP@Ag.

A possible explanation for this result would rely on a better permeation of the HeLa cell membrane to both MNP@Ag nanoparticles and mitoxantrone, induced by hyperthermia. Intracellular absorption of drugs has been reported to be enhanced by hyperthermia by increasing cell membrane permeability [33]. Thus, Berrios et al. [34] measured the cell membrane fluidity through direct and indirect methods, providing compelling evidence that both hot-water hyperthermia (HWH) and magnetic hyperthermia (MH) can increase membrane permeability, with the latter method inducing significantly greater effects as compared with HWH [33]. Therefore, the oscillations of MNP@Ag nanoparticles, while not affecting the cell viability by themselves, can induce a higher mobility of the antitumor drug molecules, which consequently can enter the tumor cells in a larger amount. Furthermore, due to the increased permeability induced by hyperthermia [33], the amount of drug transferred into the tumor cells is further enhanced. Therefore, cell death is mainly caused by the antitumor drug whose concentration in tumor cells was increased by MNP@Ag oscillations and hyperthermia-induced membrane permeation.

In our case, hyperthermia conditions seem to help MNP@Ag to affect cellular functions to a greater extent in the case of HeLa cells. In addition, once inside the cells, the MNPs could provide better and faster distribution of the antitumor drug.

However, the obtained results must be verified over periods longer than three days of treatment, until the total destruction of the tumor cells, also evaluating the longer-term damage to the fibroblasts.

4. Conclusions

To our knowledge, this is the first report focused on the combinatorial effect of multifactorial triggers (silver-coated magnetic nanoparticles, non-magnetic hyperthermia, magneto-mechanical actuation and antitumor drug (mitoxantrone) in significantly reducing short-term tumor cell viability in vitro compared to slight, non-significant reduction of normal fibroblast cells in similar conditions. Of the nine interventions in tumor and normal cells, respectively, only one combination of the four triggers led to a strong antitumor effect while maintaining the normal cells in a safe condition in the short-term. The results need to be further investigated in terms of long-term viability and the metabolic pathways potentially involved in different cell viabilities in order to determine the possible relevance for setting up a combinatorial approach in treating malignancies.

Author Contributions: Conceptualization, D.D.H.; methodology, D.D.H. and C.-M.Z.-D.; formal analysis, D.D.H. and C.-M.Z.-D.; investigation, C.-M.Z.-D., G.B., A.-E.M., C.S., G.A., G.S., M.G., M.L., O.-G.D.-P. and M.T.; writing—original draft preparation, D.D.H., L.L. and C.-M.Z.-D.; writing—review and editing, D.D.H., L.L.,C.-M.Z.-D. and A.-E.M.; visualization, D.D.H., L.L, C.-M.Z.-D., H.C. and N.L.; supervision, D.D.H.; project administration, D.D.H. and C.-M.Z.-D.; funding acquisition, D.D.H., L.L. and H.C. All authors have read and agreed to the published version of the manuscript.

Funding: This work was supported by a grant of the Ministry of Research, Innovation and Digitization CNCS–UEFIS-CDI, project number PN-III-P4-PCE-2021-1081 within PNCDI III (Contract no. 75/2022). Financial support by Romanian Ministry of Research, Innovation and Digitization, CNCS/CCCDI-UEFISCDI, Contract no. PCE20/2021 (PN-III-P4-ID-PCE-2020-2381) is also acknowledged. Financial support by Romanian Ministry of Research, Innovation and Digitization, CNCS/CCCDI-UEFISCDI, project number ERANET-EURONANOMED-3-OASIs, within PNCDI III is also acknowledged.

Institutional Review Board Statement: Not applicable.

Informed Consent Statement: Not applicable.

Data Availability Statement: Not applicable.

Conflicts of Interest: The authors declare no conflict of interest.

References

1. Soares, S.; Sousa, J.; Pais, A.; Vitorino, C. Nanomedicine: Principles, properties, and regulatory issues. *Front. Chem* **2018**, *6*, 360.
2. Blanco, E.; Hsiao, A.; Mann, A.P.; Landry, M.G.; Meric-Bernstam, F.; Ferrari, M. Nanomedicine in cancer therapy: Innovative trends and prospects. *Cancer Sci.* **2011**, *102*, 1247–1252.
3. Santucci, C.; Carioli, G.; Bertuccio, P.; Malvezzi, M.; Pastorino, U.; Boffetta, P.; Negri, E.; Bosetti, C.; La Vecchia, C. Progress in cancer mortality, incidence, and survival: A global overview. *Eur. J. Cancer Prev.* **2020**, *29*, 367–381.
4. Sung, H.; Ferlay, J.; Siegel, R.L.; Laversanne, M.; Soerjomataram, I.; Jemal, A.; Bray, F. Global cancer statistics 2020: GLOBOCAN estimates of incidence and mortality worldwide for 36 cancers in 185 countries. *CA Cancer J. Clin.* **2021**, *71*, 209–249.
5. Sambi, M.; Haq, S.; Samuel, V.; Qorri, B.; Haxho, F.; Hill, K.; Harless, W.; Szewczuk, M.R. Alternative therapies for metastatic breast cancer: Multimodal approach targeting tumor cell heterogeneity. *Breast Cancer* **2017**, *9*, 85–93.
6. Moore, J.A.; Chow, J.C.L. Recent progress and applications of gold nanotechnology in medical biophysics using artificial intelligence and mathematical modeling. *Nano Express* **2021**, *2*, 022001.
7. Zhang, H.; Liu, X.L.; Zhang, Y.F.; Gao, F.; Li, G.L.; He, Y.; Peng, M.L.; Fan, H.M. Magnetic nanoparticles based cancer therapy: Current status and applications. *Sci. China Life Sci.* **2018**, *61*, 400–414.
8. Siddique, S.; Chow, J.C.L. Recent advances in functionalized nanoparticles in cancer theranostics. *Nanomaterials* **2022**, *12*, 2826. [CrossRef]
9. Flores-Rojas, G.G.; López-Saucedo, F.; Vera-Graziano, R.; Mendizabal, E.; Bucio, E. Magnetic nanoparticles for medical applications: Updated review. *Macromol* **2022**, *2*, 374–390.
10. Saji, U.; Sang, J.L.; Kondareddy, C.; Chong-Su, C.; In-Kyu, P. Polysaccharide-coated magnetic nanoparticles for imaging and gene therapy. *BioMed Res. Int.* **2015**, *2015*, 959175.
11. El-Gendy, A.A. Core/Shell magnetic nanoparticles for biomedical applications. In *Micro and Nano Technologies, Magnetic Nanostructured Materials*; El-Gendy, A.A., Barandiaran, J.M., Hadimani, R.L., Eds.; Elsevier: Amsterdam, The Netherlands, 2018; pp. 41–58. [CrossRef]
12. Shin, T.H.; Manavalan, B.; Lee, D.Y.; Basith, S.; Seo, C.; Paik, M.J.; Kim, S.W.; Seo, H.; Lee, J.Y.; Kim, J.Y.; et al. Silica-coated magnetic-nanoparticle-induced cytotoxicity is reduced in microglia by glutathione and citrate identified using integrated omics. *Part. Fibre Toxicol.* **2018**, *18*, 42.
13. León Félix, L.; Sanz, B.; Sebastián, V.; Torres, T.E.; Sousa, M.H.; Coaquira, J.A.H.; Ibarra, M.R.; Goya, G.F. Gold-decorated magnetic nanoparticles design for hyperthermia applications and as a potential platform for their surface-functionalization. *Sci. Rep.* **2019**, *9*, 4185.
14. Haribabu, V.; Girigoswami, K.; Girigoswami, A. Magneto-silver core–shell nanohybrids for theragnosis. *Nano-Struct. Nano-Objects* **2021**, *25*, 100636.
15. Rashidi Huyeh, M.; Balouchzehi, S.; Shafiee Afarani, M.; Khajegi, P. Magnetite-Silver core–shell nanoparticles: Synthesis, characterizes, and optical properties. *Plasmonics* **2022**, *17*, 2385–2390.
16. Sharaf, E.M.; Hassan, A.; AL-Salmi, F.A.; Albalwe, F.M.; Albalawi, H.M.R.; Darwish, D.B.; Fayad, E. Synergistic antibacterial activity of compact silver/magnetite core-shell nanoparticles core shell against Gram-negative foodborne pathogens. *Front. Microbiol.* **2022**, *13*, 929491.
17. Ramirez-Acosta, C.M.; Cifuentes, J.; Cruz, J.C.; Reyes, L.H. Patchy Core/Shell, Magnetite/Silver nanoparticles via green and facile synthesis: Routes to assure biocompatibility. *Nanomaterials* **2020**, *10*, 1857.
18. Park, S.B.; White, S.B.; Steadman, C.S.; Pechan, T.; Pechanova, O.; Clemente, H.J.; Thirumalai, R.V.; Willard, S.T.; Ryan, P.L.; Feugang, J.M. Silver-coated magnetic nanocomposites induce growth inhibition and protein changes in foodborne bacteria. *Sci. Rep.* **2019**, *9*, 17499.
19. Anderson, S.D.; Gwenin, V.V.; Gwenin, C.D. Magnetic functionalized nanoparticles for biomedical, drug delivery and imaging applications. *Nanoscale Res. Lett.* **2019**, *14*, 188.
20. Zhang, X.F.; Liu, Z.G.; Shen, W.; Gurunathan, S. Silver nanoparticles: Synthesis, characterization, properties, applications, and therapeutic approaches. *Int. J. Mol. Sci.* **2016**, *17*, 1534.

21. Sofi, M.A.; Sunitha, S.; Sofi, M.A.; Pasha, S.K.; Choi, D. An overview of antimicrobial and anticancer potential of silver nanoparticles. *J. King Saud Univ. Sci.* **2022**, *34*, 101791.
22. Liu, X.; Chen, J.L.; Yang, W.Y.; Qian, Y.C.; Pan, J.Y.; Zhu, C.N.; Liu, L.; Ou, W.B.; Zhao, H.X.; Zhang, D.P. Biosynthesis of silver nanoparticles with antimicrobial and anticancer properties using two novel yeasts. *Sci. Rep.* **2021**, *11*, 15795.
23. Haque, S.; Norbert, C.C.; Acharyya, R.; Mukherjee, S.; Kathirvel, M.; Patra, C.R. Biosynthesized silver nanoparticles for cancer therapy and in vivo bioimaging. *Cancers* **2021**, *13*, 6114.
24. Lopez, S.; Hallali, N.; Lalatonne, Y.; Hillion, A.; Antunes, J.C.; Serhan, N.; Clerc, P.; Fourmy, D.; Motte, L.; Carrey, J.; et al. Magneto-mechanical destruction of cancer-associated fibroblasts using ultra-small iron oxide nanoparticles and low frequency rotating magnetic fields. *Nanoscale Adv.* **2022**, *4*, 421–436.
25. Wong, S.H.D.; Xu, X.; Chen, X.; Xin, Y.; Xu, L.; Lai, C.H.N.; Oh, J.; Wong, W.K.R.; Wang, X.; Han, S.; et al. Manipulation of the nanoscale presentation of integrin ligand produces cancer cells with enhanced stemness and robust tumorigenicity. *Nano Lett.* **2021**, *21*, 3225–3236.
26. Chiriac, H.; Radu, E.; Ţibu, M.; Stoian, G.; Ababei, G.; Lăbuşcă, L.; Herea, D.D.; Lupu, N. Fe-Cr-Nb-B ferromagnetic particles with shape anisotropy for cancer cell destruction by magneto-mechanical actuation. *Sci. Rep.* **2018**, *8*, 11538.
27. Herea, D.D.; Labusca, L.; Radu, E.; Chiriac, H.; Grigoras, M.; Panzaru, O.D.; Lupu, N. Human adipose-derived stem cells loaded with drug-coated magnetic nanoparticles for in-vitro tumor cells targeting. *Mater. Sci. Eng. C* **2019**, *94*, 666–676.
28. Chiriac, H.; Lupu, N.; Lostun, M.; Ababei, G.; Grigoraş, M.; Dănceanu, C. Low TC Fe-Cr-Nb-B glassy submicron powders for hyperthermia applications. *J. Appl. Phys.* **2014**, *115*, 17B520.
29. Herea, D.D.; Danceanu, C.; Radu, E.; Labusca, L.; Lupu, N.; Chiriac, H. Comparative effects of magnetic and water-based hyperthermia treatments on human osteosarcoma cells. *Int. J. Nanomed.* **2018**, *13*, 5743–5751.
30. Herea, D.D.; Chiriac, H.; Lupu, N.; Grigoras, M.; Stoian, G.; Stoica, B.A.; Petreus, T. Study on iron oxide nanoparticles coated with glucose-derived polymers for biomedical applications. *Appl. Surf. Sci.* **2015**, *352*, 117–125. [CrossRef]
31. Ashraf, J.M.; Ansari, M.A.; Khan, H.M.; Alzohairy, M.A.; Choi, I. Green synthesis of silver nanoparticles and characterization of their inhibitory effects on AGEs formation using biophysical techniques. *Sci. Rep.* **2016**, *6*, 20414.
32. Sastry, M.; Mayya, K.S.; Bandyopadhyay, K. pH Dependent changes in the optical properties of carboxylic acid derivatized silver colloidal particles. *Colloids Surf. A* **1997**, *127*, 221–228.
33. Liu, X.; Zhang, Y.; Wang, Y.; Zhu, W.; Li, G.; Ma, X.; Zhang, Y.; Chen, S.; Tiwari, S.; Shi, K.; et al. Comprehensive understanding of magnetic hyperthermia for improving antitumor therapeutic efficacy. *Theranostics* **2020**, *10*, 3793–3815.
34. Alvarez-Berrios, M.P.; Castillo, A.; Mendez, J.; Soto, O.; Rinaldi, C.; Torres-Lugo, M. Hyperthermic potentiation of cisplatin by magnetic nanoparticle heaters is correlated with an increase in cell membrane fluidity. *Int. J. Nanomed.* **2013**, *8*, 1003–1013.

coatings

MDPI

Article

Tunnel Magnetoresistance-Based Sensor for Biomedical Application: Proof-of-Concept

Crina Ghemes *, Oana-Georgiana Dragos-Pinzaru *, Mihai Tibu, Mihaela Lostun, Nicoleta Lupu and Horia Chiriac

National Institute of Research and Development for Technical Physics, 700050 Iasi, Romania
* Correspondence: cghemes@phys-iasi.ro (C.G.); odragos@phys-iasi.ro (O.-G.D.-P.)

Abstract: The aim of this work was to investigate and prove the possibility of the real-time detection of magnetic nanoparticles (MNPs) distributed in solid material by using a tunnel magnetoresistance-based (TMR) sensor. Following the detection tests of FeCrNbB magnetic nanoparticles distributed in transparent epoxy resin (EPON 812) and measuring the sensor output voltage changes at different particle concentrations, the detection ability of the sensor was demonstrated. For the proposed TMR sensor, we measured a maximum magnetoresistance ratio of about 53% and a sensitivity of 1.24%/Oe. This type of sensor could facilitate a new path of research in the field of magnetic hyperthermia by locating cancer cells.

Keywords: tunnel magnetoresistance; sensor; magnetic particles; sensitivity

1. Introduction

Features such as cost-effective production, small size, low power consumption, and especially high values of sensitivity and detectivity, recommend magnetoresistive (MR) sensors as the most suitable for biomedical applications [1,2]. Although the most studied magnetoresistive sensors are based on the giant magnetoresistance (GMR) effect, especially due to their high linear response to the applied field as well as the simplification of the technological process, tunnel magnetoresistance (TMR)-based sensors have better sensitivity and resolution due to the higher MR ratio, properties that are very important for biosensor applications [3]. Several methods have been used to improve the linearity and sensitivity of TMR sensors such as optimizing the magnetoresistive structure [4], integrating multiple sensors into the Wheatstone bridge [5], or integrating magnetic flux concentrators (MFCs) [6].

Due to their special magnetic properties and biocompatibility, magnetic nanoparticles can be used in medical applications such as human cell labeling and imaging [7,8] or cancer diagnosis and treatment [9,10]. Currently, the detection of the magnetic responses of MNPs can be performed by magnetorelaxometric imaging (MRX), magnetic resonance imaging (MRI), or magnetic particle imaging (MPI) [8,11,12], but there is a need to improve the cost and efficiency of magnetic nanoparticle localization.

In recent years, different solutions have been proposed for the use of magnetoresistive sensors to detect magnetic particles and evaluate their concentration in ferrofluids [13–15] as well as for the indirect detection of biomolecules or bacteria by detecting the magnetic field generated by the magnetic particles adhering to them [16–18]. On the other hand, the detection of magnetic nanoparticles by scanning human tissue with a highly sensitive sensor would be of great importance in medicine for use in magnetic hyperthermia, where a major problem is to direct the heat only to the diseased tissue without affecting the surrounding healthy tissue.

In this work, we propose a simple model of a tunnel magnetoresistance (TMR) sensor for the real-time detection of MNPs dispersed in solid material, with potential applications in magnetic hyperthermia by detecting the magnetic particles in tissues. To test the detection performance of the sensor, transparent epoxy resin (EPON 812) was used as a solid material in which the magnetic particles were dispersed. No magnetic flux concentrators or permanent magnets were used to operate the proposed sensor, so no additional steps are required in the fabrication process. The small size of the magnetoresistive device (400×250 μm^2) allows for the realization of a large number of sensors (100) on a support of 18×18 mm^2, which means a simplification of the technological process. The fabrication of the magnetoresistive sensing device and its magnetic properties are described in this work as well as the detection limits of the Fe–Cr–Nb–B magnetic nanoparticles dispersed in transparent epoxy resin (EPON 812). The ferromagnetic FeCrNbB nanoparticles (MNPs) prepared by high-energy ball milling from melt-spun ribbons (MSRs) [19] were submicron powders with a glassy structure, whose peculiarity is that the Curie temperature range of 15–50 °C can be changed depending on the Cr content [20]. This makes the magnetic FeCrNbB nanoparticles suitable for biomedical applications, especially hyperthermia applications.

2. Materials and Methods

2.1. TMR-Based Sensor Microfabrication

For the structure of the TMR-based sensors studied in this work, a thin layer of magnesium oxide (MgO) was used as the separating layer and $Co_{40}Fe_{40}B_{20}$ as the free ferromagnetic layer. To achieve higher magnetoresistance values with better sensitivity, the thicknesses of the component layers were optimized, resulting in the final configuration of: Ta (5 nm)/Ru (20 nm)/Ta (5 nm)/CoFe (3 nm)/IrMn (20 nm)/CoFe (2.5 nm)/Ru (0.85 nm)/CoFeB (3 nm)/MgO (1.8 nm)/CoFeB (3 nm)/Ta (10 nm), where the values in parentheses indicate the thickness of each layer, as shown schematically in Figure 1. The thin films were deposited on a Si/SiO_2 substrate (18×18 mm^2) using the ATC 2200/AJA International deposition system (Scituate, MA, USA), which can reach a base pressure of 5×10^{-8} Torr and allows both magnetron sputter deposition and electron beam evaporation. Except for the MgO barrier layer, which was deposited by electron beam evaporation, all other thin films were prepared by magnetron sputtering at a working pressure of 3 mTorr. During deposition, a magnetic field of 100 Oe was applied parallel to the plane of the thin films to define the easy axis of the ferromagnetic layers.

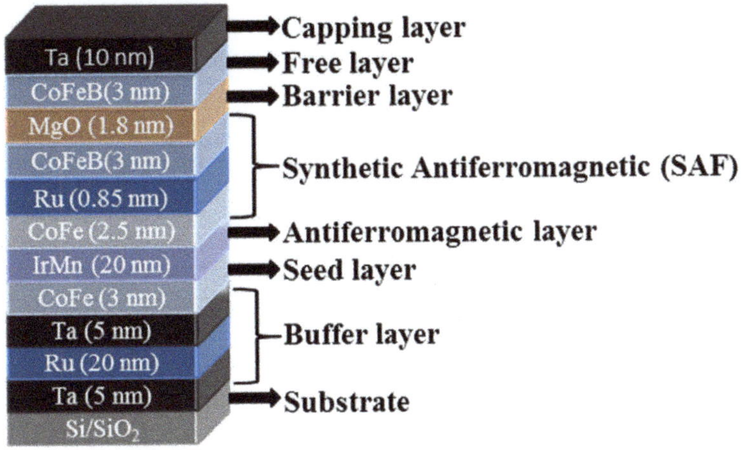

Figure 1. Schematic representation of the TMR multilayer structure.

Magnetoresistive sensors used in MNP detection applications must exhibit a linear change in electrical resistance with the applied magnetic field. It has been shown that in order to achieve this goal, it is advantageous to design the detection elements in rectangular or elliptical shapes (i.e., to have a high aspect ratio). However, as the ratio increases, the sensitivity of the sensor decreases [21]. In this work, we present a detection element structure with a low aspect ratio, with the shape of a cylindrical "pillar" and a diameter of 4 μm. The total size of the device (detection element and contacts) was 400 × 250 μm^2, so it is possible to fabricate 100 sensors on a single support. The sensors were divided into four groups of 25 sensors, marked with numbers and letter symbols so that the distinction during the measurements was facilitated (Figure 2k). Being a simple technological process, this type of sensor has the advantage of low cost and short manufacturing time. The microfabrication process of the sensor includes the following steps:

(1) Definition of a TMR structure with a rectangular shape with a size of 40 × 80 μm^2 by electron beam lithography (EBL) and ion beam etching (IBE). The detection element and the bottom contact are later defined on this rectangular structure (Figure 2a).

(2) Definition of the bottom contacts by EBL and IBE and resist stripping (Figure 2b,c).

(3) Definition of the detection element by EBL and IBE (Figure 2d). At this stage, the surface of the bottom contacts is protected by e-resist in the same lithography step as that of the pillar.

(4) A 60 nm thick SiO_2 insulating layer is deposited on the chip surface by sputtering (Figure 2e). The deposition of the insulating layer is followed by the lift-off process, which exposes only the surfaces covered by the e-resist (Figure 2f). The purpose of the insulating layer is to prevent electrical contact with the bottom contact and the top surface of the MTJ.

(5) Definition of the electrical contacts by laser beam lithography and deposition of Ta (5 nm)/Cu (200 nm). For the measurements, the CPP (current perpendicular to plane) configuration was used, which consists of the electric current flowing perpendicular to the plane of the thin films through the electrical contacts, which have a coplanar waveguide geometry (ground–signal–ground,) Figure 2g).

2.2. Magnetic Annealing

Crystallization of the CoFeB layer by magnetic annealing of the MTJ stack with highly (001)-oriented MgO barriers is crucial for maximizing the tunneling magnetoresistance. In magnetic annealing, some properties such as a strongly pinned layer or perpendicular magnetic anisotropy can also be achieved.

The values of temperature, magnetic field, and annealing time can lead to changes in the crystalline structure of the magnesium oxide layer. Therefore, the annealing parameters were studied to choose the appropriate values for each of them.

After we studied the annealing parameters, the magnetic tunnel junction multilayer stack was magnetically annealed for one hour under vacuum conditions at 320 °C and a magnetic field of 5 kOe. During annealing, the magnetic field was applied in the plane of the thin films in the direction of the anisotropy induced during deposition and kept at the same value until the sample reached room temperature. The sample was cooled down naturally to room temperature in about 20 min. Although the conditions for obtaining the magnetoresistive structure are still under investigation and improvements are needed, the images taken with the electron transmission microscope LIBRA®200MC/Carl Zeiss GmbH (Jena, Germany) and shown in Figure 3 demonstrate the transition of the CoFeB/MgO interface from the amorphous state before annealing (Figure 3a) to the crystalline state after annealing (Figure 3b).

2.3. Magnetic Nanoparticles

In order to investigate the detection capability of the sensor, magnetic FeCrNbB nanoparticles (MNPs) with a diameter of 50–500 nm, obtained by high-energy ball milling from melt-spun ribbons (MSRs) precursors [19], were dispersed in transparent epoxy resin (EPON 812) at different concentrations expressed in mass percent. The total mass of epoxy resin and NP mixture was fixed and the mass of NPs was gradually increased to obtain concentrations of 2%, 4%, 6%, and 8%. FeCrNbB particles exhibited high saturation magnetization and soft magnetic properties, which make them a very good candidate for magnetic hyperthermia.

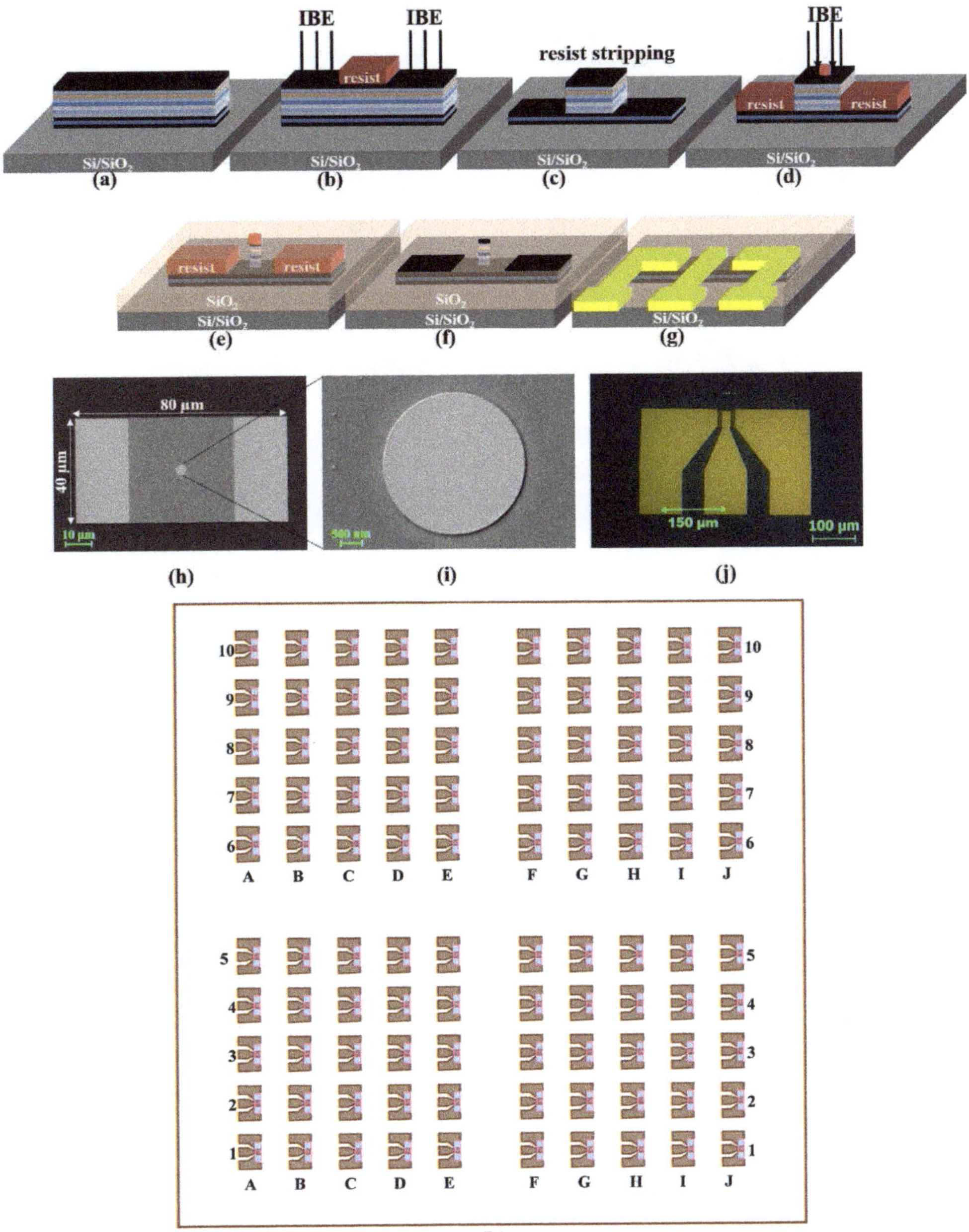

Figure 2. (**a–g**) Schematic representation of the sensor microfabrication process. (**h**) SEM image of the magnetoresistive structure of 40 × 80 µm^2 and (**i**) pillar detection element. (**j**) Optical microscopy image of the final device with coplanar waveguide geometry of the contacts. (**k**) Schematic representation of a chip with 100 MTJ sensor devices.

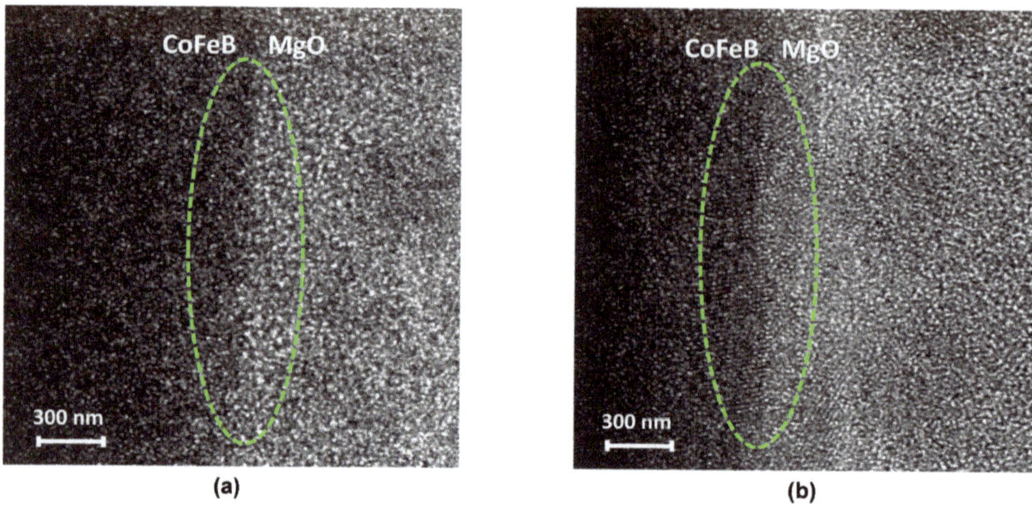

Figure 3. (**a**) TEM image of the CoFeB/MgO interface before annealing; (**b**) TEM image of the CoFeB/MgO interface after annealing.

Since epoxy resin has physical and chemical properties similar to human tissue [22], a mold of EPON 812 was prepared for the detection tests in which areas without particles alternated with areas with magnetic particles (Figure 4). Due to its similarity to tissue, epoxy resin is used in a variety of experiments that require the replacement of living organisms [23,24], thus avoiding the ethical problems that the use of human or animal tissue could lead to. Therefore, even in our case, this structure of resin and particles had the task of simulating a human tissue sample in which magnetic particles had been distributed.

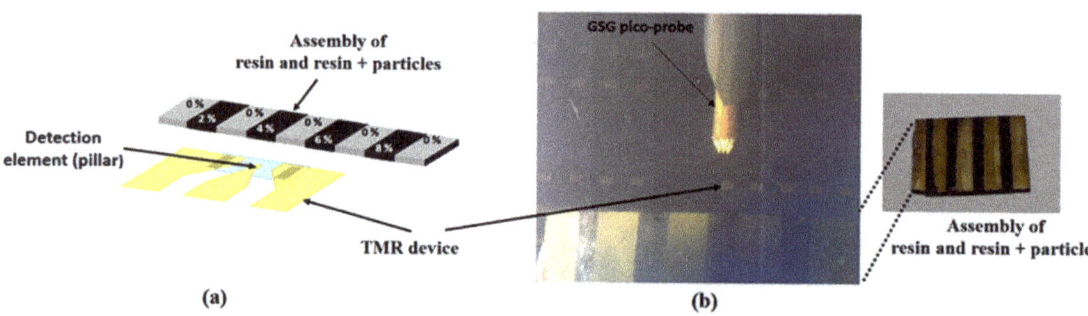

Figure 4. (**a**) Schematic representation of the resin/resin + magnetic particle assembly with different concentrations and of the magnetoresistive device; (**b**) Optical image of the sensor and the resin/resin + particle assembly, before being placed in the detection area.

3. Results

3.1. Sensor Performance

To characterize the performances of the TMR-based sensor, the magnetoresistance curve was drawn by measuring the electrical resistance of the sensor as a function of the applied magnetic field generated by a Helmholtz coil system connected to a bipolar source (Kepko BOP 100–10 MG) and measured with a Gaussmeter equipped with a Hall probe. The magnetic field was applied in the plane of the thin layers of the TMR structure in the direction of the anisotropy induced during deposition. A ground–signal–ground (GSG) probe was used to determine the electrical resistance of the TMR sensor. The system used a DC current source (Keithley 2400, Tektronix, Beaverton, OR, USA) and a multimeter (Keithley 2000, Tektronix, Beaverton, OR, USA) to measure the voltage (Figure 5). Data acquisition and system control were performed using a program developed in LabView.

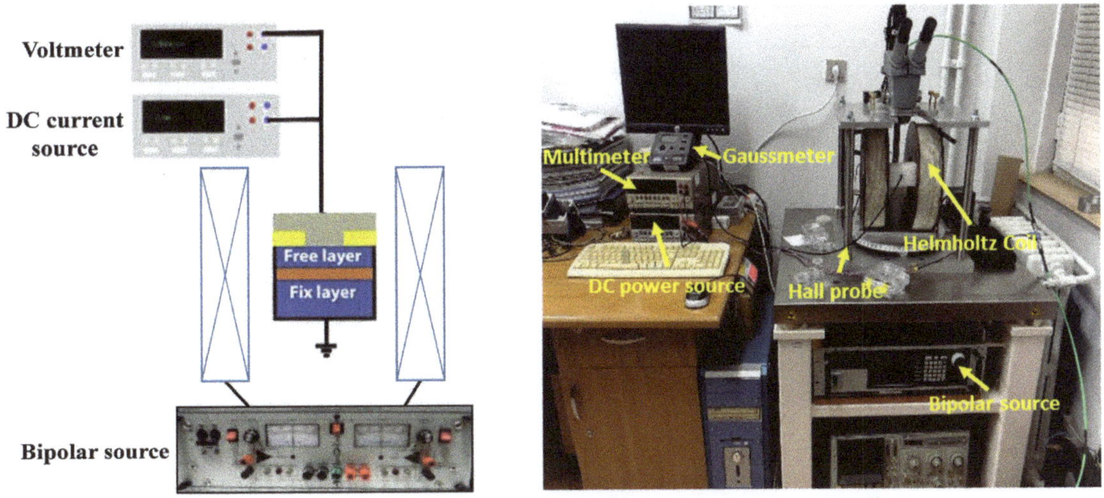

Figure 5. Schematic diagram and image of the experimental setup used for the TMR measurements.

Measurements of the sensor electrical resistance changes as a function of the applied magnetic field were performed with a bias current of 1 μA flowing through the MTJ sensor. During the measurements, the magnetic field was applied from −400 to +400 Oe in the direction of the easy axis of the pinned layer. From the sensor transfer curve (Figure 6), the proposed TMR sensor had a maximum magnetoresistance ratio of about 53% and a sensitivity of 1.24%/Oe.

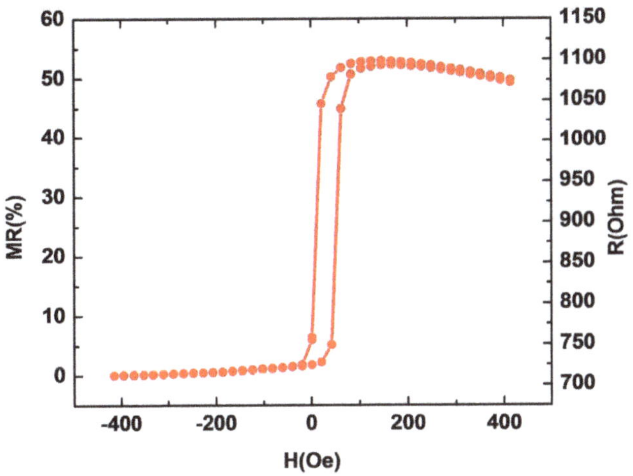

Figure 6. Sensor transfer curve.

3.2. Detection of Magnetic Nanoparticles

By using the presented sensor prototype, we demonstrated in this work the possibility of detecting different concentrations of magnetic particles distributed in a given volume. Furthermore, the design needs to be improved so that it can be easily used for magnetic sensing, especially for MPN detection. The chip structure containing the sensors will be encapsulated in a chip carrier package that prevents physical damage and corrosion, as shown schematically in Figure 7. A chip wire bonding machine was used to connect the chip to the package.

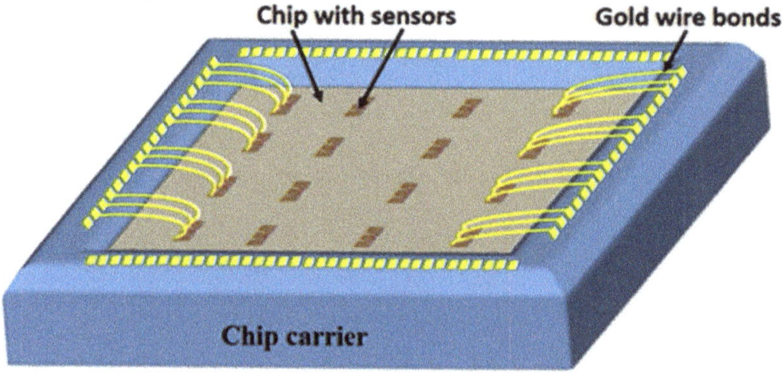

Figure 7. Schematic of the housing in which the chip structure is bonded to the chip carrier.

To magnetize the MNPs in order to obtain a detectable stray magnetic field, an external magnetic field of 20 Oe was applied perpendicular to the film plane of the magnetoresistive sensor. The sensing direction of the sensor coincides with the direction of the easy axis of the pinned layer, therefore, the sensitivity of the sensor is not affected by the applied magnetizing field for the particles.

For the detection measurements, the magnetoresistive sensor was brought to the region of maximum sensitivity of the magnetoresistance curve and then the output voltage of the sensor was measured for a few seconds (Figure 8a) for particle concentrations of 2%, 4%, 6%, and 8%. By calculating the average values of the sensor output voltages measured at the time intervals corresponding to each particle concentration, the graph in Figure 8b was obtained. It can be seen that the sensor voltage decreased linearly with increasing particle concentration from 0% to 6%. The voltage value increased slightly at 8%, which may indicate that the sensor is able to quantify particle concentrations up to a 6% concentration, while it reaches a saturation zone at higher values.

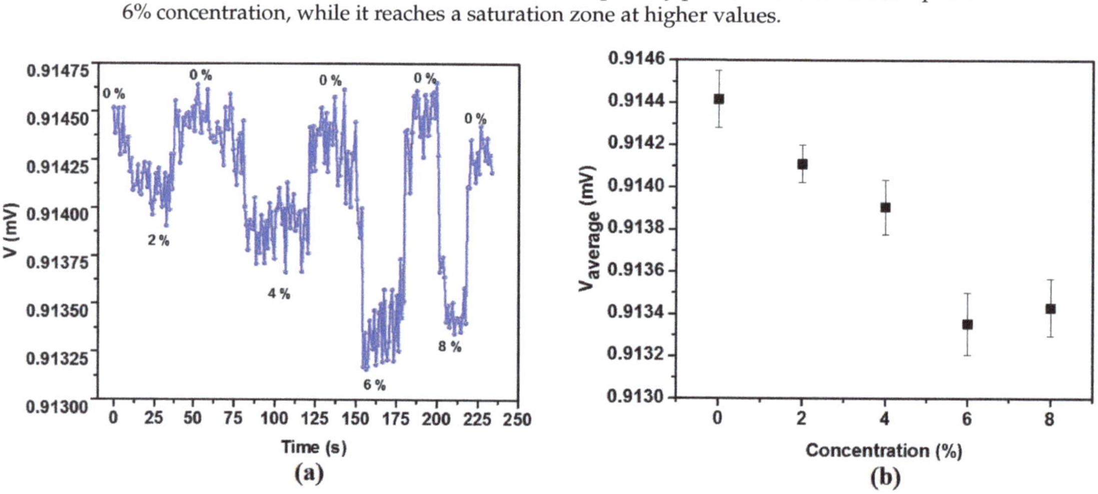

Figure 8. (**a**) Variation in the sensor output voltage in time for different particle concentrations; (**b**) the average value of the output voltages as a function of particle concentration.

4. Conclusions

We investigated the possibility of detecting magnetic nanoparticles in real-time using a simple TMR-based sensor model. Therefore, we proved the capability of measuring various concentrations of MNPs dispersed in a specific volume by utilizing the sensor prototype that described above. The presented sensor had a relatively high sensitivity and its fabrication did not require additional technological steps as in the case of using different configurations involving permanent magnets (Wheatstone bridge or flux concentrators). It has been shown that the sensor is able to detect in real-time and quantify these concentrations up to a saturation value. We believe that the proposed concept of a simple tunnel magnetoresistance (TMR) sensor for the real-time detection of MNPs could have significant potential applications in localizing cancer cells by detecting the magnetic particles in tissue and in cancer therapy by magnetic hyperthermia.

Author Contributions: Conceptualization, C.G., O.-G.D.-P., M.T., N.L. and H.C.; Methodology, C.G., O.-G.D.-P., M.T. and M.L.; Investigation, C.G., O.-G.D.-P. and M.T.; Writing—original draft preparation, C.G.; Writing—review and editing, C.G, O.-G.D.-P. and M.T; Visualization, N.L. and H.C.; Supervision, C.G.; Project administration, C.G.; Funding acquisition, C.G. All authors have read and agreed to the published version of the manuscript.

Funding: This work was supported by a grant from the Ministry of Research, Innovation, and Digitization, CCCDI—UEFISCDI, project number PN-III-P2-2.1-PED-2021-2739, within PNCDI III.

Institutional Review Board Statement: Not applicable.

Informed Consent Statement: Not applicable.

Data Availability Statement: Not applicable.

Conflicts of Interest: The authors declare no conflict of interest.

References

1. Lei, Z.Q.; Li, L.; Li, G.J.; Leung, C.W.; Shi, J.; Wong, C.M.; Lo, K.C.; Chan, W.K.; Mak, C.S.K.; Chan, S.B.; et al. Liver cancer immunoassay with magnetic nanoparticles and MgO-based magnetic tunnel junction sensors. *J. Appl. Phys.* **2012**, *111*, 07E505. [CrossRef]
2. Sun, X.; Zhi, S.; Lei, C.; Zhou, Y. Investigation of contactless detection using a giant magnetoresistance sensor for detecting prostate specific antigen. *Biomed Microdevices* **2016**, *18*, 60. [CrossRef] [PubMed]
3. Nagamine, Y.; Maehara, H.; Tsunekawa, K.; Djayaprawira, D.D.; Watanabe, N.; Yuasa, S.; Ando, K. Ultralow resistance-area product of 0.4 $\Omega(\text{Mm})^2$ and high magnetoresistance above 50% in CoFeB/MgO/CoFeB magnetic tunnel junctions. *Appl. Phys. Lett.* **2006**, *89*, 162507. [CrossRef]
4. Cardoso, S.; Leitao, D.C.; Gameiro, L.; Cardoso, F.; Ferreira, R.; Paz, E.; Freitas, P.P. Magnetic tunnel junction sensors with pTesla sensitivity. *Microsyst. Technol.* **2014**, *20*, 793–802. [CrossRef]
5. Zhang, Y.; He, G.; Zhang, X.; Xiao, G. Magnetotransport and electronic noise in superparamagnetic magnetic tunnel junctions. *Appl. Phys. Lett.* **2019**, *115*, 022402. [CrossRef]
6. Franco, F.; Cardoso, S.; Freitas, P.P. Reconfigurable spintronics wheatstone bridge sensors with offset voltage compensation at wafer level. *IEEE Trans. Magn.* **2019**, *55*, 4400705. [CrossRef]
7. Reddy, L.H.; Arias, J.L.; Nicolas, J.; Couvreur, P. Magnetic nanoparticles: Design and characterization, toxicity and biocompatibility, pharmaceutical and biomedical applications. *Chem. Rev.* **2012**, *112*, 5818–5878. [CrossRef]
8. Weissleder, R.; Cheng, H.C.; Bogdanova, A.; Bogdanov, A., Jr. Magnetically labeled cells can be detected by MR imaging. *J. Magn. Reson. Imaging* **1997**, *7*, 258–263. [CrossRef]
9. Ahmed, M.; Douek, M. The role of magnetic nanoparticles in the localization and treatment of breast cancer. *Biomed Res. Int.* **2013**, 281230. [CrossRef]
10. Revia, R.A.; Zhang, M. Magnetite nanoparticles for cancer diagnosis, treatment, and treatment monitoring: Recent advances. *Mater. Today* **2016**, *19*, 157–168. [CrossRef]
11. Liebl, M.; Steinhoff, U.; Wiekhorst, F.; Haueisen, J.; Trahms, L. Quantitative imaging of magnetic nanoparticles by magnetorelaxometry with multiple excitation coil. *Phys. Med. Biol.* **2014**, *59*, 6607–6620. [CrossRef]
12. Gleich, B.; Weizenecker, J. Tomographic imaging using the nonlinear response of magnetic particles. *Nature* **2005**, *435*, 1214–1217. [CrossRef]
13. Mukhopadhyay, S.C.; Chomsuwan, K.; Gooneratne, C.P.; Yamada, S. A novel needle-type SV-GMR sensor for biomedical applications. *IEEE Sens. J.* **2007**, *7*, 401–408. [CrossRef]

14. Loureiro, J.; Fermon, C.; Pannetier-Lecoeur, M.; Arrias, G.; Ferreira, R.; Cardoso, S.; Freitas, P.P. Magnetoresistive detection of magnetic beads flowing at high speed in microfluidic channels. *IEEE Trans. Magn.* **2009**, *45*, 4873–4876. [CrossRef]
15. Li, L.; Mak, K.Y.; Leung, C.W.; Ng, S.M.; Lei, Z.Q.; Pong, P.W.T. Detection of 10-nm superparamagnetic iron oxide nanoparticles using exchange-biased GMR sensors in wheatstone bridge. *IEEE Trans. Magn.* **2013**, *49*, 4056–4059. [CrossRef]
16. Kokkinis, G.; Jamalieh, M.; Cardoso, F.; Cardoso, S.; Keplinger, F.; Giouroudi, I. Magnetic-based biomolecule detection using giant magnetoresistance sensors. *J. Appl. Phys.* **2015**, *117*, 17B731. [CrossRef]
17. Yang, S.-Y.; Lien, K.-Y.; Huang, K.-J.; Lei, H.-Y.; Lee, G.-B. Micro flow cytometry utilizing a magnetic bead-based immunoassay for rapid virus detection. *Biosens. Bioelectron.* **2008**, *24*, 855–862. [CrossRef]
18. Herea, D.-D.; Labusca, L.; Radu, E.; Chiriac, H.; Grigoras, M.; Panzaru, O.D.; Lupu, N. Human adipose-derived stem cells loaded with drug-coated magnetic nanoparticles for in-vitro tumor cells targeting. *Mater. Sci. Eng. C* **2019**, *94*, 666–676. [CrossRef]
19. Chiriac, H.; Lupu, N.; Lostun, M.; Ababei, G.; Grigoraş, M.; Dănceanu, C. Low T_C Fe-Cr-Nb-B glassy submicron powders for hyperthermia applications. *J. Appl. Phys.* **2014**, *115*, 17B520. [CrossRef]
20. Chiriac, H.; Whitmore, L.; Grigoras, M.; Ababei, G.; Stoian, G.; Lupu, N. Influence of Cr on the nanoclusters formation and superferromagnetic behavior of Fe-Cr-Nb-B glassy alloys. *J. Appl. Phys.* **2015**, *117*, 17B522. [CrossRef]
21. Luz Ventosa, T.L.L. High Performance Magnetic Tunnel Junctions for Magnetic Scanning. Master's Thesis, Instituto Superi-or Tecnico, Lisboa, Portugal, May 2017.
22. Edris, A.; Choi, B.; Aguilar, G.; Nelson, J.S. Measurements of laser light attenuation following cryogen spray cooling spurt termination. *Laser Surg. Med.* **2003**, *32*, 143–147. [CrossRef] [PubMed]
23. White, D.R.; Martin, R.J.; Darlison, R. Epoxy resin based tissue substitutes. *Br. J. Radiol.* **1977**, *50*, 814–821. [CrossRef] [PubMed]
24. Yemby, H.T.; Arnold, M.P.; Giancarlo, A.V.; Felipe, C.; José, V.R. Construction and characterization of materials equivalent to the tissues and organs of the human body for radiotherapy. *Radiat. Phys. Chem.* **2019**, *159*, 70–75. [CrossRef]

coatings

MDPI

Article

Synthesis and Characterization of Gold-Shell Magnetic Nanowires for Theranostic Applications

Oana-Georgiana Dragos-Pinzaru [1], Gabriela Buema [1,*], Dumitru-Daniel Herea [1,*], Horia Chiriac [1], Nicoleta Lupu [1], Anca Emanuela Minuti [1], George Stoian [1], Daniel Shore [2], Valerie C. Pierre [3], Ibro Tabakovic [2] and Bethanie J. H. Stadler [2]

1 National Institute of R&D for Technical Physics, 700050 Iasi, Romania
2 ECE Department, University of Minnesota, Minneapolis, MN 55455, USA
3 Department of Chemistry, University of Minnesota, Minneapolis, MN 55455, USA
* Correspondence: gbuema@phys-iasi.ro (G.B.); dherea@phys-iasi.ro (D.-D.H.)

Abstract: Increasing interest has been given in recent years to alternative physical therapies for cancer, with a special focus on magneto-mechanical actuation of magnetic nanoparticles. The reported findings underline the need for highly biocompatible nanostructures, along with suitable mechanical and magnetic properties for different configurations of alternating magnetic fields. Here, we show how the biocompatibility of magnetic nanowires (MNWs), especially CoFe, can be increased by gold coating, which can be used both in cancer therapy and magnetic resonance imaging (MRI). This study provides a new approach in the field of theranostic applications, demonstrating the capabilities of core–shell nanowires to be used both to increase the cancer detection limit (as T2 contrast agents) and for its treatment (through magneto-mechanical actuation). The MNWs were electrodeposited in alumina templates, whereas the gold layer was electroless-plated by galvanic replacement. The gold-coated CoFe nanowires were biocompatible until they induced high cellular death to human osteosarcoma cells via magneto-mechanical actuation. These same MNWs displayed increased relaxivities (r1, r2). Our results show that the gold-coated CoFe nanowires turned out to be highly efficient in tumor cell destruction, and, at the same time, suitable for MRI applications.

Keywords: gold-coated magnetic nanowires; electrodeposition; MRI contrast agents; cancer therapy

1. Introduction

Numerous applications in nanomedicine, electronics, and catalysis utilize metallic materials in the form of nanowires, nanorods, or nanotubes. The physical characteristics of nanostructured materials are substantially influenced by the particle size and geometry, and are distinct from those of the same materials in bulk or single-crystal form [1–8]. The increased surface/volume ratio [9,10], the particular shapes of the nanostructured materials [11], their aspect ratio [12], and the preparation conditions are all responsible for the change in physical and chemical characteristics [13,14].

The surface chemistry, the geometry of the nanoparticles, and the nature of the magnetic material may greatly affect the magnetic behavior of the nanoparticles. Magnetic nanoparticles have found a series of biomedical applications including hyperthermia, drug delivery systems, tissue engineering, biomaterial/device coatings, theranostic platforms, lab on a chip, and magnetic separation due to their specific chemical and physical properties [15,16]. Moreover, tumor cell destruction by magneto-mechanical actuation of magnetic particles has been proven as a new promising physical approach in the field. This technique utilizes mechanical forces to kill the cancer cells through vibrations/rotations of magnetic particles, including magnetic nanowires (MNWs) [17]. Due to their increased potential in biomedical areas, such as live cell manipulation, cancer treatment, nanowarming, and/or contrast agents, noble metal and transition-metal nanowires are among the most investigated nanowires [18–29].

According to the World Health Organization, cancer is the cause of death for 8.2 million people each year, representing approximately 13% of all deaths, and the diagnosis rate increased by 4.5% from 2014 to 2018 [30]. In order to improve cancer outcome and survival, early cancer detection remains the priority for both patients and clinicians. In this context, the development of new useful materials for cancer detection in the earliest stages is imperative. Moreover, the combination of the diagnosis and therapy provided by a single-nanomaterial-based platform comes with a series of advantages, such as improved diagnosis, targeted delivery of antitumor drugs, lower cytotoxic effects on healthy tissues, etc. [31].

However, a major drawback of uncoated nanoparticles/nanowires in biomedical applications arises from their limited stability over time due to their propensity to clump together. By using 1D nanostructures, such as nanowires, this drawback may be greatly overcome since the agglomeration phenomena are less likely to occur in one-dimensional structures [32]. The cytotoxicity of uncoated magnetic nanoparticles and the fact that these materials might quickly deteriorate in the human body are other specific drawbacks. To avoid or diminish the toxicity and to prevent rapid degradation, the surface of the nanomaterials is often covered with organic molecules, such as chitosan, polyethylene glycol, or dextran [33,34]. This encapsulation of magnetic nanoparticles in a layer of organic compounds increases the duration of circulation in the body and affords good biocompatibility. However, due to the high reactivity of organic compounds in the blood, the coating can still be degraded by the human body. In order to avoid the rapid biodegradation of nanomaterials and to reduce their toxicity, a new method has been developed which involves coating magnetic nanoparticles with noble metals, because they are known to be nontoxic and stable in the human body [35]. The new smart materials prepared following this route are multifunctional, meaning they can be used both for diagnosis and treatment. The advantages of using noble metal (e.g., Au) to coat magnetic 1D nanoparticles arise from their high biocompatibility, high corrosion resistance, high chemical and physical stability, and increased capability to be functionalized via a thin, biocompatible layer of the noble metal, which creates an ideal interface to bind other biofunctional molecules via thiol groups [36].

A versatile method for metal plating is electroless deposition, representing a spontaneous or in situ reduction of metal ions to their metallic state in the absence of an external power source [37]. This particular method can be used to prepare core–shell nanostructures, in order to improve the mechanical, physical, and chemical properties of the nanomaterials [38,39].

The aim of this work was to prepare a new nanocomposite material, based on core–shell nanowires, to be used in theranostic applications, especially in oncology. To this end, we have developed an easy two-step method based on the electrodeposition of magnetic material into AAO (Anodic Aluminum Oxide) templates, followed by a galvanic displacement reaction to prepare core–shell MNWs as follows: NiFe@Au, CoPt@Au, CoFe@Au. The most biocompatible core–shell nanowires were tested for rapid in vitro destruction of cancer cells by magneto-mechanical actuation as well as for their MRI capabilities. Remarkably, r2 transverse relaxivities up to 10.64 $mM^{-1}\ s^{-1}$ (at 1.5 T), together with a higher biocompatibility, were achieved when CoFe@Au core–shell nanowires were used. This capability allows highly accurate cancer cell detection. At the same time, the as-prepared core–shell nanowires manifested an efficient antitumor effect when the AMF (Alternating Magnetic Fields) was applied. Our study showed for the first time the ability of core–shell nanowires to be used for theranostic applications. More specifically, we prepared magnetic core nanowires based on NiFe, CoPt, and CoFe alloys via electrodeposition inside AAO templates, followed by gold-shell electroplating. The as-prepared core–shell nanowires were analyzed from a morphological, compositional, and structural point of view using scanning electron microscopy, energy-dispersive X-ray spectroscopy, and X-ray diffraction, respectively. Next, the biocompatibility of the gold-shell nanowires was evaluated using an MTT (3-(4,5-dimethylthiazol-2-yl)-2,5-diphenyltetrazolium bromide) assay. The antitumor effect of the NW-mediated magneto-mechanical actuation was evaluated, and the relaxation times were determined on a Bruker Minispec mq60 NMR analyzer at 1.5 T. The results show that the as-synthetized gold-shell nanowires can be successfully used for theranostic applications.

2. Materials and Methods

The MNW core was synthesized by pulse electrodeposition in a closed three-electrode cell from an aqueous electrochemical bath containing the ions of the core magnetic materials (Ni, Fe, Co, and Pt) into the nanopores of the AAO templates (with pore diameters of 200 nm and a template thickness of 40 μm) provided by Whatman Ltd. (Chalfont St. Giles, England). The applied potential was controlled through a Bipotentiostat/Galvanostat HEKA PG 340 (HEKA, Ludwigshafen/Rhein, Germany). After performing the electrodeposition, the magnetic core nanowires were liberated from the template by dissolution of the alumina in an aqueous 3 M NaOH solution. Finally, a gold shell was plated from a tetra-Chloroauric (III) acid aqueous solution. All reagents for synthesis and analysis were of 99.98% purity and procured from AlfaAesar, (Karlsruhe, Germany). The length and morphology of the as-prepared nanowires were observed by HR-SEM (high-resolution—scanning electron microscopy) using a CrossBeam System Carl Zeiss NEON40EsB (Carl Zeiss, Jena, Germany). The elemental composition was determined by EDX (energy-dispersive X-ray spectroscopy) measurements. The crystalline structure of the electrodeposited nanowires was studied by X-ray diffraction (XRD) by means of a Bruker AXS D8-Advance X-Ray Diffractometer (Bruker, Brno, Czech Republic)with parallel optical geometry using Cu Kα radiation (λ = 1.5406 Å). The relaxation times (longitudinal—T1 and transverse—T2) were measured on a Bruker Minispec mq60 NMR Analyzer (Bruker, Woonsocket, USA) at 1.5 T (60 MHz) at 25 °C using the inversion recovery sequence and the Carr–Purcell–Meiboom–Gill sequence, respectively. Prior to the measurements, the core–shell nanowire surfaces were functionalized using poly(ethylene glycol) 2-mercaptoethyl ether acetic acid, SH-PEG-COOH.

The in vitro biocompatibility of the core–shell NiFe, CoFe, and CoPt nanowires and their core–shell counterparts was tested on human fibroblast cells while the potential antitumor effect of the most biocompatible nanowires, i.e., CoFe@Au, was tested on human osteosarcoma cells. The cell viability was evaluated using the MTT assay, a test that consists of the cell reduction of 3-(4,5-dimethylthiazol-2-yl)-2,5-diphenyltetrazolium bromide to formazan. Briefly, cells were seeded in 96-well plates until confluency, when the nanowires dispersed in cell media were added. For biocompatibility, the MTT assay was performed 24 h after co-incubation of cells with nanowires, while for the magneto-mechanically actuated ones, the samples were first treated in the magnetic field and evaluated after another 24 h.

All the experiments were performed in triplicate. Statistical analysis was performed using Microsoft Excel 2013 to calculate the mean $\pm$ SD and Origin 2019 software to test for differences between means through ANOVA analysis with a post hoc Tukey's test.

3. Results

3.1. Core–Shell Nanowire Preparation

The main steps of preparing nanowires with a magnetic core and Au shell, using electrodeposition inside AAO (anodic aluminum oxide) templates, followed by Au shell coating using electroless plating, are shown in Figure 1.

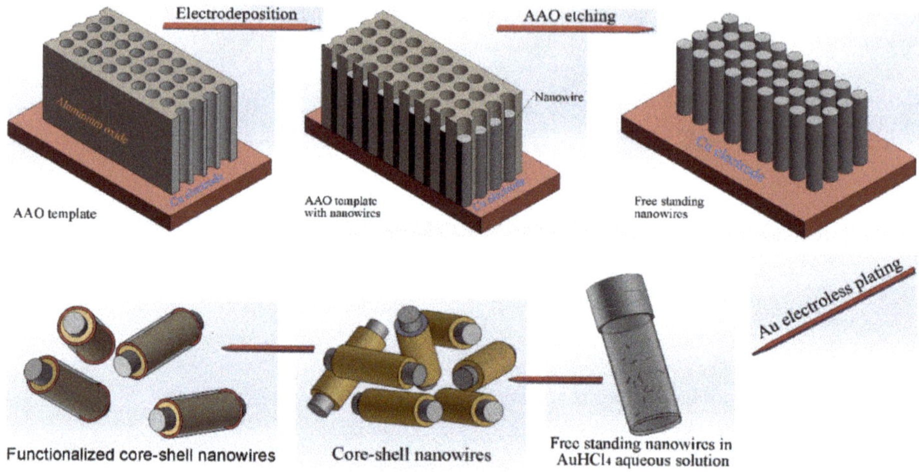

Figure 1. Schematic diagram of preparation of core–shell nanowires. The nanowires were functionalized with Thiol-PEG-Carboxyl (SH-PEG-COOH).

The synthesis of the core–shell nanocomposite materials involved several steps. In the first step, the synthesis of MNWs was conducted by electrodeposition of the magnetic material inside the nanopores of AAO template. In the second step, the nanowires were released from AAO by dissolution of the AAO template in an aqueous 3 M NaOH solution. Finally, a thin layer of Au was deposited on the surface of the MNWs by galvanic displacement.

Magnetic NiFe, CoPt, and CoFe nanowires, used as core magnetic materials, were grown by electrodeposition inside the AAO template procured from Whatman International Ltd. AAO templates with diameters of 25 mm, thicknesses of 40 μm, and nominal pore diameters of 200 nm were used. Prior to the alloy electrodeposition, thin layers of Ti (10 nm) and Cu (300 nm) were deposited by sputtering onto one side of the AAO template, covering the pores completely, and serving as a working electrode during the electrodeposition. The role of the 10 nm Ti layer was to increase the adherence of the Cu layer to the AAO. Each magnetic material was electrodeposited from a different electrochemical bath containing the ions of the metals that are found in the composition of the alloy.

The electrodeposition of NiFe alloy nanowires was carried out in a sulfate/chloride plating solution containing 0.4 M H_3BO_3, 0.3 M NH_4Cl, 0.2 M $NiSO_4{\cdot}6H_2O$, 0.02 M $FeSO_4{\cdot}7H_2O$, 5 mM saccharin as a Na salt, and 0.3 mM sodium lauryl sulfate (Na-LS). The electrolyte solution was adjusted to pH 3.0. These relatively high concentrations of Fe^{2+} and Ni^{2+} salts were used following recently proposed concentration optimization in the case of NiFe nanowire electrodeposition [5]. The potential was pulsed between −1.0 V/SCE during a "time-on" period of 2.5 s and a "rest" potential of −0.7 V/SCE during a time-off period of 1 s.

The electrolyte used for the electrodeposition of the CoPt alloy was a stable hexachloroplatinate CoPt solution with pH = 5.5 containing 0.4 M H_3BO_3, 0.3 M NH_4Cl, 0.1 M $CoSO_4{\cdot}7H_2O$, and 0.00386 M H_2PtCl_6, with saccharine as an additive. The solution pH was adjusted using an aqueous solution of 0.1 M NaOH [40]. The CoPt alloys were electrodeposited by pulsing the potential between −0.8 V/SCE during the time-on period of 2.5 s and −0.1 V/SCE during the time-off period of 1 s.

The CoFe nanowires were electrodeposited by pulsing the potential between −1.15 V/SCE during the time-on period of 2.5 s and −0.7 V/SCE during the time-off period of 1 s, using a sulfate/chloride plating solution containing 0.4 M H_3BO_3, 0.3 M NH_4Cl, 0.1 M $CoSO_4{\cdot}7H_2O$, 0.2 M $FeSO_4{\cdot}7H_2O$, 1.0 mM malonic acid as an organic additive (MA), and

0.3 mM sodium lauryl sulfate (Na-LS) as a wetting agent. The electrolyte solution was adjusted to pH 3.0 [11].

The pulse electrodeposition was performed in a three-electrode cell with a platinum wire as a counter electrode, an AAO template working electrode, and a saturated calomel electrode (SCE) reference. The cell volume used for the experiments was 100 mL. A schematic representation of the experimental process is presented in Figure 2.

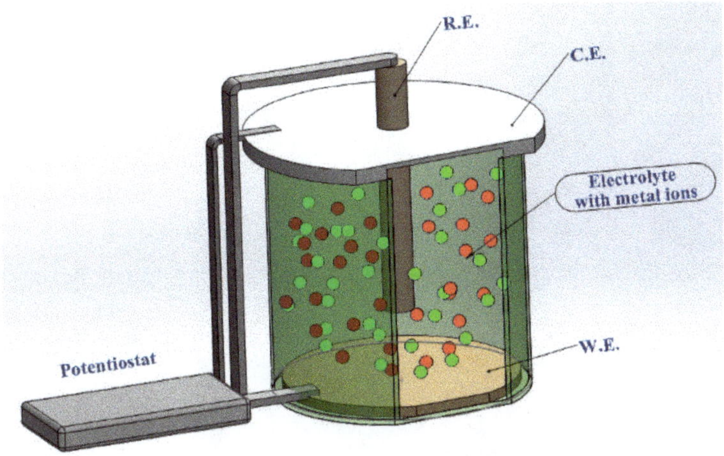

Figure 2. Schematic diagram of the experimental setup. The electrolyte contained different positive and negative metal ions: Co^{2+}, Pt^{2+}, Fe^{2+}, Ni^{2+}, Cl^{-}, HO^{-}.

The crystallinity and the purity of the as-prepared nanowires were checked with the XRD analysis, with the data being presented in the Supplementary Materials, Figures S1–S3. Our data show that NiFe as well as CoFe presented a cubic structure while CoPt had a hexagonal structure. The diffraction patterns presented only the reflections of the NiFe, CoPt, and CoFe alloys (from the electrodeposited nanowires), with no other peaks being visible on the diffractograms. After the preparation of the magnetic core in the nanopores of the AAO template, the AAO membrane was dissolved in a 3 M NaOH aqueous solution in order to release the nanowires. After template dissolution, the MNWs were rinsed several times with distilled water and magnetically separated. In the next step, a thin coating of Au was chemically deposited by electroless plating on the surface of the MNWs. In order to cover the MNWs' surface with Au, the washed nanowires were immersed in Au plating solution for 5 min. The plating solution contained 0.5 mM $HAuCl_4$ and 5M H_3BO_3. The experiments were carried out at room temperature. In order to measure the relaxation times, the surfaces of the CoFe@Au core–shell nanowires were functionalized according to the procedure presented previously by Shore et al. [41] as follows: 1 mL of the SH-PEG-COOH solution was poured onto the nanowires, mixed, and left overnight to functionalize.

3.2. *Morphological and Compositional Characterization*

The properties of nanowires are significantly impacted by their shape, aspect ratio, and composition. Hence, it is crucial to evaluate these characteristics. In this context, extremely precise control over the synthesis parameters is required to produce alloys with very low batch-to-batch variations. After electrodeposition, the cross-section of the AAO template with nanowires within was examined by HR-SEM. All samples had an identical distribution of electrodeposited material inside the AAO template as seen in the HR-SEM pictures. The length of the electrodeposited nanowires was 1 ± 0.2 μm.

Figure 3 shows a typical HR-SEM image of an AAO template cross-section filled with electrodeposited MNWs, including NiFe nanowires (Figure 3a,d), CoPt nanowires (Figure 3b,e), and CoFe nanowires (Figure 3c,f). The measured diameter of the nanowire was 200 nm.

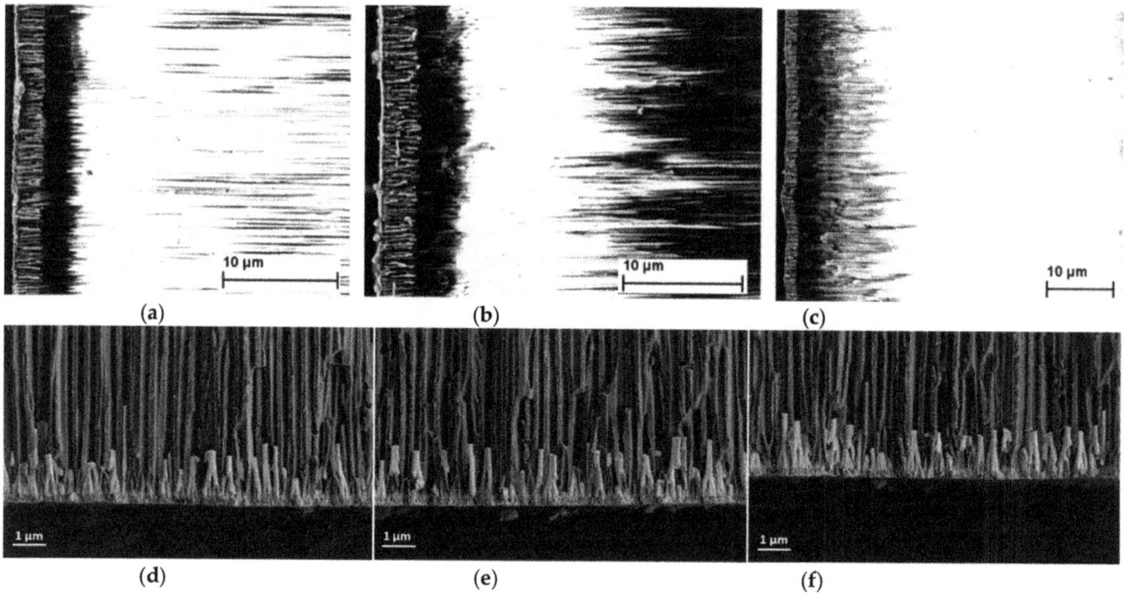

Figure 3. HR-SEM image of the AAO template cross-section filled with electrodeposited MNWs, including (**a**,**d**) NiFe nanowires, (**b**,**e**) CoPt nanowires, and (**c**,**f**) CoFe nanowires. Note: the AAO is dark in SEM along the Debye length (where electrons are conducted to the MNWs), above which it is bright due to charging.

The preparation process of core–shell nanowires was characterized, after each synthesis step, by scanning electron microscopy (SEM). Figure 4a shows an SEM microscopy image of CoFe nanowires prepared and released from the AAO template, connected via the Cu working electrode. In order to remove the Cu electrode, prior to the Au coating process, the nanowires were rinsed with an aqueous 0.1 M HNO_3 solution. In this way, the thin Cu layer was dissolved and the CoFe magnetic core nanowires were detached from the Cu electrode. Figure 4b (as well as Figure S4 in the Supplementary Materials) shows the SEM picture of the core–shell nanowires with a magnetic CoFe core and Au coating.

The SEM microscopy images show that the nanowires were completely released from the AAO template during the dissolution step and the fact that the nanowires' surface was covered with a thin layer of Au. This is also clear from EDX analysis. Figure 5 shows the SEM images of the magnetic core–shell nanowires of CoPt and NiFe with Au sheaths, respectively, together with the obtained EDX spectra. The amount of Au in the two samples, determined using EDX analysis, was 6%. The compositional analysis indicates that the prepared samples contained the following elements: NiFe and Au (for the NiFe@Au core–shell nanowires), Co, Pt, and Au (for the CoPt@Au core–shell nanowires), and Co, Fe, and Au (for the CoFe@Au core–shell nanowires). These elements are also the ones we expected to find; nickel, cobalt, platinum, and iron came from the electrodeposited magnetic core, while gold came from the coating made of noble metal.

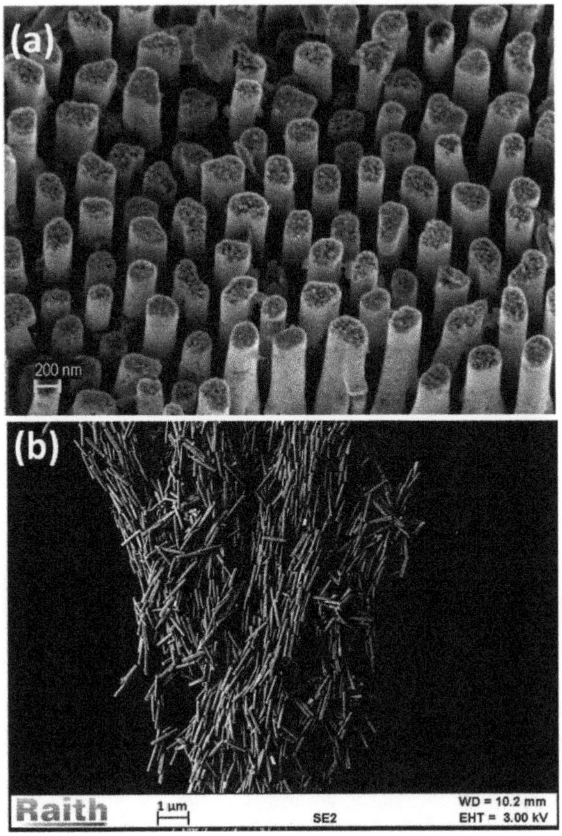

Figure 4. (**a**) SEM micrograph image of CoFe nanowires released from the nanoporous alumina membrane; (**b**) SEM micrograph image of core–shell nanowires with CoFe magnetic core and Au coating.

3.3. Evaluation of the Gold-Shell MNWs for Theranostic Applications

In order to determine the plausibility of their use for biomedical applications, we analyzed the "in vitro" biocompatibility of the core–shell NWs and their potential to be used for tumor cell destruction. The ability of artificial materials not to produce a negative reaction from the host's immune system, especially an inflammatory one, in a specific application, preserving its designed functionality related to a biomedical treatment, is generally known as biocompatibility. Testing a medical device's biocompatibility is mandatory to guarantee patient security.

Biocompatibility Evaluation of the Core–Shell Nanowires

The biocompatibility of the as-prepared core–shell nanowires was evaluated using human fibroblast cells after co-incubation for 40 h. Cell viability was determined using the MTT assay (5-dimethylthiazol-2-yl-2, 5-diphenyltetrazolium bromide), according to the supplier's instructions. Dimethyl sulfoxide was used as the dissolving agent, whereas the absorbance was read at 570 nm (Synergy HTX Multi-Mode Reader—Bioteck, Santa Clara, USA). Cell viability (CV), expressed by optical density (OD), was calculated using the formula $CV = 100 \times (ODs - ODb)/(ODc - ODb)$, where ODs = OD of the cells incubated with nanowires; ODb = OD of blank (media only); ODc = OD of untreated (control) cells.

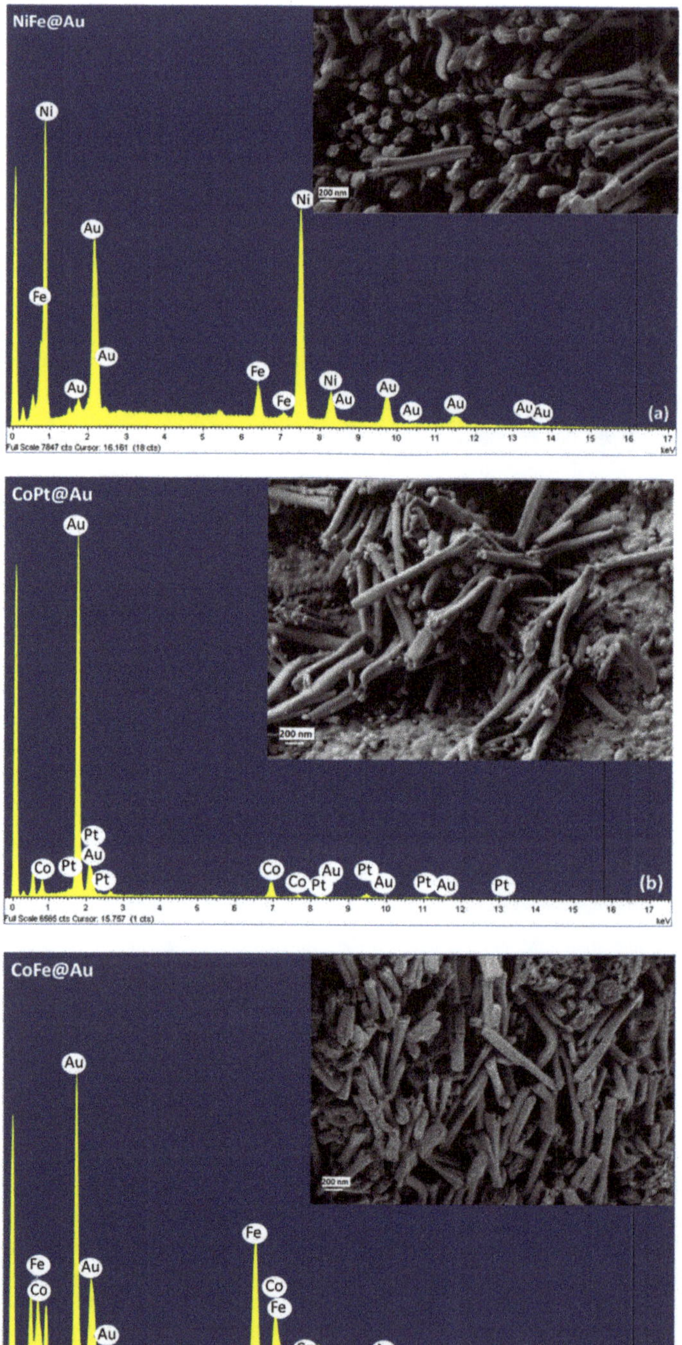

Figure 5. EDX spectra and SEM micrograph images of as-prepared core–shell nanowires: (**a**) NiFe@Au; (**b**) CoPt@Au; (**c**) CoFe@Au.

The results are shown in Figure 6. Bare and gold-coated CoPt nanowires were found to be the most cytotoxic to fibroblast cells. At the opposite pole, CoFe@Au nanowires were found to be highly biocompatible, as were NiFe and NiFe@Au nanowires. Statistical analysis indicated that there was a significant difference between CoFe@Au NWs and CoPt@Au NWs, but not between NiFe@Au NWs and CoPt@Au NWs. Additionally, we did not observe a statistical difference between CoFe@Au NWs and NiFe@Au NWs. However, similar viabilities would have been expected from the coated nanowires regardless of the metallic core. This difference between the coated nanowires may simply be the result of the different quality properties of the gold coatings during electroless plating induced by the different galvanic potentials of the metallic substrates, for an identical reaction time. Therefore, an optimized electroless plating would most likely lead to similar viability results for all gold-coated nanowires, regardless of core composition. However, by corroborating the statistical and the absolute viability results, we considered only CoFe@Au nanowires in testing antitumor capabilities via magneto-mechanical actuation.

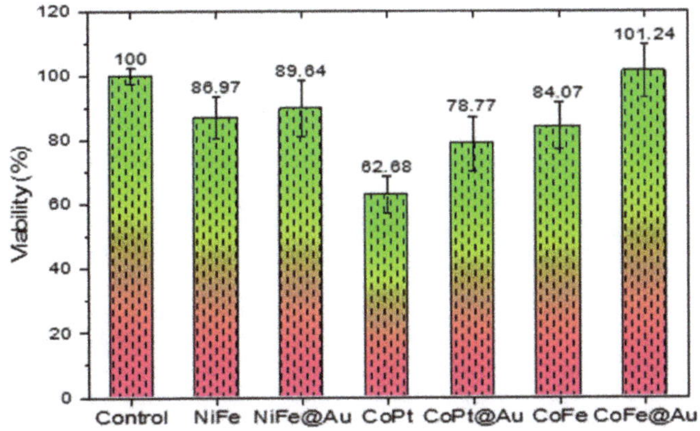

Figure 6. Cell viability of bare NiFe, CoPt, CoFe, and gold-coated (@Au) nanowires for 0.2 mg/mL concentration.

In order to increase the efficiency of the magneto-mechanic process, magnetic nanomaterials with improved magnetic susceptibilities, such as nickel, cobalt, or NiFe alloys, coated with a gold layer to ensure biocompatibility, may be required [17]. A collateral advantage of a gold layer on nanowires is the possibility of grafting organic molecules via self-assembly of thiolates [17], which are known to quickly adsorb on gold surfaces.

Magneto-Mechanical Destruction of Tumor Cells Incubated with CoFe@Au Nanowires

The antitumor effect of the NW-mediated magneto-mechanical actuation was evaluated on a culture of human osteosarcoma cells incubated with NWs in 96-well plates, at a density of 1×10^5 cells/well, and 0.2 mg NWs/mL, in an alternating magnetic field. Each well contained 0.2 mL of culture medium. The nanowire solution was prepared by weighing the 0.2 mg of dried CoFe@Au core–shell nanowires, followed by their release in the culture medium. In order to generate and control the magnetic field, a Helmholtz coil system and proprietary software were used. The well plate was positioned at the center of the system, and the magnetic field (85 Oe) was rotated with a frequency of 2 Hz.

The results show a decrease in the tumor cell viability down to 20% when the AMF was applied for 30 min (Figure 7). Given the reduced time of AMF application and one-shot treatment, the decrease in the cell viability is very significant. Most probably, serial applications of the AMF for longer time periods induce a more intense cytotoxic effect on such tumor lines. Compared to other reported data [17], the synthesized CoFe@Au nanowires afforded one of the best results in terms of "in vivo" antitumor efficacy, given the treatment time period. Consequently, we assume that this type of magnetic gold-coated nanowire could be considered for magneto-mechanical therapy of solid tumors. Regarding the magnetic properties of the CoFe nanowires, a detailed analysis can be found in our previous reported work [11], including data about the easy axis of the magnetization that was found to be parallel to the longitudinal axis of the nanowire.

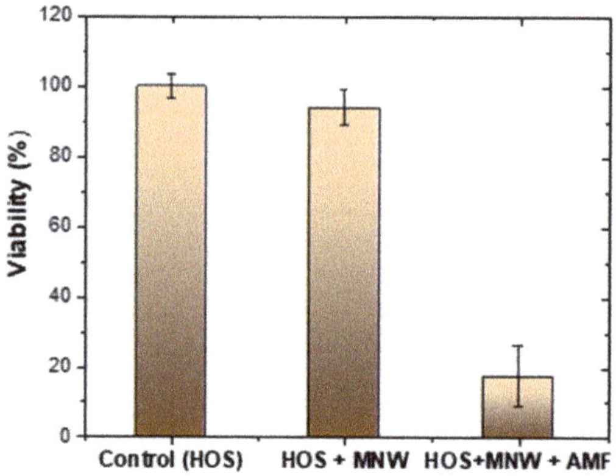

Figure 7. Viability of human osteosarcoma cells incubated with nanowires, in alternating magnetic field (AMF).

Evaluation of the CoFe@Au NWs for Their Use as T2 Contrast Agents

The relaxation times (longitudinal (T1) and transverse (T2)) of the core–shell MNWs in deionized water were determined on a Bruker Minispec mq60 NMR analyzer at 1.5 T (60 MHz) at 25 °C using the inversion recovery sequence and the Carr–Purcell–Meiboom–Gill sequence, respectively. Prior to the experiments, the CoFe@Au core–shell nanowires were subjected to SH-PEG-COOH functionalization, as previously described. The measurements showed that the r1 and r2 relaxivity values were 0.38 and 10.64 mM^{-1} s^{-1}, respectively, while the r2/r1 ratio was 28. The r2/r1 ratio value indicates if the compounds can be used as T1-dominated contrast agents (compounds showing $r2/r1 < 5$) or as T2-dominated contrast agents (compounds showing $r2/r1 > 8$). Our data clearly showed that the as-prepared CoFe@Au core–shell nanowires can be used as T2-dominated contrast agents.

4. Conclusions

In this work, gold-coated MNWs (with magnetic cores formed by NiFe, CoPt, or CoFe) were successfully prepared by combining electrodeposition inside AAO templates with electroless plating. The microstructure and the chemical composition of the as-prepared magnetic core, as well as the core–shell nanowires, were analyzed by SEM and EDX techniques, which showed nanowires of about 1 µm in length and a uniform Au shell over the magnetic core. The amount of Au determined by EDX was about 6%, regardless of the nature of the magnetic core. The biocompatibility of the as-prepared nanocomposite materials were also evaluated, with the results showing that the CoPt@Au core–shell nanowires manifest the poorest biocompatibility, while the CoFe@Au core–shell nanowires manifest the highest biocompatibility value. Accordingly, the CoFe@Au core–shell nanowires were further tested for a theranostic application, i.e., the capacity for tumor destruction through a magneto-mechanical process, as well as the evaluation of their relaxation times. The obtained data showed a decrease in the tumor cell viability of about 80% after 30 min of magnetic actuation in an alternating magnetic field. At the same time, the relaxation time measurements showed that the r2 value of the CoFe@Au core–shell nanowires was 10.64, while the r2/r1 ratio was 28. The results show that the as-prepared magnetic core–shell nanowires can also be successfully used as T2-dominated contrast agents. In conclusion, in accord with the aim of this work, we have successfully demonstrated that CoFe@Au nanocomposite materials in shape of nanowires can be used for theranostic applications, with the materials manifesting good biocompatibility, tumor destruction capacities, and high relaxation time values. Future work will be carried out in order to evaluate the influence of the core–shell nanowires' aspect ratio (e.g., ratio between nanowires' diameter and length) on their theranostic application potential. In this regard, CoFe@Au core–shell nanowires with different lengths (ranging from 0.3 to 1.5 μm) and different diameters (45, 75, and 200 nm) will be prepared, and the biocompatibility, the capacity for tumor destruction, and the relaxation time will be determined.

Supplementary Materials: The following supporting information can be downloaded at: https://www.mdpi.com/article/10.3390/coatings12111755/s1, Highlight section; Figure S1: XRD patterns of the NiFe nanowires arrays; Figure S2: XRD patterns of the CoPt nanowires arrays; Figure S3: XRD patterns of the CoFe nanowires arrays; Figure S4: SEM picture of the CoFe@Au core–shell nanowires.

Author Contributions: Conceptualization, O.-G.D.-P., D.-D.H., B.J.H.S. and N.L.; methodology, I.T., D.-D.H., B.J.H.S. and O.-G.D.-P.; formal analysis, O.-G.D.-P.; investigation, G.B., A.E.M., G.S., H.C., D.S. and V.C.P.,; resources, O.-G.D.-P. and D.-D.H.; writing—original draft preparation, O.-G.D.-P. and D.-D.H.; writing—review and editing, O.-G.D.-P., G.B., V.C.P., I.T. and B.J.H.S.; visualization, N.L.; supervision, O.-G.D.-P. and B.J.H.S.; project administration, O.-G.D.-P. and D.-D.H.; funding acquisition, O.-G.D.-P. and D.-D.H. All authors have read and agreed to the published version of the manuscript.

Funding: This work was supported by a grant of the Ministry of Research, Innovation and Digitalization, CNCS-UEFISCDI, project number PN-III-P4-PCE-2021-1395/GreenEn, within PNCDI III (contract No. PCE 110/2022); by a grant of the Ministry of Research, Innovation and Digitization CNCS–UEFISCDI, project number PN-III-P4-PCE-2021-1081 within PNCDI III (contract No. 75/2022); and by (UEFISCDI) contract no. PCE20/2021 (PN-III-P4-ID-PCE-2020-2381).

Institutional Review Board Statement: Not applicable.

Informed Consent Statement: Not applicable.

Data Availability Statement: The data presented in this study are available on request from the corresponding author.

Conflicts of Interest: The authors declare no conflict of interest.

References

1. Buema, G.; Herea, D.-D.; Dragos-Pinzaru, O.-G. Special issue: Ceramic and metallic biomaterials nanoparticles for applications in medical sciences. *Coatings* **2022**, *12*, 998. [CrossRef]
2. Zhang, J.; Shi, W.; Ma, Q.; Cui, H.; Zhang, L. Application of nanotechnology in immunity against infection. *Coatings* **2021**, *11*, 430. [CrossRef]
3. Aldawsari, H.M.; Alhakamy, N.A.; Padder, R.; Husain, M.; Md, S. Preparation and characterization of chitosan coated PLGA nanoparticles of resveratrol: Improved stability, antioxidant and apoptotic activities in H1299 lung cancer cells. *Coatings* **2020**, *10*, 439. [CrossRef]
4. Haque, S.T.; Karim, M.E.; Abidin, S.A.Z.; Othman, I.; Holl, M.M.B.; Chowdhury, E.H. Fe/Mg-modified carbonate apatite with uniform particle size and unique transport protein-related protein corona efficiently delivers doxorubicin into breast cancer cells. *Nanomaterials* **2020**, *10*, 834. [CrossRef] [PubMed]
5. Dragos, O.; Chiriac, H.; Lupu, N.; Grigoras, M.; Tabakovic, I. Anomalous codeposition of fcc NiFe nanowires with 5–55% Fe and their morphology, crystal structure and magnetic properties. *J. Electrochem. Soc.* **2016**, *163*, D83. [CrossRef]
6. Budi, S.; Manaf, A. The effects of saccharin on the electrodeposition of NiCoFe films on a flexible substrate. *Mater. Res. Express.* **2021**, *8*, 086513. [CrossRef]
7. Krimpalis, S.; Dragos, O.; Moga, A.E.; Lupu, N.; Chiriac, H. Magnetization processes in electrodeposited NiFe/Cu multilayered nanowires. *J. Mater. Res.* **2011**, *26*, 1081. [CrossRef]
8. Schneegans, O.; Moradpour, A.; Dragos, O.G.; Franger, S.; Dragoe, N.; Pinsard-Gaudart, L.; Chrétien, P.; Revcolevschi, A. Na_xCoO_2: A new opportunity for rewritable media? *J. Am. Chem. Soc.* **2007**, *129*, 7482. [CrossRef]
9. Harja, M.; Buema, G.; Lupu, N.; Chiriac, H.; Herea, D.D.; Ciobanu, G. Fly ash coated with magnetic materials: Improved adsorbent for Cu (II) removal from wastewater. *Materials* **2021**, *14*, 63. [CrossRef]

10. Aricò, A.S.; Bruce, P.; Scrosati, B.; Tarascon, J.M.; Schalkwijk, W.V. Nanostructured materials for advanced energy conversion and storage devices. *Nat. Mater.* **2005**, *4*, 366–377. [CrossRef]
11. Ghemes, A.; Dragos-Pinzaru, O.; Chiriac, H.; Lupu, N.; Grigoras, M.; Shore, D.; Stadler, B.; Tabakovic, I. Controlled electrodeposition and magnetic properties of $Co_{35}Fe_{65}$ nanowires with high saturation magnetization. *J. Electrochem. Soc.* **2017**, *164*, D13. [CrossRef]
12. Meneses, F.; Bran, C.; Vazquez, M.; Bercoff, P.G. Enhanced in-plane magnetic anisotropy in thermally treated arrays of Co-Pt nanowires. *Mater. Sci. Eng. B* **2020**, *261*, 114669. [CrossRef]
13. Franger, S.; Berthet, P.; Dragos, O.G.; Baddour-Hadjean, R.; Bonville, P.; Berthon, J. Large influence of the synthesis conditions on the physico-chemical properties of nanostructured Fe_3O_4. *J. Nanoparticle Res.* **2007**, *9*, 389–402. [CrossRef]
14. Tabakovic, I.; Qiu, J.-M.; Dragos-Pinzaru, O. Electrodeposition of thin CoPt films with very high perpendicular anisotropy from hexachloroplatinate solution: Effect of saccharin additive and electrode substrate. *J. Electrochem. Soc.* **2016**, *163*, D287. [CrossRef]
15. Cardoso, V.F.; Francesko, A.; Ribeiro, C.; Banobre-Lopez, M.; Martins, P.; Lanceros-Mendez, S. Advances in magnetic nanoparticles for biomedical applications. *Adv. Healthc. Mater.* **2018**, *7*. [CrossRef] [PubMed]
16. Herea, D.D.; Labusca, L.; Radu, E.; Chiriac, H.; Grigoras, M.; Dragos-Panzaru, O.; Lupu, L. Human adipose-derived stem cells loaded with drug-coated magnetic nanoparticles for in-vitro tumor cells targeting. *Mat. Sci. Eng. C* **2019**, *94*, 666–676. [CrossRef]
17. Naud, C.; Thebault, C.; Carriere, M.; Hou, Y.; Morel, R.; Berger, F.; Dieny, B.; Joisten, H. Cancer treatment by magneto-mechanical effect of particles, a review. *Nanoscale Adv.* **2020**, *2*, 3632–3655. [CrossRef]
18. Meng, X.; Seton, H.C.; Lu, L.T.; Prior, I.A.; Thanh, N.T.K.; Song, B. Magnetic CoPt nanoparticles as MRI contrast agent for transplanted neural stem cells detection. *Nanoscale* **2011**, *3*, 977–984. [CrossRef]
19. San, B.H.; Lee, S.; Moh, S.H.; Park, J.G.; Lee, J.H.; Hwang, H.Y.; Kim, K.K. Size-controlled synthesis and characterization of CoPt nanoparticles using protein shells. *J. Mater. Chem. B* **2013**, *1*, 1453. [CrossRef]
20. Choi, D.; Fung, A.; Moon, H.; Ho, D.; Chen, Y.; Kan, E.; Rheem, Y.; Yoo, B.; Myung, N. Transport of living cells with magnetically assembled nanowires. *Biomed. Microdevices* **2007**, *9*, 143–148. [CrossRef]
21. Czechowicz, A.; Weissman, I. Purified hematopoietic stem cell transplantation: The next generation of blood and immune replacement. *Immunol. Allergy. Clin. N. Am.* **2010**, *30*, 159–171. [CrossRef] [PubMed]
22. Pavlath, G.K.; Rando, T.A.; Blau, H.M. Transient immunosuppressive treatment leads to long-term retention of allogeneic myoblasts in hybrid myofibers. *J. Cell. Biol.* **1994**, *127*, 1923–1932. [CrossRef] [PubMed]
23. Hultgren, A.; Tanase, M.; Chen, C.S.; Meyer, G.J.; Reich, D.H. Cell manipulation using magnetic nanowires. *J. Appl. Phys.* **2003**, *93*, 7554. [CrossRef]
24. Gao, N.; Wang, H.; Yang, E.H. An experimental study on ferromagnetic nickel nanowires functionalized with antibodies for cell separation. *Nanotechnology* **2010**, *21*, 105107. [CrossRef]
25. Lin, W.S.; Lin, H.M.; Chen, H.H.; Hwu, Y.K.; Chiou, Y.J. Shape effects of iron nanowires on hyperthermia treatment. *J. Nanomater.* **2013**, *2013*, 237439. [CrossRef]
26. Gunther, A.; Bender, P.; Tschope, A.; Birringer, R. Rotational diffusion of magnetic nickel nanorods in colloidal dispersions. *J. Phys. Condens. Matter* **2011**, *23*, 325103. [CrossRef]
27. Choi, D.S.; Park, J.; Kim, S.; Gracias, D.H.; Cho, M.K.; Kim, Y.K.; Fung, A.; Lee, S.E.; Chen, Y.; Khanal, S.; et al. Hyperthermia with magnetic nanowires for inactivating living cells. *J. Nanosci. Nanotechnol.* **2008**, *8*, 2323–2327. [CrossRef]
28. Contreras, M.F.; Sougrat, R.; Zaher, A.; Ravasi, T.; Kosel, J. Non-chemotoxic induction of cancer cell death using magnetic nanowires. *Int. J. Nanomedicine* **2015**, *10*, 2141–2153. [CrossRef]
29. Shore, D.; Ghemes, A.; Dragos-Pinzaru, O.; Gao, Z.; Shao, Q.; Sharma, A.; Um, J.; Tabakovic, I.; Bischof, J.C.; Stadler, B.J.H. Nanowarming using Au-tipped $Co_{35}Fe_{65}$ ferromagnetic nanowires. *Nanoscale* **2019**, *11*, 14607–14615. [CrossRef]
30. Available online: http://www.who.int/cancer/en/ (accessed on 20 July 2022).
31. Palekar-Shanbhag, P.; Jog, S.V.; Chogale, M.M.; Gaikwad, S.S. Theranostics for cancer therapy. *Curr. Drug Deliv.* **2013**, *10*, 357–362. [CrossRef]
32. Banadaki, A.D.; Kajbafvala, A. Recent advances in facile synthesis of bimetallic nanostructures: An overview. *J. Nanomater.* **2014**, *2014*, 985948.
33. Josephson, L.; Tung, C.H.; Moore, A.; Weissleder, R. High-efficiency intracellular magnetic labeling with novel superparamagnetic-tat peptide conjugates. *Bioconjug. Chem.* **1999**, *10*, 186–191. [CrossRef] [PubMed]
34. Yu, M.; Huang, S.; Yu, K.J.; Clyne, A.M. Dextran and polymer polyethylene glycol (PEG) coating reduce both 5 and 30 nm iron oxide nanoparticle cytotoxicity in 2D and 3D cell culture. *Int. J. Mol. Sci.* **2012**, *13*, 5554–5570. [CrossRef] [PubMed]
35. Qin, J.; Laurent, S.; Jo, Y.S.; Roch, A.; Mikhaylova, M.; Bhujwalla, Z.M.; Muller, R.N.; Muhammed, M. A high-performance magnetic resonance imaging T2 contrast agent. *Adv. Mater.* **2007**, *19*, 1874–1878. [CrossRef]
36. Chatterjee, K.; Sarkar, S.; Jagajjanani Rao, K.; Paria, S. Core/shell nanoparticles in biomedical applications. *Adv. Colloid. Interface Sci.* **2014**, *209*, 8–39. [CrossRef] [PubMed]
37. Tang, J.; Ou, Q.; Zhou, H.; Qi, L.; Man, S. Seed-mediated electroless deposition of gold nano-particles for highly uniform and efficient SERS enhancement. *Nanomaterials* **2019**, *9*, 185. [CrossRef] [PubMed]

38. Bruno, L.; Urso, M.; Shacham-Diamand, Y.; Priolo, F.; Mirabella, S. Role of substrate in Au nanoparticle decoration by electroless deposition. *Nanomaterials* **2020**, *10*, 2180. [CrossRef]
39. Mashentseva, A.A.; Ibragimova, M.A.; Akhmetova, S.B.; Kozlovskiy, A.L.; Zdorovets, M.V.; Zhanerke, T. Amirkhanova. Synthesis, radical scavenging, and antimicrobial activities of core–shell Au/Ni microtubes. *Chem. Pap.* **2020**, *74*, 2189–2199. [CrossRef]
40. Dragos-Pinzaru, O.-G.; Stoian, G.; Borza, F.; Chiriac, H.; Lupu, N.; Tabakovic, I.; Stadler, B.J.H. CoPt nanowires with low Pt content for the catalytic methanol oxidation reaction (MOR). *ACS Appl. Nano Mater.* **2022**, *5*, 8089–8096. [CrossRef]
41. Shore, D.; Pailloux, S.L.; Zhang, J.; Gage, T.; Flannigan, D.J.; Garwood, M.; Pierre, V.C.; Stadler, J.H. Electrodeposited Fe and Fe–Au nanowires as MRI contrast agents. *Chem. Commun.* **2016**, *52*, 12634–12637. [CrossRef]

coatings

MDPI

Review

Developments in Dental Implant Surface Modification

Bożena Łosiewicz *, Patrycja Osak, Delfina Nowińska and Joanna Maszybrocka

Faculty of Science and Technology, Institute of Materials Engineering, University of Silesia in Katowice, 75 Pułku Piechoty 1A, 41-500 Chorzów, Poland; patrycja.osak@us.edu.pl (P.O.); delfina.nowinska09@gmail.com (D.N.); joanna.maszybrocka@us.edu.pl (J.M.)
* Correspondence: bozena.losiewicz@us.edu.pl; Tel.: +48-32-3497-527

Abstract: The development of dental implants has significantly advanced due to technological innovations aimed at improving their performance and patient outcomes. This work presents key factors influencing the success of dental implants, including osseointegration, which is the direct connection between living bone and the implant surface, and the various surface modifications that enhance this process. This review highlights the importance of surface roughness, chemical composition, and the use of bioactive coatings to promote better integration with surrounding bone tissue. Innovations such as nanotechnology, 3D printing, and smart surfaces are paving the way for more effective and personalized dental implant solutions. This review underscores the importance of ongoing research and development to improve success rates, enhance patient comfort, and reduce healing times. It focuses on creating cost-effective, reliable methods that integrate multiple functions, such as combining antibacterial and osteoconductive properties to improve overall implant performance.

Keywords: biomaterials; dental implants; long-term use; osseointegration; peri-implantitis; surface modification; titanium

1. Introduction

The development of dental implants is driven by technological advancements [1–3], patient demand [4,5], aging population [4,6–8], cost considerations [9–12], very high success rates between 90% and 98% [13–15], market growth—which is expected to expand at a compound annual growth rate (CAGR) of 9.8% from 2024 to 2030, reaching USD 9.62 billion by 2030 [16]—and patient expectations [17–20]. These factors collectively underscore the importance of ongoing innovation in the field of dental implants.

Dental implant surface modification is a critical area of research aimed at enhancing the osseointegration and longevity of dental implants. This field has seen significant advancements in recent years, driven by the need to improve the biocompatibility, stability, and overall success rates of dental implants [1,13–15,21–28]. A graphical representation of data from a Scopus database search was created to illustrate the trend in publications related to dental implant surface modifications from 1975 to 2024 (Figure 1). The data indicate a timeline of publications spanning 49 years, highlighting the growth in research on this topic over time. The results shown in Figure 1 were the basis for a large analysis of trends in dental implant research, particularly concerning surface modifications, driven by the potential to improve implant performance and patient outcomes.

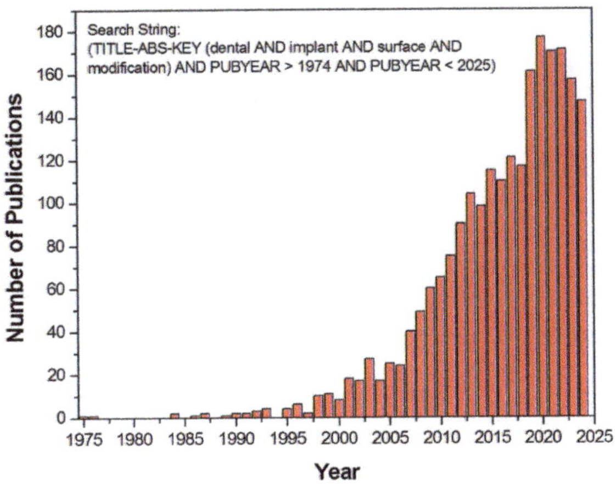

Figure 1. The number of publications on dental implant surface modifications indexed in the Scopus database from 1975 to 2024 (collected on 28 October 2024).

The analyzed studies highlight the need for more comprehensive long-term studies on dental implant surface modifications [29–41]. It was found that bioactive modifications, such as collagen-based coatings and combinations with bone morphogenetic protein-2 (BMP-2), aim to enhance osseointegration and implant longevity by improving biological properties for better bone integration [29,34–39]. Various biomolecular coatings, including bone morphogenetic proteins, growth factors, peptides, and extracellular matrix molecules, have shown early-stage benefits for bone formation and osseointegration [31,40,41]. The addition of bioactive molecules to titanium surfaces is a novel research area, with promising short-term results, but long-term clinical studies are needed to validate these benefits [31].

The reported results also show that laser modification is rapidly evolving as a physicochemical surface process for dental implants [42–48]. This technique creates complex surface topographies that enhance osseointegration and reduce bacterial colonization without affecting bulk properties [42–44,46]. Key parameters like repetition rate, pulse energy, scanning speed, and smoothness must be controlled to achieve the desired surface characteristics. Laser-modified surfaces promote better osseointegration, minimize bone loss, and allow the direct attachment of periodontal ligaments, acting as a barrier against bacterial invasion [42,43]. Additionally, laser modification improves optical, frictional, and biological properties, making implants more suitable for clinical applications [43]. Recent studies indicate that laser-modified surfaces significantly enhance early peri-implant bone healing and overall implant stability, increasing interest in their clinical use [44,45].

Previous research emphasizes that nanoengineering techniques modify dental implant surfaces at the nano-scale to influence osseointegration by altering surface features for better interaction with bone cells [30,37,49–51]. Ongoing research aims to develop new surface treatments, combining physical, chemical, and biological modifications to enhance implant surface properties [51].

This review article provides a comprehensive overview of current advancements and techniques in modifying dental implant surfaces. It details various surface treatments, including acid etching, plasma spraying, and coating applications, and their effects on biocompatibility, osseointegration, and clinical performance. The review updates readers on strategies to improve success rates and healing processes, emphasizing the importance of surface roughness and other properties in enhancing implant–bone interaction and reducing bacterial colonization.

2. Mechanism of the Osseointegration Process

The success of implant treatment highly depends on osseointegration, confirmed by implant immobility and the absence of inflammation [27,31,48,52–59]. Osseointegration is the direct morphological and functional connection between living bone tissue and the implant surface [59]. Key factors of osseointegration include the implantation technique, the primary stabilization of the implant in the embedded bone, bone quality, implant surface type, and loading method. Primary stabilization, influenced by bone density, implant shape, and surface type, is crucial for osseointegration [60–63].

Increased implant stability can be achieved by increasing the length or diameter of the implant, especially in lower-density bones [64–69]. Increasing implant diameter helps prevent bone resorption by reducing stress around the implant neck and better distributing forces on the alveolar bone. Biomechanically, threaded screw implants with various thread pitches optimize force transfer [70]. Implant surface topography also affects stabilization [70]. Rough surfaces with numerous micro-hooks improve anchoring and force distribution compared to smooth surfaces. They facilitate easier force transfer acting on the bone through the prosthetic structure and promote bone growth on the implant surface [71–77]. Despite the high clinical success of titanium dental implants, the cellular and molecular mechanisms behind osseointegration are not fully understood, mainly due to limited methodological tools in this field [78–82]. Titanium is now seen as an immunomodulatory biomaterial, as its implantation triggers a transient inflammatory state that may activate the immune pathways involved in osseointegration [83–88].

Research on osseointegration primarily focuses on surface modifications of titanium endosseous implants to enhance fusion with bone tissue [78,89,90]. In vitro studies examine surface topography, structure, and chemical composition to improve bone cell differentiation and mineralization, but are limited by the intrinsic characteristics of cell cultures, which do not fully simulate in vivo interactions. Preclinical evaluations are often conducted in animals with robust skeletal structures, such as minipigs and dogs, which can mimic human craniofacial bone architecture. However, these models have limitations, including size, weight, and a lack of specific experimental tools, making it challenging to understand the biological basis of osseointegration [66,91–101].

The osseointegration process of a titanium dental implant in the edentulous space between the maxillary right first molar and incisor along the alveolar crest in mice was studied at 3, 7, 14, and 21 days post-implantation [94], as shown in Figure 2. Molecular tests were conducted on 48 one-week-old male wild-type C57Bl/6 mice with an average weight of 25 g. Osseointegration kinetics identified the key elements responsible for cell migration, proliferation, mesenchymal stem cell (MSC) deposition, maturation, angiogenesis, bone formation, and bone remodeling at the implant–bone interface. Biguetti et al. [94] found that osseointegration in C57Bl/6 mice involves multiple stages, including the host's immune inflammatory response, bone cell differentiation, new bone formation and maturation on the titanium surface, and bone remodeling, which overlap during the process.

Bone healing and osseointegration are complex processes involving inflammation, cell migration and differentiation, matrix formation and mineralization, and remodeling, ultimately leading to stable implant anchorage [94–126]. After surgical trauma, bone healing begins with an acute inflammatory reaction, characterized by the release of inflammatory mediators like histamine and prostaglandins, increasing vascular permeability and attracting neutrophils to the site [94,102–104]. This phase includes blood clot formation, which serves as a scaffold for healing.

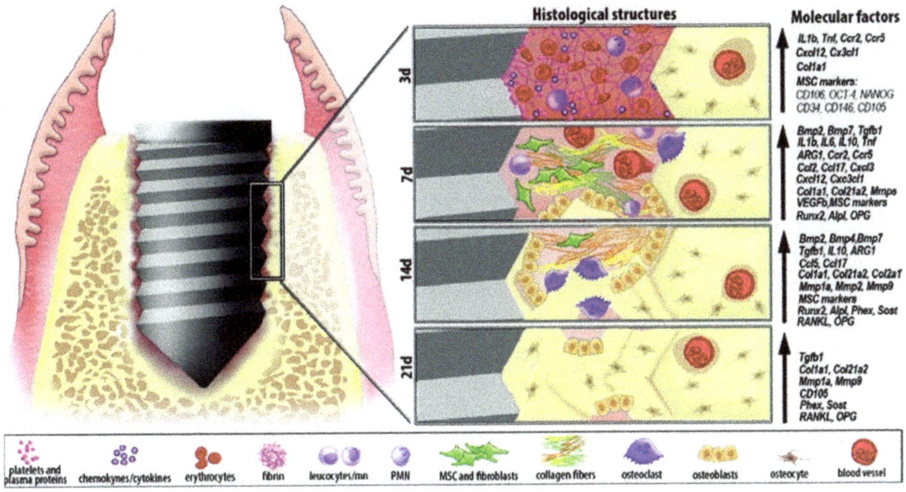

Figure 2. Osseointegration process of titanium dental implant 3, 7, 14, and 21 days after implantation in mice [94].

The disruption of blood vessels and bone nutrition leads to local necrosis and inflammation. Micro-cracks from surgery stimulate bone healing by releasing factors like TGF-β, VEGF-B, and various cytokines [94,105–109]. Bone morphogenetic proteins (BMPs) are crucial, inducing the differentiation of cells into osteoblasts and initiating bone formation. A temporary granulation tissue rich in growth factors forms, and the pH around the implant initially decreases but returns to normal [94,110–112].

The blood clot organizes into a protein network, forming young connective tissue on the implant surface. Platelet activation influences the expression of growth factors and cytokines, promoting new blood vessel formation, cell migration, and differentiation into osteoblasts [94,113–116]. Approximately 7 days post-surgery, bone-forming cells secrete an osteoid matrix, which matures and mineralizes over time, replacing the initial repair bone with mature lamellar bone [94].

The remodeling phase involves bone resorption and apposition, strengthening the repair bone and increasing the bone–implant contact (BIC) coefficient [117–119]. Osteoclasts and precursor cells become active, regulated by the RANKL/RANK/OPG signaling pathway [94,120–123]. In challenging cases, stem cells and bone-forming agents like BMPs and platelet-rich plasma (PRP) can accelerate healing and enable immediate implant loading [94,124–126].

Surface modifications of dental implants are proposed to accelerate their osseointegration and increase their biological activity [28,127–132]. The roughness of the implant surface plays a crucial role in the initial stability and long-term success of the implant. While moderate roughness can enhance osseointegration by promoting better bone–implant contact, excessive roughness can increase the risk of bacterial colonization and subsequent infection.

3. Types of Dental Implant Surface

3.1. Machined Surface

In the 1970s, titanium dental implants with a smooth (machined) surface were introduced to the dental market (Figure 3 [133]). The machined surface of dental implants is characterized by several key features that influence their clinical performance and integration with bone tissue. However, machined surfaces are relatively smooth compared to other types of implant surfaces, and due to the low development of the surface topography, show the lowest degree of osseointegration [129,134–140]. This smoothness is achieved through mechanical processes such as turning or milling, which result in a uniform and polished surface finish [135,136]. The roughness of machined surfaces is typically low, which can be quantified using parameters such as Ra (average roughness). This low roughness can affect the initial interaction between the implant and bone tissue, influencing the rate and quality of osseointegration [137–139]. The arithmetic mean deviation of the roughness profile (Ra) for implants with a machined surface obtained by machining ranges from 0.8 to 1.2 μm. It should be noted that the values of surface roughness parameters like Ra and Rq depend on the measurement length (or area) and the technique used. It is a common misconception to treat these values as "absolute numbers" without specifying the measurement conditions, which can lead to misunderstandings. To ensure accurate interpretation and comparison, always include the length or area over which the roughness parameters were measured and the specific method or equipment used. Different lengths can yield different values, and comparing data obtained with different techniques can lead to incorrect conclusions. Including this essential information when presenting roughness data helps us to understand the context and limitations of the provided values [141].

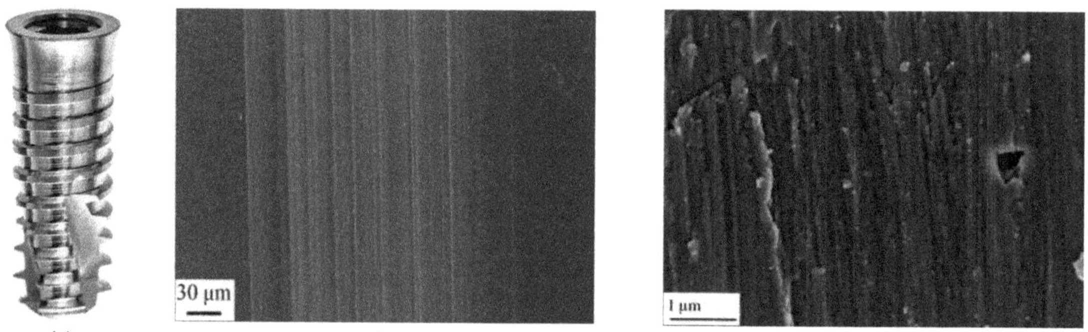

(a) (b) (c)

Figure 3. (**a**) Machined WINSIX dental implant by BioSAF IN [133]; (**b**,**c**) SEM image of a titanium implant with machined surface [134].

The machined surface of titanium implants is isometric, anisotropic, and characterized by the presence of parallel grooves with a width of about 10 μm [129]. The thickness of the TiO_2 layer on the machined surface is approximately 17 nm. Machined implants have the lowest BIC of approximately 15%, which is associated with an increased risk of implant failure [139]. The wettability of machined surfaces can vary, but they generally exhibit moderate hydrophilicity. This property is important as it affects the initial blood clot formation and subsequent cellular attachment, which are crucial steps in the osseointegration process [138]. While machined surfaces are generally smooth, they can still possess micro- and nano-topographical features that can influence cell behavior and tissue integration. These features, though subtle, can modulate the early biological response to the implant. Machined surfaces are typically made of biocompatible materials such as titanium or titanium alloys, which are known for their good biocompatibility and corrosion resistance. This ensures that the implant does not cause adverse reactions within the body [140]. Machined surfaces have been used in dental implants for many years and have shown good clinical performance. They are often used in situations where a smooth surface is preferred, such as in areas with high aesthetic demands or in patients with specific medical conditions [142].

Examples of machined titanium dental implants include the Brånemark Standard Implants system by Nobel Biocare, the Restore Machined Implants system by Lifecore Dental, and the Machined WINSIX Implants system by BioSAF IN [143,144]. The Brånemark

Standard Implants system is known for its parallel-walled design, which is recommended for all bone qualities. The implants feature a machined collar of 0.8 mm for various sizes, such as NP 3.3 and RP 3.75 [143]. The Restore Machined Implants system utilizes machined titanium implants designed to provide a smooth surface that promotes bone integration and stability [144]. The Machined WINSIX Implants system also employs machined titanium implants, focusing on achieving optimal bone-to-implant contact and long-term success [144].

3.2. Titanium Plasma Surface

The first modification of the surface of titanium implants performed in 1976 was their coating with titanium plasma (TPS), produced by the thermal spraying of titanium powder in a plasma stream [145,146]. The TPS surface of dental implants is characterized by several key features that enhance their biological performance and integration with bone tissue. The plasma treatment creates a roughened surface with micro- and nano-topographical features. The roughness of Ra is typically in the range of 20 to 30 μm, which is achieved by depositing a porous titanium coating using plasma spray technology. The thickness of the TPS layer is from 30 to 40 μm. The TPS surface is highly porous, which increases the surface area available for bone tissue interaction. This porosity is crucial for promoting osseointegration, as it allows for better bone ingrowth and mechanical interlocking with the implant [78,145–149]. The TPS surface is characterized by numerous undercuts and intrusions (Figure 4 [148]). Implants covered with a TPS layer increase the BIC coefficient by 6 times in comparison with implants with a machined surface [150]. The increase in the development of the implant surface with the TPS surface improves the implant–bone tissue connection, and the roughness being 10 times greater than in the case of the machined surface causes faster bone resorption around the implant during long-term use. The plasma treatment can improve the wettability of the titanium surface, making it more hydrophilic. This property is beneficial for the initial blood clot formation and subsequent cellular attachment, which are critical steps in the osseointegration process [151]. Titanium and its alloys are inherently biocompatible, and the plasma treatment further enhances this property. The resulting surface is able to bond with osteoblasts, promoting bone formation and integration with the surrounding bone tissue [151]. The plasma-coated surface can improve the mechanical properties of the implant, such as its wear resistance and fatigue strength. This is particularly important for ensuring the long-term stability and durability of the implant [152]. Some plasma treatments can incorporate antibacterial elements, such as silver nanoparticles, which can reduce the risk of infection and improve the overall success rate of the implant [153].

Titanium dental implants with a TPS surface are manufactured by Densply Friadent in the IMZ TPS system [154], Straumann Institute in the Bonefit system [155], Lifecore Dental in the Restore TPS system [156], Nobel Biocare in the Steri-Oss TPS system [157], and Spotimplant in the IMZ Original system [148]. The implants in the IMZ TPS system are designed to mimic the shape of natural tooth roots, which helps in achieving a more natural and secure fit during the implantation process. This system includes various types of implants, such as cylinder and screw implants, which can be used in different clinical scenarios to meet the specific needs of patients. The implants in the IMZ TPS system feature an intermobile element that acts as a stress buffer between the implant and the surrounding bone, which helps in distributing the forces more evenly and reducing the risk of implant failure. The Bonefit system was introduced in 1986 as a two-piece implant system with a smooth transgingival shoulder, which was a significant advancement in soft-tissue-level implant design [155]. The Straumann Research Institute has been involved in the development and continuous improvement of the Bonefit system. The Restore

TPS system is a well-documented and FDA-approved dental implant system known for its titanium plasma spray coating, external hexagon connection, and compatibility with various abutments [156]. It has been evaluated positively in long-term studies. The Restore TPS system is a bone-level implant, meaning the implant is placed at the bone level, and the abutment is connected to the implant after the healing period. This system is compatible with various abutments, including temporary abutments with a textured surface for acrylic, available in locking or non-locking options for single and multiple-unit restorations. The Steri-Oss TPS system includes both bone-level and tissue-level implants. These implants can be non-threaded or threaded, and they come in various configurations, such as Hex-Loc Cylindrical Non-threaded and Original Non-threaded. The TPS coating promotes bone growth and integration with the implant, which is crucial for the long-term success of dental implants. Steri-Oss was founded in 1985 and was shut down in 1997. However, the company was acquired by Nobel Biocare in 1998, and the Steri-Oss TPS system continued to be developed and marketed under Nobel Biocare's brand [157]. The IMZ Original system by Spotimplant is a standard dental implant designed for tissue-level placement [148]. The connection between the implant and the abutment is internal, and it has a non-anti-rotational shape, which means the abutment fits inside the implant without any mechanism to prevent rotation. The implant body is straight and does not have threads. It includes a shock-absorbing intermobile element, which helps in distributing forces and reducing stress on the bone. IMZ Original is a versatile and reliable option for dental implant procedures, suitable for both partially and totally edentulous arches.

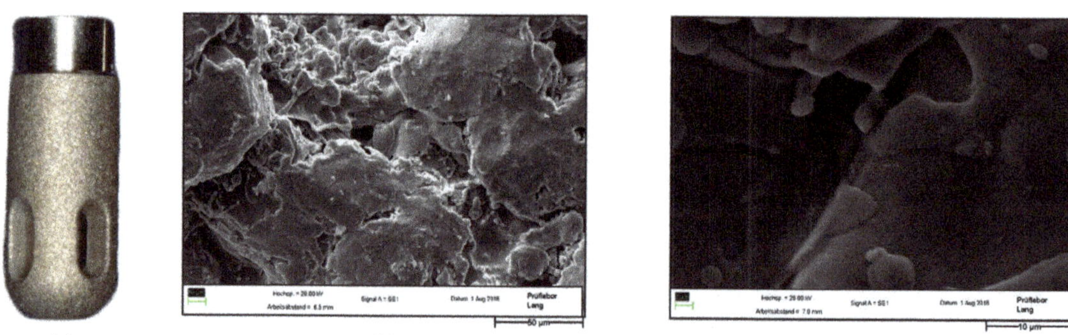

(**a**) (**b**) (**c**)

Figure 4. (**a**) IMZ Original dental implant by Spotimplant [148]; (**b**,**c**) SEM image of a titanium implant with a TPS surface after 9 years of implant production, with a typical structure resembling molten metal splashes [149].

3.3. *Sandblasted Surface*

In a significant group of dental implants, a sandblasted surface is used to enhance osseointegration by increasing surface roughness, promoting better bone-to-implant contact, improving mechanical stability, and reducing the risk of implant failure [28,53,54,74,90,129,134,158–169]. During the sandblasting process, the implant surface is bombarded with abrasive particles, most often noble corundum with grain sizes from 25 to 250 µm. It has been shown that the optimal surface roughness of Ra 1–3 µm, which provides the best osseointegration effects, is obtained when sandblasting with Al_2O_3 particles with a size of 25–75 µm [129,159]. The type and size of the abrasive are important, in addition to the shape of its particles [160]. The highest BIC coefficient is obtained using Al_2O_3 particles with a sharp shape. After the sandblasting process using noble corundum, a small amount of residual Al_2O_3 is observed on the titanium surface, which may hinder bone healing. On the other hand, recent progress in dental implant research has shown that negative static charges are generated on the surface of titanium implants with embedded Al_2O_3 particles, which support the osseointegration process as a result of the selective activation of osteoblasts and the inhibition of fibroblasts [161]. However, the static charges accumulated on the titanium surface during sandblasting disappear over time. The challenge is to find ways to maintain the stability of these charges after quantifying the desired level of negative charges needed to stimulate osteoblast activity in the process of osseointegration around dental implants. Alumina does not dissolve in acidic solutions, such as nitric acid baths used for passivation. However, blasting residues on the surface can be mostly removed when using acids that dissolve titanium, like hydrofluoric acid, after the sandblasting process. A blasting material made from phosphate ceramicis is called Resorbable Blasting Medium (RBM) [162]. This means that any particles left on the surface after blasting can be dissolved by acidic baths, making it easy to remove the blasting residues. In the sandblasting process, TiO_2 is used with grain sizes from 10 to 125 μm. Hydroxyapatite (HA), or other forms of calcium phosphate, e.g., tricalcium phosphate (β-tricalcium phosphate) [162], is also used. The increase in the grain size of the abrasive affects the increase in surface roughness. Sandblasted titanium implants are characterized by an irregular, isotropic surface, the development of which is about 34% greater in comparison with the machined surface [163]. The thickness of the TiO_2 layer on the surface of sandblasted titanium implants is about 2–5 nm [161,164].

Figure 5a shows a sandblasted dental implant of the Prima Plus 4.1 (RD) model by Lifecore Dental, which features a tapered implant body with threaded features, including reverse buttress, V-shaped, and square threads [165].

Figure 5. (**a**) Sandblasted Prima Plus 4.1 dental implant by Lifecore Dental [165]; (**b**,**c**) SEM image of a titanium implant with a sandblasted surface [134].

The Prima system is a versatile and flexible dental implant system that offers a range of implant body shapes, thread types, and connection options. It includes both one-piece and two-piece implants, making it suitable for various clinical applications. This system's compatibility with different prosthetic attachments further enhances its utility in dental treatments.

Currently, titanium implants with a surface sandblasted with noble Al_2O_3 are manufactured by the Polish company Osteoplant Research and Development in the Standard and Hex systems [166], using TiO_2 by Astra Tech Dental in the TiOblast system [167,168], and using HA by Lifecore Dental in the Renova [169] and Prima [165] systems.

3.4. *Hydroxyapatite Surface*

With the development of implant prosthetics, titanium implants with an HA surface covered with a 50 μm thick HA layer and a roughness of Ra of about 8.2–10.2 μm have appeared on the global market [53,54,74,90,134,162,170–181]. Anisotropic HA layers on the surface of titanium implants bonding with living bone can be obtained by electrochemical deposition, involving the deposition of HA through an electrochemical process [170,173]; the electrophoretic deposition method, using an electric field to deposit HA particles onto the titanium surface [173,174]; the sol–gel method, where a gel-like precursor is formed and then converted into a solid material [171,172]; pyroprocessing and hydroprocessing, where hydroprocessing involves the deposition of HA on a titanium substrate using a hydrothermal process [175]; and biomimetic deposition, which mimics the natural process of bone formation to deposit HA on the titanium surface, enhancing biocompatibility and osseointegration [176].

Titanium implants covered with an HA layer enable faster implant–bone tissue connection compared to implants with a machined, sandblasted, or TPS surface; however, the high temperature necessary to sinter the ceramic layer with the metal substrate causes changes in the HA phase. These changes lead to the degradation of the HA layer in the body and the reduced biocompatibility of implants [177]. Thermal treatment also causes changes in the titanium structure and a decrease in the bond strength of the implant with the HA layer. Currently, HA layers obtained in amorphous or crystalline form are sought, which will not require sintering with the substrate and will reduce the risk of bone tissue loss during osseointegration [28].

The dental implant type 3i T3 by BIOMET 3i shown in Figure 6a utilizes Discrete Crystalline Deposition (DCD) technology, which is a sophisticated surface modification technique that enhances the osseointegration of dental and medical implants by depositing nanometer-sized HA crystals (Figure 6c) [162]. These features are intended to help retain blood clots along the threaded part of the implant [178]. The discrete nature of the crystal deposition allows for precise control over the surface topography, which can be tailored to specific biomechanical requirements. The DCD process can be combined with other surface modification techniques such as microblasting, acid etching, and anodization to further enhance the implant's performance.

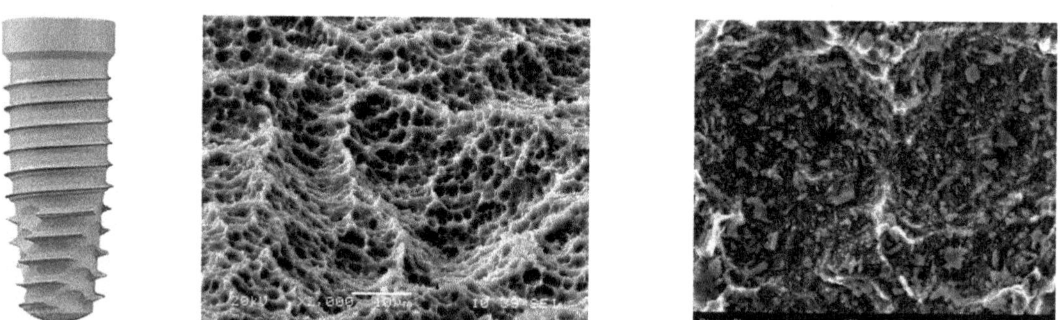

(a) (b) (c)

Figure 6. (**a**) Dental implant type 3i T3 by BIOMET 3i [178]; (**b**,**c**) SEM image of a titanium implant with HA surface [162].

Another example of titanium dental implants with an HA surface is the IMZ HA system by Densply Friadent, which incorporates an intramobile element (IME) designed to simulate the viscoelasticity of the periodontal ligament, thereby reducing stress on the surrounding bone and improving the long-term outcome of the implant. This system is known for its ability to mimic natural tooth movement and enhance the stability and longevity of dental implants [179]. The Nobel Replace External Hex (Steri-Oss) is a standard dental implant covered with an HA layer produced by Nobel Biocare, featuring a tissue-level implant with an external hexagon connection and a tapered body with buttress threads [181].

3.5. Double-Etched Surface

The double-etched (DE) surface of titanium implants is obtained by two acid etching steps with a mixture of HCl + H_2SO_4 or HF + HNO_3 acids, and is characterized by several key features that enhance their performance and integration with bone tissue [90,134,182–191]. The DE process creates a micro-rough surface on the titanium implant, which is crucial for improving osseointegration. The Ra of the DE surface is typically in the range of 1–2 μm, which has been shown to enhance bone–implant contact and mechanical interlocking with the surrounding bone [182–185]. The micro-rough surface generated by the DE process promotes better osseointegration by increasing surface roughness, enhancing cell adhesion and proliferation, mimicking natural bone environments, improving biocompatibility, and accelerating bone formation and healing [183–185]. The DE process creates a micro-rough surface with a hierarchical micro- and submicron-scale structure. This increased roughness provides a larger surface area for cell attachment and proliferation, which is crucial for the initial stages of osseointegration. The micro-rough surface has been shown to enhance the adhesion, proliferation, and differentiation of osteoblasts (bone-forming cells). This is because the rough surface mimics the natural extracellular matrix, providing a more favorable environment for cell attachment and growth. The micro-rough surface can also simulate the natural bone environment, which helps in guiding the formation of the bone matrix. This biomimetic property accelerates the bone healing process and promotes a stronger bond between the implant and the bone. The DE process not only increases surface roughness, but also modifies the surface chemistry of the titanium implant. This can enhance the biocompatibility of the implant, reducing the risk of adverse reactions and improving the overall integration with the bone. The micro-rough surface facilitates quicker bone formation and healing by providing an optimal environment for osteoblast activity. This results in a faster and more robust osseointegration process, allowing for earlier loading and functional use of the implant [192]. Interestingly, the BIC coefficient of the DE surface is higher both in comparison to less-rough machined surfaces and rougher surfaces, such as TPS, HA, and sandblasted. This means that the characteristic microstructure of the DE surface has the strongest osteoconductive properties and influences the increased adhesion of bone-forming cells, protein adsorption, and the stimulation of angiogenesis, facilitating the binding of fibrin clots [182].

However, the DE surface is obtained as a result of etching with a mixture of aggressive acids, which can cause fluoride ions to remain in the resulting cavities. Fluoride ions are retained on the surface of a dental implant through a combination of surface adsorption, ion exchange reactions, chemical reactions with surface atoms, surface modifications, and the formation of insoluble fluoride compounds. These processes are facilitated by the chemical reactivity of the surface and the nature of the etching process. Acid etching can modify the surface of the material, making it more reactive. This increased reactivity can facilitate the retention of fluoride ions through enhanced chemical reactions. The etching process can create a more porous surface with a higher surface area, which can trap fluoride ions through physical adsorption and chemical bonding. The retention of fluoride ions on the surface of titanium dental implants during a DE process is generally considered a positive phenomenon due to the resulting enhanced surface properties, improved corrosion resistance, promotion of osseointegration, antibacterial properties, and beneficial surface modification. However, careful control of the process is necessary to avoid potential negative effects associated with excessive fluoride, which can potentially lead to toxicity issues [186].

The DE surface treatment results in the improved mechanical stability of implants. This is particularly beneficial in challenging clinical scenarios where the bone quality is poor, as it helps in achieving better initial and secondary stability [183,185]. Some studies have indicated that the DE surface can also exhibit antibacterial properties, which can help in reducing the risk of infection around the implant. This is an additional benefit that contributes to the overall success of the implant [186]. The DE process creates submicron and nanometer-scale cavities on the surface. These cavities mimic the bone structure, further enhancing the biological response and integration with the bone [184]. Implants with a DE surface have shown excellent clinical success rates, with high survival rates and minimal marginal bone loss over extended periods. This has been observed in long-term clinical evaluations, demonstrating the effectiveness of the DE surface treatment [185].

Figure 7a shows the Osseotite dental implant system developed by Biomet 3i, part of Zimmer Biomet, now belonging to ZimVie [187]. The Osseotite implant features a DE surface that is designed to improve clot/implant attachment, potentially increasing platelet activation and red blood cell adherence. The SEM image of this surface presented in Figure 7b reveals its isotropic nature and the presence of numerous elevations and depressions, which facilitate bone ingrowth [188]. Figure 7c shows an enhanced microscopy image of the Osseotite surface with platelet activation [189].

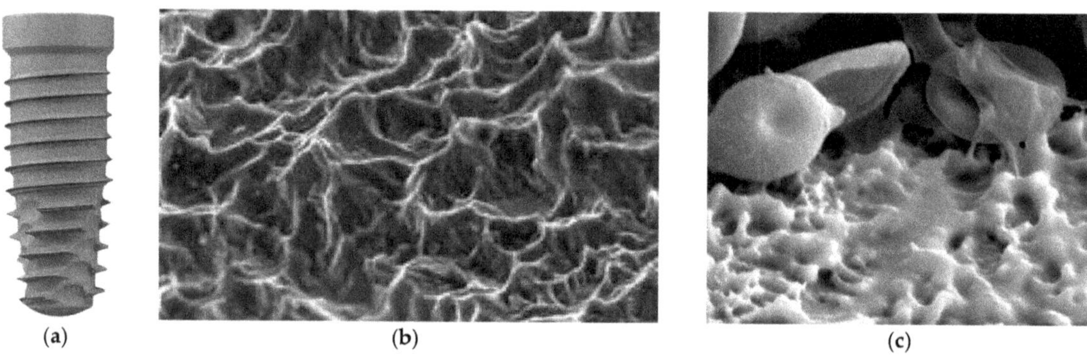

Figure 7. (**a**) The Osseotite dental implant system from Biomet 3i [187]; (**b**) SEM image of the Osseotite titanium implant with DE surface at 20,000× magnification [188]; (**c**) enhanced microscopy image of the Osseotite surface showing platelet activation [189].

The Osseotite implant is one of the most well-researched dental implants, with studies showing cumulative success rates of up to 98%. It has demonstrated effectiveness in various clinical scenarios, including immediate loading, and human histologic data confirm enhanced osteoconduction and contact osteogenesis, promoting bone healing and integration. Suitable for single-tooth, multiple-tooth, and full-arch replacements, its design ensures adequate primary stability and appropriate occlusal loading, which are essential for successful integration [187–189].

Another example of a titanium dental implant with a DE surface is the Steri-Oss Etched implant by Nobel Biocare [190,191]. The Steri-Oss Etched implant is suitable for various dental applications, including single-tooth replacements, multiple-tooth replacements, and full-arch restorations. During the healing phase, titanium healing abutments are used to prevent soft tissue from closing over the implant and to record the height and position of each healing abutment [191].

3.6. Sandblasted and Etched Surface

The sandblasted, large-grit, acid-etched (SLA) surface is a type of surface treatment commonly used in dental implants to enhance osseointegration [78,90,134,193–203]. SLA is a registered trademark of the Straumann Institute, and the SLA surface was introduced to the market by Straumann in 1997 [204]. The SLA process involves two main steps: large-grit sandblasting followed by acid etching. This combination creates a surface with both macro- and micro-roughness. The sandblasting step uses large particles to create a macro-rough surface with large pores and sharp edges, which increases the surface area for bone contact. In the sandblasting process of the SLA-type surface, typically Al_2O_3 (250–500 μm) [194], TiO_2, or HA are used as abrasives [193–197].

The acid etching step further refines the surface, creating micro-pits and a micro-rough texture that enhances the surface's ability to retain bone cells and promote bone growth. The resulting surface topography includes large dips, sharp edges, and small micro-pits. This complex structure provides an ideal environment for bone cells to adhere and proliferate [193]. The Ra values typically range around 1.5 μm, which indicates a higher roughness than machined and DE surfaces, and has been shown to improve osseointegration [78,193]. The SLA surface has a thicker oxide layer than machined and DE surfaces, consisting mainly of TiO_2, TiO, and Ti_2O_3 [193]. The developed and differentiated SLA surface has increased protein adsorption, and the BIC coefficient increases compared to the machined surface [159]. The process significantly improves the implant's surface hydrophilicity, which is crucial for bone cell adhesion and growth. SLA surfaces have been widely used and studied in clinical settings, demonstrating improved implant stability and reduced early failure rates compared to smoother surfaces [198]. The surface characteristics allow for the early loading of implants, which can reduce overall treatment time and improve patient outcomes.

Figure 8a shows a dental implant of the Standard Implants type by the Straumann Institute. This is a type of dental implant with an SLA surface designed for tissue-level placement [198].

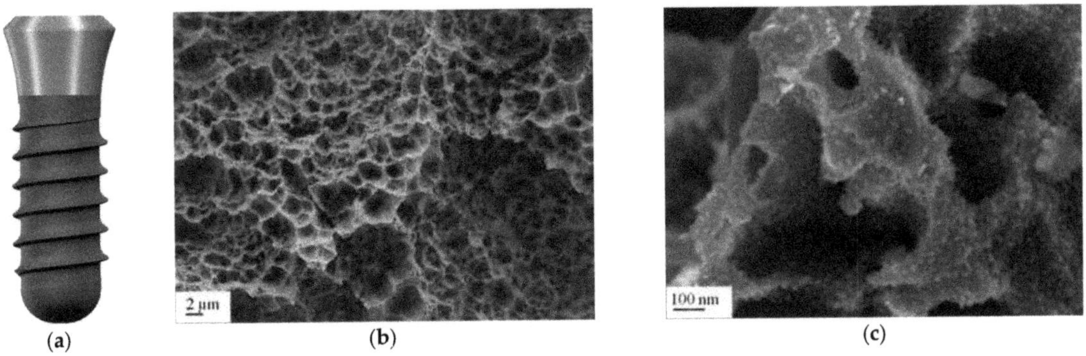

(a) (b) (c)

Figure 8. (**a**) Dental implant of the Standard Implants type with SLA surface by Straumann Institute [198]; (**b**,**c**) SEM image of a titanium implant with SLA surface [134].

Straumann Standard implants are particularly recommended for use in patients whose prosthetic solution will be based on a retention system, such as a bar, ball locks, or clasps. In such a situation, the aesthetics of the soft tissue around the implant are not crucial, and the high neck makes it easier for the patient to maintain hygiene around the implant. The implant collar has an optimized conical shape that allows for the use of a one-step clinical protocol. The implant is placed intragingivally and closed with a healing screw immediately after implantation, which allows avoiding the procedure of re-exposing the implant. The optimal thread shape ensures optimal primary and secondary stability [198]. The SLA surface of dental implants is isotropic and has a characteristic micro-topography, with wide craters of 20–40 μm in diameter (Figure 8b,c [134]). The craters created in the sandblasting process contain microcavities of 0.5–3 μm in diameter and nanofeatures that are formed as a result of etching [78,134]. The Al_2O_3 sandblasting process and etching with a HCl + H_2SO_4 mixture are used in the production of Standard Implants, Standard Plus Implants, and Tapered Effect Implants by the Straumann Institute [134,198,200], as well as SPI [201] and DFI [202] systems by Alpha Bio.

3.7. Hydrophilic Surface

The hydrophilic surface (SLActive) is obtained in the same way as the SLA surface, but after the etching process, the implants are dried in a nitrogen atmosphere and then kept in a physiological saline solution to protect from hydrocarbon adsorption [78,90,134,205–211]. The surface hydrophilicity of SLActive is primarily due to its unique surface treatment process. This process preserves the high surface energy and chemical purity of the implant, resulting in superhydrophilic properties. The hydrophilic nature allows for the rapid attachment of blood and proteins, which accelerates the osseointegration process and reduces the healing time to 3–4 weeks [212]. A clinical study comparing conventional SLA and SLActive implants found that there is generally no significant difference in clinical outcomes between the two types of implants. The cumulative survival rate was 99.4% overall, with 99.1% for SLA implants and 100% for SLActive implants. Additionally, the mean marginal bone resorption for all implants was similar, indicating comparable clinical performance between the two types of implants [213]. In the case of the SLA surface, the water contact angle is 139.90°, which indicates its ultrahydrophobic properties and lack of bioadhesion (Figure 9a) [207]. The reflections visible under a drop of water on the SLA surface result from the presence of air bubbles trapped between the water and the implant surface. The initial water wetting angle for the SLActive surface is 0°, which indicates its superhydrophilic properties and very strong bioadhesion (Figure 9b) [207]. After immersing SLA and SLActive dental implants in water, a meniscus is visible at the water–air–implant interface only in the case of the SLActive surface, due to its superhydrophilicity (Figure 9b) [207].

In the osseointegration process, two key factors are crucial for the success of dental implants: primary stability and secondary stability (Figure 9c) [207]. Primary stability refers to the mechanical stability of the implant immediately after placement. It is influenced by the quality of the bone where the implant is inserted and is determined by the specific design features of the implant system, such as thread pitch, the precision of the implant's dimensions, and accurate sizing, ensuring a snug fit in the bony socket's corresponding drills. Secondary stability relates to the biological healing process that occurs after the implant is placed. Unlike primary stability, secondary stability is not directly controlled by the clinician. It develops over time as the bone integrates with the implant, which is influenced by various factors, the most critical being the speed of osseointegration. The surface texture and composition of the implant can significantly enhance the rate at which this integration occurs. The combination of primary and secondary stability is known as total stability. It is important to note that a delay in the healing process can lead to a notable decrease in total stability, particularly between weeks 2 and 4 post-implantation. This period, often referred to as the "stability dip," is critical for the success of the osseointegration process, as compromised stability during this time can affect the long-term success of the implant. Increasing the hydrophilicity of the SLActive surface and maintaining high surface energy until the moment of implantation affects the stability of the implant after implantation (Figure 9d) [207]. The SLActive implant surface is chemically active and reaches maximum secondary stabilization twice as fast as the implant with the SLA surface, which minimizes the risk of implant loss or loosening. The increase in secondary stabilization is due to the fact that in the early stages of bone tissue regeneration, the adsorption of proteins on the SLActive surface subjected to constant moistening is facilitated, which accelerates the process of new bone formation and makes it possible to place a prosthetic reconstruction even on the day of implantation. In the case of other types of surfaces, the placement of the proper crown usually takes place about eight weeks after the procedure, when about 75% of implant anchorage is achieved [207,208]. The non-activated SLA surface is less stable, which makes it susceptible to strong mechanical stress and post-implantation complications, or loosening of the implant during mechanical loading. The chemical composition of the SLActive surface shows more than twice the carbon content of the SLA surface [208]. An increased BIC coefficient has also been shown for the SLActive surface [207,208]. Considering the fact that the SLActive and SLA surfaces do not differ in terms of roughness, the acceleration of osseointegration in the case of the SLActive surface is due to its increased hydrophilicity.

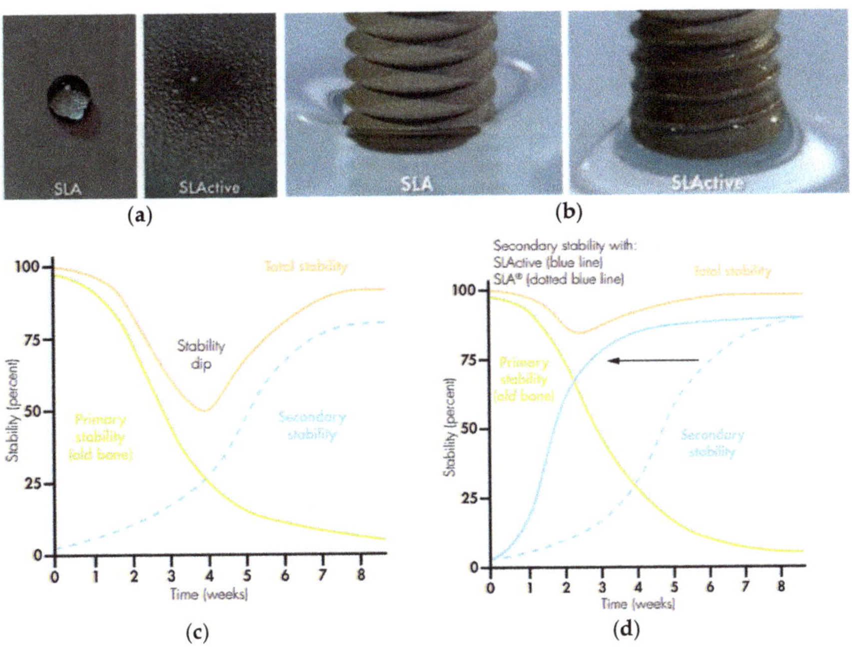

Figure 9. (**a**) Demonstration of the ultrahydrophobic properties of SLA surface and the superhydrophilic properties of SLActive surface. (**b**) Immersion of dental implants with SLA and SLActive surfaces in water. (**c**) Implant stability after implantation, with reduction in stability combined with gradual increase in secondary stability leading to a decline in overall stability (stability dip) during the period between weeks 2 and 4. (**d**) Acceleration of osseointegration between weeks 2 and 4 as a result of using an implant with SLActive surface [207].

It should be stressed that the wettability of dental implant surfaces is a critical factor that influences their interaction with water or body fluids. This property is not constant and is primarily determined by the surface chemistry of the dental implants, which are most often made of titanium. Titanium and its self-passive oxide layer are high-energy surfaces, inherently making them hydrophilic, similar to sandblasted or acid-etched surfaces. However, the absorption of hydrocarbons from the atmosphere can alter the surface properties, causing them to transition from hydrophilic to hydrophobic. This change in surface wettability can significantly impact the osseointegration process. Hydrophilic surfaces tend to adsorb plasma proteins more rapidly, which promotes clot formation, initiates healing, and supports bone regeneration. Consequently, it is suggested that dental implants be stored under controlled conditions, such as in oxygen-free atmospheres or liquids, to mitigate the biological aging process and maintain their desired surface properties. The wettability of dental implant surfaces is crucial for their performance and integration with the bone. Ensuring that these surfaces remain hydrophilic through proper storage and handling can enhance the overall success of dental implants [207,208].

Dental implant systems with SLActive surfaces, such as Bone Level Implants with extensive healing potential, Standard Implants, Standard Plus Implants and Tapered Effect Implants with SLActive surfaces, were introduced to the dental market by the Straumann Institute [207–211]. In the SEM image of the Roxolid® SLActive® surface shown in Figure 10c, clearly shaped nanostructures can be observed, which are not present in the Roxolid® SLA® surface topography visible in Figure 10b [210]. Recent in vitro studies have demonstrated that the enhanced formation of fibrin networks on the Roxolid® SLActive® surface with nanostructures is significant [211]. SEM imaging reveals that after just 15 min of contact with human whole blood, these nanostructures contribute to improved fibrin network development (Figure 10e) [211]. Interestingly, while hydrophilicity has been recognized for its role in promoting osseointegration, it does not fully account for the accelerated integration observed with Roxolid® SLActive® surfaces.

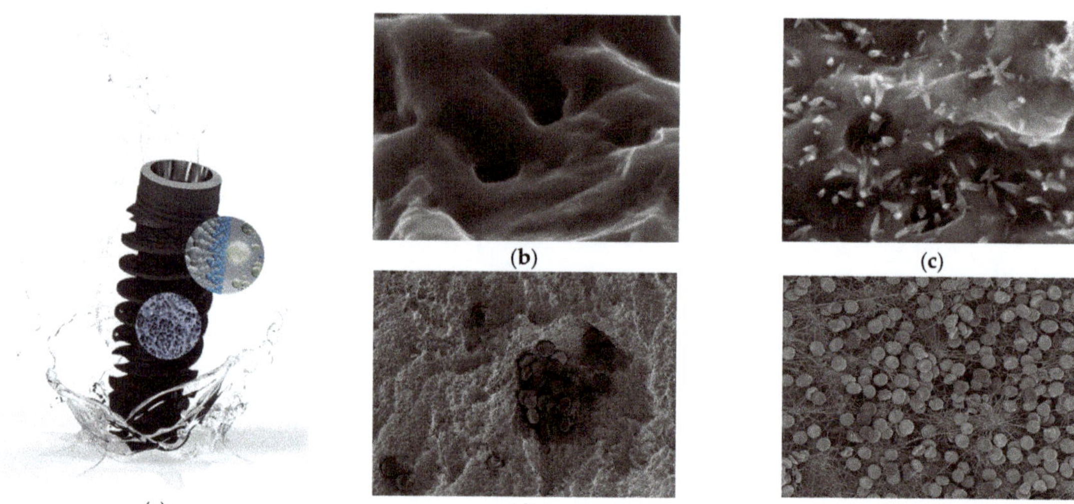
(a) (b) (c) (d) (e)

Figure 10. (**a**) Dental implant with Straumann® SLActive® surface (Basel, Switzerland) [209]; (**b**) SEM image of Roxolid® SLA® surface [210]; (**c**) SEM image of Roxolid® SLActive® surface [210]; (**d**) SEM image of Roxolid® SLActive® surface without nanostructures [211]; (**e**) SEM image of Roxolid® SLActive® surface with nanostructures [211].

The latest findings indicate that the presence of nanostructures on the Roxolid® SLActive® surface not only enhances fibrin network formation, but also supports the mineralization process of bone cells. This results in more effective early-stage osseointegration compared to the Roxolid® SLActive® surfaces lacking these nanostructures.

3.8. Oxidized Surface

Oxidized (anodized) surfaces are obtained via an electrochemical oxidation process carried out at a constant current density or voltage in the range of 1–300 V in electrolyte solutions, most often based on H_2SO_4 and H_3PO_4, with the addition of HF, which increases the current density [21–23,78,90,134,214–226]. The oxidized surface was introduced to the dental market in 2001 as TiUnite [215]. Depending on the electrochemical oxidation parameters, a continuous oxide layer [216], a porous oxide layer [217], or a layer of oxide nanotubes [21–23,218] of different thickness, structure, and surface morphology can be produced on the anode surface. The anodizing process is carried out in the optimal temperature range of 18–38 °C [219]. With the increase in the electrolyte temperature, the oxide layer thickens, which increases the resistance of titanium implants to wear and corrosion and accelerates osseointegration in the initial period after implantation [218]. The oxidized surface is rough, with an Ra value above 2 μm, and the thickness of the oxide layer ranges from 1–2 to 7–10 μm [128,220].

The TiUnite Dental Implant by Nobel Biocare is a highly advanced dental implant system with an anodized surface, known for its superior performance and clinical success (Figure 11a) [220]. The characteristic feature of the oxidized surface is isotropy and flowery morphology, with pores of dimensions ranging from 1 to 10 μm, along with smaller pores that have diameters less than 1 micrometer (Figure 11b) [134]. This results in shorter healing times and improved stability [222]. Such a microscopically porous surface enhances bone cell attachment and integration [223,224]. Hemostasis by the newly formed fibrin matrix on the TiUnite surface is shown in Figure 11c [224]. This fibrin matrix plays a crucial role in blood clot formation and subsequent hemostasis. The anodized surface of titanium, which has these specific pore dimensions, significantly enhances thrombogenicity and blood clot formation. The well-structured and spread transitory matrix formed by the fibrin clot facilitates the colonization of osteoprogenitor cells, thereby promoting osseointegration and overall hemostatic efficiency [223,224].

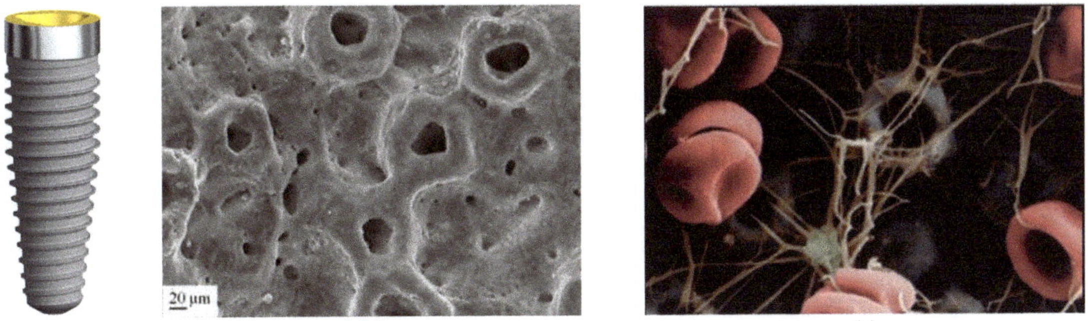

(a) (b) (c)

Figure 11. (**a**) TiUnite® Dental Implant by Nobel Biocare [220]; (**b**) SEM image of the anodized TiUnite® implant with a porous surface [134]; (**c**) hemostasis by the newly formed fibrin matrix on the TiUnite surface [224].

The results of clinical studies indicate a higher degree of osseointegration of implants with an oxidized surface than titanium implants with a machined or TPS surface [134,220,222,224,225]. However, surface treatment via anodizing may cause titanium brittleness. The TiUnite implant has demonstrated a high success rate, with a whole-patient group success rate of 98.2%, including a 100% success rate for implants with an oxidized surface [223,226]. Studies have shown that TiUnite implants have a remarkably low early failure rate and support long-term clinical survival. Early implant- and patient-level survival rates exceed 99% at one year, and the late implant-level survival rate is estimated at 95.1% after 10 years [223]. The TiUnite surface has been proven to support peri-implant health and bone maintenance, with low rates of peri-implantitis and successful soft tissue outcomes [220,223]. This surface is used in various Nobel Biocare implant systems, including the NobelActive and Brånemark System Mk III, making it suitable for different bone types and clinical scenarios [226]. It allows for immediate function of the implant, meaning patients can start using their new teeth soon after the procedure, which is beneficial for both patient comfort and clinical efficiency [220,223]. The TiUnite surface has been extensively researched, with over 465 publications featuring more than 89,500 implants and 22,600 patients, providing a robust body of evidence supporting its efficacy [223].

The latest development in anodizing technology is plasma electrolytic oxidation (PEO), also known as micro arc oxidation (MAO), which is an electrochemical surface treatment process that allows for the production of porous and hard oxide layers on titanium [227–230]. The PEO method involves high-voltage electrochemical oxidation at voltages of several hundred V, during which plasma microdischarges are generated at the titanium–electrolyte interface in the form of sparks, causing structural changes to the anode surface. Oxide layers obtained by the PEO method are about 1.3 μm thick, and their surface contains pores with a diameter of over 1.3 μm, which indicates a roughness similar to that of a machined surface [230]. Titanium implants with the TiUnite surface—Replace, Perfect, and Direct, manufactured by Nobel Biocare (Kloten, Switzerland)—are available on the dental market [220,225,226]. Oxide layers formed on the anode surface in the PEO process can be additionally enriched with Mg, Ca, S, and P ions, which provides a mechanical and biochemical connection between the implant surface and the bone tissue [231]. An example of a surface with the addition of Ca and P ions is the Biomimetic Advanced Surface developed by Avinent (Barcelona, Spain) for the Coral and Ocean dental systems [232]. These systems are characterized by optimal micro-porosity and macro-roughness, which provide high primary stabilization, and the hydrophilic properties of their surface support the osseointegration process by facilitating the migration of bone-forming and osteogenic cells.

3.9. Biologically Active Surface

The continuous development of technology has led to the creation of titanium implants with a biologically active surface, which are still in the experimental phase [36,42–45,47,48,78,90,139,150,171,193,215,217,233–244]. Biologically active surfaces on dental implants refer to surfaces that have been modified or coated with bioactive materials to enhance their interaction with biological tissues, particularly bone. These surfaces are designed to improve the implant's integration with the surrounding bone, enhance its mechanical properties, and provide additional functionalities such as antibacterial properties and improved tissue healing. The biologically active surface of titanium implants is achieved through various surface modification techniques that enhance their interaction with bone tissue, improving osseointegration and overall implant performance. Open-pored coatings are designed to improve bone ingrowth and enhance the mechanical strength of the bone–implant interface. The open-pored structure allows for better integration with the surrounding bone tissue [90,150,171,217]. Multilevel micro-pit structures are created through a series of surface treatments, including sandblasting, acid etching, and glow discharge. This process forms a hierarchical microstructure that enhances the biocompatibility and osseointegration of the titanium implant [90,215]. Micro–nano composite structures are created using laser processing and multiple acid etching steps, which improve cell behavior and bone integration [36,42–45,47,48,193]. The biological modification of titanium implants based on the loading of specific bioactive substances, such as growth factors, peptides, proteins, and drugs, has been proposed [233–235,237–243]. These substances directly interact with bone cells and the surrounding tissues, promoting cell proliferation, differentiation, and mineralization. Methods of biological modification include physical adsorption, nanotechnology, chemical bonding, self-assembly, and nucleic acid-related technologies. These methods ensure the stability of the surface morphology and create a drug delivery system that releases bioactive substances over time [233–235]. To address the issue of bacterial adhesion and postoperative infections, antibacterial coatings are applied to titanium implants. These coatings can release antibiotics or other antibacterial agents, reducing the risk of infection and promoting bone healing [78,153,241–243]. Osteoconductive coatings are designed to enhance bone growth and integration. They often include materials like hydroxyapatite and bioactive glass, which are known for their osteoconductive properties [28,41,44,170–177,218,245]. Advanced surface treatment as anodization forms a TiO_2 nanotube layer on the titanium surface, which improves biocompatibility and antibacterial properties. The TiO_2 nanotubes enhance osseointegration and reduce the risk of infection [21–23,218]. Additive manufacturing technique allows for the creation of complex surface structures with controlled porosity, which can be tailored to enhance bone integration and mechanical properties [112].

Research is also being conducted on fluoride-enriched surfaces [134,139,244]. An example is the nanotextured surface of the OsseoSpeed implant by Astra Tech, which, thanks to the presence of incorporated fluoride ions, stimulates the bone formation process and accelerates osseointegration, ensuring stable implant placement in the jaw bone (Figure 12a) [244]. The OsseoSpeed dental implant is a high-performance dental implant designed for immediate loading and enhanced osseointegration. The implant features a microthreaded neck that increases the surface area for bone contact, thereby enhancing stability and osseointegration. The implant surface is treated with a unique surface technology that promotes rapid bone growth and integration. This technology is designed to improve the initial stability and long-term success of the implant. The SEM image shown in Figure 12b,c depicts a titanium dental implant that has a surface enriched with fluoride [134]. This unique surface treatment accelerates the osseointegration process, leading to faster healing and a quicker return to normal oral function. The OsseoSpeed dental implant offers a combination of advanced surface technology, immediate loading capability, and versatile design, making it a reliable choice for dental implant procedures.

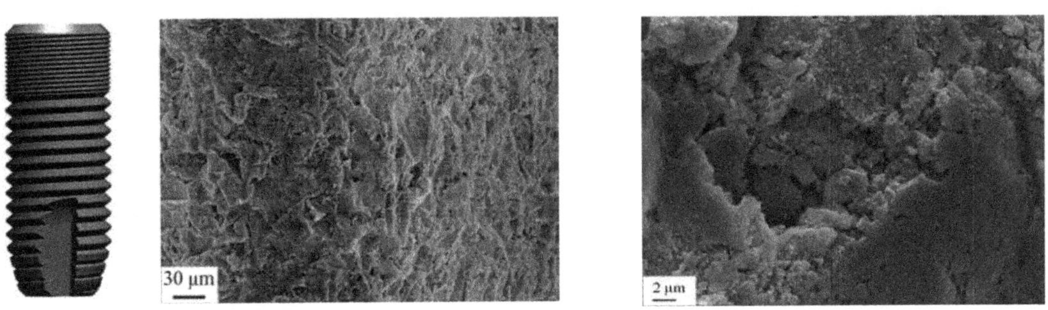

(a) (b) (c)

Figure 12. (**a**) OsseoSpeed dental implant by Astra Tech [244]; (**b**,**c**) SEM image of a titanium implant with fluoride-enriched surface [134].

The use of metal–organic frameworks (MOFs) in dental implant surface modification is a novel approach. MOFs can be used to create bioactive surfaces that improve the stability and biocompatibility of the implants, as well as provide additional functionalities [246–249].

MOFs can be used to modify the surface of titanium implants to improve osseointegration. This is achieved by creating a more bioactive surface that promotes bone cell adhesion and proliferation. For example, the use of UiO-66/AgNPs nanocomposite coatings has shown promise in enhancing osseointegration and preventing bacterial colonization, which is crucial for the long-term success of dental implants [246,247]. MOFs can incorporate antibacterial agents, such as silver nanoparticles (AgNPs), to create surfaces that resist bacterial colonization. This is particularly important in preventing peri-implantitis, a common complication in dental implantology. The UiO-66/AgNP nanocomposite, for instance, has been shown to have effective antibacterial properties [246]. Certain MOFs, like Bio-MOF-1, are designed with bio-derived constituents that are inherently biofriendly. These materials can enhance the bioactivity of the implant surface, promoting bone regeneration and integration with the surrounding tissue [248]. MOFs can be engineered to act as drug delivery systems, releasing bioactive molecules or antibiotics in a controlled manner. This can help in managing post-surgical infections and promoting healing. The use of MOFs as a confinement matrix for drug delivery systems is an emerging area of research in dental implantology [246,247]. MOFs can be used to create complex surface topographies and roughness at the micro- and nano-scale. This enhances the mechanical interlocking of the implant with the bone, further improving osseointegration. Techniques such as anodization and laser surface treatment can be combined with MOF coatings to achieve these effects [247,249]. MOFs can be tailored to improve the biocompatibility and osteoconductivity of the implant surface. This is achieved by incorporating bioactive materials like hydroxyapatite or other bone-like minerals into the MOF structure, which can mimic the natural bone environment and promote bone growth [247,248]. The field of dental implant surface modification is dynamic and continuously evolving. Advances in bioactive coatings, laser treatments, nanoengineering, and the use of MOFs are contributing to the development of more effective and durable dental implants. These innovations are expected to significantly improve patient outcomes and the overall success rates of dental implant procedures.

The addition of bioactive molecules to the surface of titanium implants induces a response from living tissue [250]. It has been shown that the biological surface supports cell adhesion to the implant surface by absorbing blood and circulating bone tissue [31]. The bioactive surface of titanium dental implants can promote bone formation around the implant, which will result in increased osseointegration in the early stages of healing. However, long-term clinical studies are required due to the fact that the results of in vivo animal studies do not fully reflect the clinical reality of the human body [28].

3.10. Hybrid Surface

Current trends in the development of dental implants with hybrid surfaces concern the search for new ways of modifying the surface of titanium implants, which will affect the acceleration of the osseointegration process and will enable the immediate functional loading of implants [27,31,48,52–63,78–85,162,173,194,208,224,237,238]. Considering the fact that too high roughness of the implant surface can cause effects opposite to those intended, surfaces with optimal micro-roughness and appropriate topography, structure, chemical composition, surface charge, wettability, and bioactivity are being developed, which will stimulate bone tissue to grow. The type of titanium implant surface is one of the key factors ensuring the success of the implantation procedure.

A hybrid surface on titanium dental implants refers to a combination of different surface treatments or coatings that aim to enhance various properties of the implant, such as osseointegration, biocompatibility, and resistance to biofilm formation [251–258]. Hybrid surfaces often combine micro- and nano-scale features. For example, a controllable hybrid micro–nano titanium model surface has been developed and contrasted with commercially available nano-featured surfaces, showcasing improved properties for dental implants [92,251]. Hybrid coatings on titanium implants can include antibacterial agents, which help in reducing the risk of infection. These coatings also promote soft tissue attachment through fibroblasts, enhancing the overall integration of the implant with the surrounding tissues [251]. Some hybrid coatings are designed to improve the aesthetic properties and wear resistance of the implant, combining the benefits of titanium and zirconia [251]. The hybrid surface design can influence biofilm formation on dental implants. Studies have evaluated the biofilm formation on hybrid titanium implants with moderately rough and turned surfaces, indicating that the hybrid surface can affect the colonization of bacteria [252]. Implants with hybrid surfaces have been evaluated for their radiological, clinical, and microbiological outcomes, particularly in patients with a history of periodontitis. These studies aim to assess the long-term success and stability of hybrid surface implants [253]. Various surface modification techniques are used to create hybrid surfaces, including chemical treatments, laser irradiation, and plasma spraying. These methods can alter the surface micro/nano-topography and composition, improving hydrophilicity, mechanical properties, and osseointegration [42,254–258].

The prototype for a hybrid implant with a micro-rough implant surface in the endosseous portion to promote faster and better osseointegration of the implant during healing and during function, and a smooth, machined surface in the neck/collar region for the transcrestal area with contact to the peri-implant sulcus was developed by Dennis Tarnow in 1993 [258]. His Hybrid Design (HD) is a significant advancement in dental implant technology, focusing on improving aesthetic outcomes and bone integration through the use of different surface textures and materials. HD has been implemented in modern IMAX NHSI hybrid dental implants, which were introduced to the medical market as standard bone-level implants by iRES, founded in Switzerland in 2014 [256], as shown in Figure 13a. The iRES implants are additionally covered with a double layer of hyaluronic acid, which is covalently bonded to the titanium surface and ensures increased hydrophilicity and the absorption of growth factors in the healing phase, as well as their better utilization by extracellular matrix proteins. The bioactive hyaluronic coating also results in a more intensive recruitment, proliferation, and differentiation of osteoblasts, and the long-term provision of the implant–bone connection. The most promising hybrid dental implants are those with machined surfaces where the arithmetic mean deviation of the surface unevenness height from the reference plane (Sa) of 0.5 μm occurs in the coronal part of the implant on 40% of its length, while the rough surface is Sa = 1.9 μm and occupies the remaining 60% (Figure 13a) [256]. The machined surface does not support the growth of bacteria; therefore, its task is to protect the implant from inflammation, resulting from the formation of an aggressive biofilm around the exposed implant neck and from bone tissue loss around the implant (peri-implantitis) (Figure 13b) [257]. The implant surface is most often exposed as a result of non-bacterial disturbance of the biological balance of tissues around the implant. Bacteria attach secondarily when the implant surface is exposed, and this favors their settlement and multiplication. On rough surfaces, bacteria form a bacterial plaque and penetrate deeper into the implant more easily.

Figure 13c presents a sequence of Wilhelmy plate measurements for such a hybrid implant covered with a bioactive hyaluronic acid nanolayer over a 1.5 s interval, with each frame taken every 0.5 s [257]. The implant, connected to a microbalance, was suspended above a beaker rising with ultrapure water. Initially, at t = 0, the implant was dry, showcasing its lower matte micro-rough texture and upper shiny machined finish. Once the apex of the implant contacted the water, capillary action rapidly occurred through the threaded connector, influenced by the implant's surface chemistry and roughness. After 0.5 s, water rose approximately halfway up the 10.5 mm micro-rough area; by 1 s, the micro-rough section was fully wetted, and at 1.5 s, the capillary rise was advancing toward the implant platform, 16 mm from the water surface. Despite being stored for a year, the implant demonstrated full wettability.

(a) (b) (c)

Figure 13. (**a**) IMAX NHSI hybrid dental implant by iRES covered with a bioactive hyaluronic acid nanolayer [256]; (**b**) the hybrid dental implant features a section labeled M that has been machined, while the section labeled A has undergone a sandblasting process followed by a double acid etching treatment [257]; (**c**) the timeframe image of the Wilhelmy plate experiment [257].

The results of the long-term wettability tests conducted on clinically available hyaluronan-coated titanium implants, which were packaged and sterile, or in their "on the shelf" condition, after one year from production, demonstrate that nanoengineering the implant surface by attaching the hydrophilic hyaluronic acid molecule provides these titanium implants with permanent wettability. This eliminates the need for wet storage, which is currently used to maintain long-term hydrophilic implant surfaces.

Figure 14 presents SEM images of the hybrid implant surface covered with a bioactive hyaluronic acid nanolayer at magnifications of 20,000× and 50,000×, showing both machined (a,b) and micro-rough (c,d) sections [257]. The SEM images are primarily characterized by the distinct features of each microarchitecture: the machined section (a,b) exhibits parallel grooves, indicative of machining tools, and the micro-rough section (c,d) shows sharp, closely packed peaks, resulting from double acid etching. At this level of magnification, the macro-roughness typical of sandblasted–acid-etched surfaces, caused by blasting, is not visible because the field of view is too small. Consequently, these images do not provide much information about the overlying hyaluronic acid layer, including its presence or homogeneity. Increasing the magnification further would not yield more information, as image contrast cannot offer vertical resolution at the nanometer scale. From an analytical perspective, the introduction of nanoengineered implant surfaces into routine clinical practice necessitates the advancement of analytical methods to the nano-scale.

In the case of developed surfaces, the time of osseointegration is shortened and a balance between the coexisting phenomena of osteogenesis and osteolysis is more easily achieved [78]. Therefore, the challenge is to design implants that are not only mechanically durable and strongly bound to the body, but that also exhibit antibacterial resistance [92].

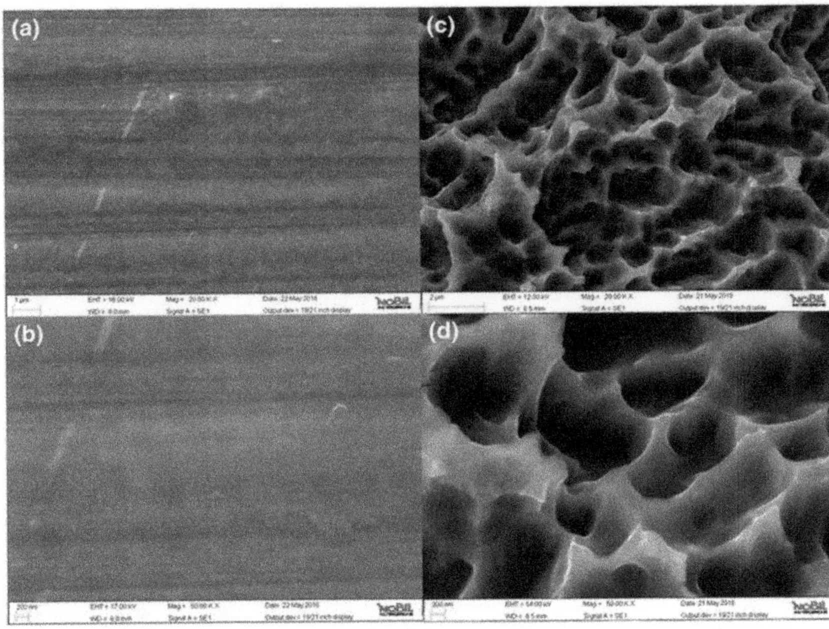

Figure 14. SEM image of the hybrid dental implant covered with a bioactive hyaluronic acid nanolayer: (**a**) machined surface in the section labeled M at 20,000× magnification; (**b**) machined surface in the section labeled M at 50,000× magnification; (**c**) sandblasted, double acid-etched surface in the section labeled A at 20,000× magnification; (**d**) sandblasted, double acid-etched surface in the section labeled A at 50,000× magnification [257].

3.11. Laser-Structured Surface

The laser-structured surface of dental implants is characterized by precise and controlled surface textures, increased roughness, enhanced osseointegration, improved bacterial resistance, material versatility, and high reproducibility [36,42–45,48,193]. Laser surface modifications can create a contamination-free titanium implant with a thick oxide layer, which promotes better osseointegration compared to conventional methods. This means the implant integrates more effectively with the surrounding bone tissue, leading to the improved stability and longevity of the implant [42,259]. Lasers can create precise and controlled surface textures on dental implants. This is achieved using different types of laser pulses, such as femtosecond, picosecond, and nanosecond pulses, which allow for the creation of micro- and nano-topographies on the implant surface. These textures can mimic the natural bone structure, enhancing the biological response and integration with the bone [260]. The micro- or nano-topography created by laser texturing can provide antimicrobial properties, which help in reducing the risk of infection around the implant. This is particularly important for preventing peri-implantitis, a common complication in dental implantology [47,261].

The controlled surface modifications can also enhance the mechanical properties of the implant, making it more durable and resistant to wear and tear [43]. Several in vivo studies have demonstrated the effectiveness of laser-structured surfaces in improving the biointegration of dental implants. These studies have shown that implants with laser-structured surfaces exhibit better osseointegration and reduced inflammation compared to those with traditionally treated surfaces [262]. Laser technology allows for the customization of implant surfaces, which can be tailored to meet specific clinical needs. This includes the ability to create surfaces that are more conducive to bone growth and integration, as well as surfaces that can be used for specific types of dental implants, such as ceramic implants [263].

Over the past few decades, the application of laser technology in dental implantology has evolved significantly. Techniques such as laser etching (Laser-Lok) have been developed and proven effective in creating a biological seal around the implant, enhancing its stability and longevity [264]. The origins of laser etching can be traced back to the invention of the laser in the early 1960s. The first laser engraver was used in 1978, and since then, the technology has evolved to become a widely used marking method in various industries [265]. Laser-Lok is a proprietary dental implant surface treatment developed by BioHorizons (Birmingham, AL, USA) [266,267].

Figure 15a shows the Tapered Internal Plus (4.5) dental implant by BioHorizons, which is a high-quality implant that offers several advantages, including predictable results, enhanced bone maintenance, improved soft tissue health, and ease of use [266]. Its design features include a 45° conical internal hex connection, color-coded platform, anatomically tapered body, and aggressive buttress threads.

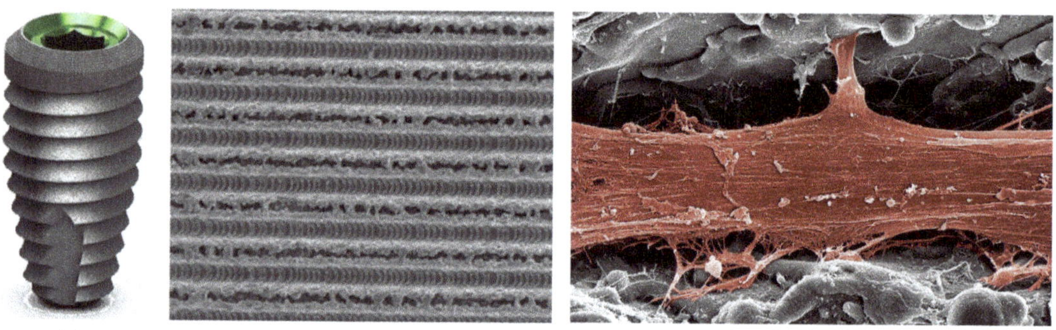

(a) (b) (c)

Figure 15. (**a**) Tapered Internal Plus (4.5) dental implant with Laser-Lok surface technology by BioHorizons [266]; (**b**) the Laser-Lok surface at 800× magnification with uniformly shaped microchannels designed to facilitate and enhance tissue growth [267]; (**c**) a colorized SEM image of a dental implant showing connective tissue that is physically attached and interwoven with the Laser-Lok surface [267].

The Laser-Lok surface of this dental implant is characterized by uniformly shaped microchannels designed to facilitate and enhance tissue growth (Figure 15b) [267]. A colorized SEM image reveals that the unique Laser-Lok surface has demonstrated its ability to trigger a biological reaction that prevents the downward growth of epithelial tissue and promotes the bonding of connective tissue. This connection creates a biological seal around the implant, which helps safeguard and preserve the health of the crestal bone (Figure 15c) [267].

BioHorizons Laser-Lok technology is a sophisticated surface treatment for dental implants and abutments that enhances the biological integration and mechanical stability of dental implants and abutments, leading to improved clinical outcomes. It consists of precision-engineered, cell-sized microchannels that are laser-machined onto the surface of dental implants and abutments. This technology is designed to enhance osseointegration and promote bone growth around the implant. The microchannels are created using a patented laser ablation process [267].

4. Future Directions

The future of dental implant surface modification is promising, with ongoing research and development aimed at improving the success rates, longevity, and patient satisfaction of dental implant procedures. Advances in materials science, nanotechnology, and bioengineering are driving innovations in implant surface modification.

Nanotechnology has already made significant inroads into dental implant surface modification, with nanostructured surfaces shown to enhance cell adhesion, proliferation, and differentiation. These surfaces can mimic the natural environment of bone, promoting better osseointegration. Researchers are exploring more sophisticated nanostructures, such as nanotubes and nanofibers, which can further optimize the surface properties of implants. These structures can be designed to incorporate bioactive molecules or drugs for controlled release, enhancing osseointegration and reducing the risk of infection.

Three-dimensional printing technology is being used to create customized dental implants with complex, porous structures that promote bone ingrowth. These implants can be designed to match the specific anatomy of the patient, improving fit and osseointegration. Advances in 3D printing materials and techniques will allow for even more precise and personalized implant designs. Researchers are also exploring the integration of nanostructured surfaces and bioactive coatings into 3D-printed implants, creating multifunctional surfaces that optimize osseointegration, mechanical stability, and antibacterial properties.

Bioactive coatings, such as hydroxyapatite and calcium phosphate, are used to enhance the biological response to dental implants. These coatings promote the formation of a calcium phosphate layer on the implant surface, mimicking the natural bone mineral. Researchers are developing more advanced biomimetic coatings that not only promote bone growth, but also mimic the complex structure and composition of natural bone. These coatings can incorporate growth factors, peptides, and other bioactive molecules to stimulate bone formation and reduce healing time.

Smart surfaces that can respond to changes in the local environment are being developed for dental implants. These surfaces can release drugs or bioactive molecules in a controlled manner, targeting specific stages of the healing process or responding to infections. The integration of smart surfaces with advanced drug delivery systems will enable more precise and effective treatment strategies. For example, surfaces that can sense bacterial infections and release antibiotics locally can reduce the risk of implant failure due to infection.

Laser and plasma surface modification techniques are used to create micro- and nano-scale patterns on implant surfaces, enhancing cell attachment and bone integration. Advances in laser and plasma technologies will allow for even more precise and controlled surface modifications. Researchers are exploring the creation of hierarchical structures that combine micro- and nano-scale features to optimize both mechanical interlocking and biological responses.

Infection is a major complication of dental implant surgery. Surface modifications that reduce bacterial adhesion and growth are being developed to minimize this risk. Researchers are exploring new materials and coatings with inherent antibacterial properties, as well as surfaces that can release antibacterial agents in a controlled manner. These surfaces can help prevent biofilm formation and reduce the risk of peri-implantitis.

Advanced imaging techniques, such as computed tomography (CT) and cone beam computed tomography (CBCT), are used to plan dental implant procedures. The integration of advanced imaging with dental implant surface modification will enable more precise and personalized treatment planning. For example, imaging data can be used to design customized implant surfaces that optimize fit and osseointegration based on the patient's specific anatomy and bone quality.

The future of dental implant surface modification is bright, with ongoing research and development focused on creating smarter, more biocompatible, and personalized implant surfaces. These innovations will not only improve the success rates of dental implant procedures, but also enhance patient comfort and reduce healing times. As these technologies advance, the field of dental implantology will continue to evolve, providing better solutions for tooth replacement and restoration. While many surface modification techniques have shown promise, they often come with high costs or inconsistent results, limiting their widespread clinical application. Ongoing research aims to develop more cost-effective and reliable methods. Future advancements will likely focus on integrating multiple functions into a single surface treatment, such as combining antibacterial and osteoconductive properties to enhance overall implant performance.

5. Conclusions

The future of dental implant surface modification is promising, with ongoing research focused on enhancing success rates, longevity, and patient satisfaction. Advances in materials science and bioengineering are driving innovations that improve osseointegration and reduce the risk of complications such as peri-implantitis. Techniques like laser surface modification and the incorporation of bioactive molecules are expected to play a crucial role in developing implants that not only integrate better with bone, but also resist bacterial colonization. As these technologies evolve, they will likely lead to more reliable and effective dental implant solutions, ultimately improving patient outcomes and satisfaction.

The development of dental implant surface modification techniques has significantly advanced the field of dental implantology. By optimizing the surface properties of implants, clinicians can achieve better osseointegration, improved mechanical stability, and a reduced risk of complications. Ongoing research continues to explore new materials and techniques to further enhance the performance and success of dental implants.

The latest surface modification techniques for dental implants improve patient outcomes by enhancing osseointegration, biocompatibility, and antibacterial properties, while also incorporating drug delivery systems and improving mechanical and durability characteristics. These advancements collectively contribute to higher success rates, reduced infection risks, and the better long-term stability of dental implants.

The most important factor for the success of the osseointegration process is the composition and type of the surface layer of the implants. The surfaces of titanium implants, which affect the biomechanical potential of the implant–bone tissue contact and the rate of protein adsorption, have been modified over the years through various methods. These modifications include physical, chemical, and biological approaches to enhance the surface properties of the implants.

Author Contributions: Conceptualization, B.Ł.; methodology, B.Ł., P.O., D.N. and J.M.; writing—original draft preparation, B.Ł., P.O. and D.N.; writing—review and editing, J.M.; visualization, B.Ł, P.O., D.N. and J.M.; supervision, B.Ł.; project administration, D.N. All authors have read and agreed to the published version of the manuscript.

Funding: This work was created partly within the framework of the Student Research Grant of the European City of Science Katowice, 2024, entitled: "Development of an innovative method for manufacturing porous oxide layers on titanium for applications in regenerative medicine" (University of Katowice, Poland).

Conflicts of Interest: The authors declare no conflicts of interest.

References

1. Waghmare, G.; Waghmare, K.; Bagde, S.; Deshmukh, M.; Kashyap, D.N.; Shahu, V.T. Materials Evolution in Dental Implantology: A Comprehensive Review. *J. Adv. Res. Appl. Mech.* **2024**, *123*, 75–100. [CrossRef]
2. Lanis, A.; Peña-Cardelles, J.F.; Negreiros, W.M.; Hamilton, A.; Gallucci, G.O. Impact of digital technologies on implant surgery in fully edentulous patients: A scoping review. *Clin. Oral Implants Res.* **2024**, *35*, 1000–1010. [CrossRef] [PubMed]
3. Shivgotra, R.; Soni, B.; Kaur, M.; Thakur, S. Advancement in Biomaterials in the Form of Implants. In *Engineered Biomaterials. Engineering Materials*; Malviya, R., Sundram, S., Eds.; Springer: Singapore, 2023. [CrossRef]
4. Nitschke, I.; Krüger, K.; Jockusch, J. Age-related knowledge deficit and attitudes towards oral implants: Survey-based examination of the correlation between patient age and implant therapy awareness. *BMC Oral Health* **2024**, *24*, 403. [CrossRef] [PubMed]
5. Zhao, Y.; Kim, B. The Effects of the Expansion of Dental Care Coverage for the Elderly. *Healthcare* **2024**, *12*, 1949. [CrossRef]
6. Kim, H.-J.; Sung, I.-Y. Analysis of Dental Prosthetic Treatment in Patients with Cancer Aged 65 Years and Older after Expanded Health Insurance Coverage: A Retrospective Clinical Study. *Medicina* **2024**, *60*, 1509. [CrossRef]
7. Gargallo-Albiol, J.; Ortega-Martínez, J.; Salomó-Coll, O.; López-Boado, A.P.; Paternostro-Betancourt, D.; Hernández-Alfaro, F. Mouth opening limitation and influence of age and surgical location for static fully guided dental implant placement: An observational, cross-sectional clinical study. *Int. J. Oral Max. Surg.* **2024**, *53*, 526–532. [CrossRef]
8. Dave, M.; Tattar, R.; Patel, N. Medical considerations in the ageing implant patient. *Oral Surgery* **2024**, *17*, 59–66. [CrossRef]
9. Karlsson, K.; Derks, J.; Wennström, J.L.; Petzold, M.; Berglundh, T. Health economic aspects of implant-supported restorative therapy. *Clin Oral Implants Res.* **2022**, *33*, 221–230. [CrossRef]
10. Brägger, U.; Krenander, P.; Lang, N.P. Economic aspects of single-tooth replacement. *Clin Oral Implants Res.* **2005**, *16*, 335–341. [CrossRef]
11. Massa, L.O.; Fraunhofer, J.A. Economics of Dental Implants. In *The ADA Practical Guide to Dental Implants*; Wiley-Blackwell: Hoboken, NJ, USA, 2021; pp. 139–143. [CrossRef]
12. Losenická, J.; Gajdoš, O.; Kamenský, V. Cost-utility analysis of an implant treatment in dentistry. *BMC Oral Health* **2021**, *21*, 433. [CrossRef]
13. Fan, Y.Y.; Li, S.; Cai, Y.J.; Wei, T.; Ye, P. Smoking in relation to early dental implant failure: A systematic review and meta-analysis. *J. Dent.* **2024**, *151*, 105396. [CrossRef] [PubMed]
14. Howe, M.S.; Keys, W.; Richards, D. Long-term (10-year) dental implant survival: A systematic review and sensitivity meta-analysis. *J. Dent.* **2019**, *84*, 9–21. [CrossRef] [PubMed]
15. French, D.; Ofec, R.; Levin, L. Long term clinical performance of 10 871 dental implants with up to 22 years of follow-up: A cohort study in 4247 patients. *Clin. Implant Dent. Relat. Res.* **2021**, *23*, 289–297. [CrossRef] [PubMed]
16. Dental Implants Market Growth & Trends. Available online: https://www.grandviewresearch.com/press-release/global-dental-implants-market (accessed on 28 October 2024).
17. O' Dwyer, S.; Riordain, R.N. The patient experience of dental implant surgery: A literature review of pertinent qualitative studies. *Ir. J. Med. Sci.* **2021**, *190*, 835–842. [CrossRef]
18. Chakaipa, S.; Prior, S.J.; Pearson, S.; van Dam, P.J. Improving Patient Experience through Meaningful Engagement: The Oral Health Patient's Journey. *Oral* **2023**, *3*, 499–510. [CrossRef]
19. Improve competitive advantage, job satisfaction and the patient experience. *Br. Dent. J.* **2022**, *232*, 347. [CrossRef]
20. Shrivastava, R.; Luxenberg, R.; Sutton, E.; Emami, E. Patients experience and satisfaction with immediate loading of implant-supported overdentures—A qualitative study. *J. Dent.* **2023**, *137*, 104644. [CrossRef]
21. Stróż, A.; Dercz, G.; Chmiela, B.; Stróż, D.; Łosiewicz, B. Electrochemical Formation of Second Generation TiO_2 Nanotubes on Ti13Nb13Zr Alloy for Biomedical Applications. *Acta Phys. Pol. A* **2016**, *130*, 1079–1080. [CrossRef]
22. Smołka, A.; Dercz, G.; Rodak, K.; Łosiewicz, B. Evaluation of corrosion resistance of nanotubular oxide layers on the Ti13Zr13Nb alloy in physiological saline solution. *Arch. Metall. Mater.* **2015**, *60*, 2681–2686. [CrossRef]
23. Smołka, A.; Rodak, K.; Dercz, G.; Dudek, K.; Łosiewicz, B. Electrochemical Formation of Self-Organized Nanotubular Oxide Layers on Ti13Zr13Nb Alloy for Biomedical Applications. *Acta Phys. Pol. A* **2014**, *125*, 932–935. [CrossRef]
24. Szklarska, M.; Dercz, G.; Rak, J.; Simka, W.; Łosiewicz, B. The influence of passivation type on corrosion resistance of Ti15Mo alloy in simulated body fluids. *Arch. Metall. Mater.* **2015**, *60*, 2687–2693. [CrossRef]
25. Łosiewicz, B.; Stróż, A.; Kubisztal, J.; Osak, P.; Zubko, M. EIS and LEIS Study on In Vitro Corrosion Resistance of Anodic Oxide Nanotubes on Ti–13Zr–13Nb Alloy in Saline Solution. *Coatings* **2023**, *13*, 875. [CrossRef]
26. Hosseini-Faradonbeh, S.A.; Katoozian, H.R. Biomechanical evaluations of the long-term stability of dental implant using finite element modeling method: A systematic review. *J. Adv. Prosthodont.* **2022**, *14*, 182–202. [CrossRef] [PubMed]
27. Xiao, L.; Zhou, Y.J.; Jiang, Y.B.; Tam, M.S.; Cheang, L.H.; Wang, H.J.; Zha, Z.G.; Zheng, X.F. Effect of Diabetes Mellitus on Implant Osseointegration of Titanium Screws: An Animal Experimental Study. *Orthop Surg.* **2022**, *14*, 1217–1228. [CrossRef]
28. Osak, P.; Maszybrocka, J.; Kubisztal, J.; Ratajczak, P.; Łosiewicz, B. Long-Term Assessment of the In Vitro Corrosion Resistance of Biomimetic ACP Coatings Electrodeposited from an Acetate Bath. *J. Funct. Biomater.* **2021**, *12*, 12. [CrossRef]

29. Shayeb, M.A.; Elfadil, S.; Abutayyem, H.; Shqaidef, A.; Marrapodi, M.M.; Cicciù, M.; Minervini, G. Bioactive surface modifications on dental implants: A systematic review and meta-analysis of osseointegration and longevity. *Clin. Oral Investig.* **2024**, *28*, 592. [CrossRef]
30. Canullo, L.; Menini, M.; Pesce, P.; Iacono, R.; Sculean, A.; Del Fabbro, M. Nano-superhydrophilic and bioactive surface in poor bone environment. Part 1: Transition from primary to secondary stability. A controlled clinical trial. *Clin. Oral Investig.* **2024**, *28*, 372. [CrossRef]
31. Meng, H.W.; Chien, E.Y.; Chien, H.H. Dental implant bioactive surface modifications and their effects on osseointegration: A review. *Biomark. Res.* **2016**, *4*, 24. [CrossRef]
32. López-Valverde, N.; Flores-Fraile, J.; Ramírez, J.M.; Macedo de Sousa, B.; Herrero-Hernández, S.; López-Valverde, A. Bioactive Surfaces vs. Conventional Surfaces in Titanium Dental Implants: A Comparative Systematic Review. *J. Clin. Med.* **2020**, *9*, 2047. [CrossRef]
33. Katić, J.; Šarić, A.; Despotović, I.; Matijaković, N.; Petković, M.; Petrović, Ž. Bioactive Coating on Titanium Dental Implants for Improved Anticorrosion Protection: A Combined Experimental and Theoretical Study. *Coatings* **2019**, *9*, 612. [CrossRef]
34. Munisamy, S.; Vaidyanathan, T.K.; Vaidyanathan, J. A bone-like precoating strategy for implants: Collagen immobilization and mineralization on pure titanium implant surface. *J. Oral Implantol.* **2008**, *34*, 67–75. [CrossRef] [PubMed]
35. de Almeida, T.C.S.; Valverde, T.M.; Martins, T.M.d.M.; Oliveira, F.d.P.; Cunha, P.d.S.; Tavares, M.A.B.; Rodrigues, E.M.; Albergaria, J.D.S.; Vieira, G.M.; Gomes, D.A.; et al. Enhanced osteogenic response by collagen type I coating on surface-modified titanium bone implants. *Mater. Today Commun.* **2024**, *39*, 108535. [CrossRef]
36. Belloni, A.; Argentieri, G.; Orilisi, G.; Notarstefano, V.; Giorgini, E.; D'Addazio, G.; Orsini, G.; Caputi, S.; Sinjari, B. New insights on collagen structural organization and spatial distribution around dental implants: A comparison between machined and laser-treated surfaces. *J. Transl. Med.* **2024**, *22*, 120. [CrossRef] [PubMed]
37. Erturk, P.A.; Altuntas, S.; Irmak, G.; Buyukserin, F. Bioinspired Collagen/Gelatin Nanopillared Films as a Potential Implant Coating Material. *ACS Appl. Bio Mater.* **2022**, *5*, 4913–4921. [CrossRef]
38. Petrović, Ž.; Šarić, A.; Despotović, I.; Katić, J.; Peter, R.; Petravić, M.; Ivanda, M.; Petković, M. Surface Functionalisation of Dental Implants with a Composite Coating of Alendronate and Hydrolysed Collagen: DFT and EIS Studies. *Materials* **2022**, *15*, 5127. [CrossRef]
39. O'Neill, L.; Twomey, B.; Tan, F.; O'Donoghue, J.; Junt, J.A. Collagen Coating of Titanium Implants Using Non-thermal Plasma. *Plasma Med.* **2020**, *11*, 63–79. [CrossRef]
40. Abdulghafor, M.A.; Mahmood, M.K.; Tassery, H.; Tardivo, D.; Falguiere, A.; Lan, R. Biomimetic Coatings in Implant Dentistry: A Quick Update. *J. Funct. Biomater.* **2024**, *15*, 15. [CrossRef]
41. Lee, S.W.; Hahn, B.D.; Kang, T.Y.; Lee, M.J.; Choi, J.Y.; Kim, M.K.; Kim, S.G. Hydroxyapatite and collagen combination-coated dental implants display better bone formation in the peri-implant area than the same combination plus bone morphogenetic protein-2-coated implants, hydroxyapatite only coated implants, and uncoated implants. *J. Oral Maxillofac. Surg.* **2014**, *72*, 53–60. [CrossRef]
42. Kolarovszki, B.; Ficsor, S.; Frank, D.; Katona, K.; Soos, B.; Turzo, K. Unlocking the potential: Laser surface modifications for titanium dental implants. *Lasers Med. Sci.* **2024**, *39*, 162. [CrossRef]
43. Saran, R.; Ginjupalli, K.; George, S.D.; Chidangil, S.; Unnikrishnan, V.K. LASER as a tool for surface modification of dental biomaterials: A review. *Heliyon* **2023**, *9*, e17457. [CrossRef]
44. Santos, A.F.P.; da Silva, R.C.; Hadad, H.; de Jesus, L.K.; Pereira-Silva, M.; Nímia, H.H.; Oliveira, S.H.P.; Guastaldi, A.C.; Queiroz, T.P.; Poli, P.P.; et al. Early Peri-Implant Bone Healing on Laser-Modified Surfaces with and without Hydroxyapatite Coating: An In Vivo Study. *Biology* **2024**, *13*, 533. [CrossRef] [PubMed]
45. The Use of Lasers for Dental Implant Surgery. Available online: https://www.deserthillsdental.com/dental-implants-and-laser-dentistry/ (accessed on 28 October 2024).
46. Papa, S.; Maalouf, M.; Claudel, P.; Sedao, X.; Maio, Y.D.; Hamzeh-Cognasse, H.; Thomas, M.; Guignandon, A.; Dumas, V. Key topographic parameters driving surface adhesion of *Porphyromonas gingivalis*. *Sci. Rep.* **2023**, *13*, 15893. [CrossRef] [PubMed]
47. Fenelon, T.; Bakr, M.; Walsh, L.J.; George, R. Effects of lasers on titanium dental implant surfaces: A narrative review. *Laser Dent. Sci.* **2022**, *6*, 153–167. [CrossRef]
48. Luczak, W.; Reiner-Rozman, C.; Muck, M.; Heitz, J.; Mitov, G.; Pfaffeneder, F.; See, C.; Hassel, A.W.; Kleber, C. Laser Treatment of Dental Implants toward an Optimized Osseointegration: Evaluation via Tapping-Mode Atomic Force Microscopy and Scanning Electron Microscopy. *Phys. Status Solidi Appl. Mater. Sci.* **2023**, *220*, 2200605. [CrossRef]
49. Alamoudi, A. Nanoengineering and Surface Modifications of Dental Implants. *Cureus* **2024**, *16*, e51526. [CrossRef]
50. Gulati, K. Nano-Engineering Solutions for Dental Implant Applications. *Nanomaterials* **2022**, *12*, 272. [CrossRef]
51. Nagamoto, K.; Nakanishi, K.; Akasaka, T.; Abe, S.; Yoshihara, K.; Nakamura, M.; Hayashi, H.; Takemoto, S.; Tamura, M.; Kitagawa, Y.; et al. Investigation of a new implant surface modification using phosphorylated pullulan. *Front. Bioeng. Biotechnol.* **2024**, *12*, 1378039. [CrossRef]

52. Karthik, K.; Thangaswamy, V. Evaluation of implant success: A review of past and present concepts. *J. Pharm. Bioallied. Sci.* **2013**, *5* (Suppl. S1), S117–S119. [CrossRef]
53. Han, W.; Fang, S.; Zhong, Q.; Qi, S. Influence of Dental Implant Surface Modifications on Osseointegration and Biofilm Attachment. *Coatings* **2022**, *12*, 1654. [CrossRef]
54. Kligman, S.; Ren, Z.; Chung, C.-H.; Perillo, M.A.; Chang, Y.-C.; Koo, H.; Zheng, Z.; Li, C. The Impact of Dental Implant Surface Modifications on Osseointegration and Biofilm Formation. *J. Clin. Med.* **2021**, *10*, 1641. [CrossRef]
55. Cooper, L.F.; Shirazi, S. Osseointegration—The biological reality of successful dental implant therapy: A narrative review. *Front. Oral Maxillofac. Med.* **2022**, *4*, 39. [CrossRef]
56. Lechner, J.; von Baehr, V.; Notter, F.; Schick, F. Osseointegration and osteoimmunology in implantology: Assessment of the immune sustainability of dental implants using advanced sonographic diagnostics: Research and case reports. *J. Int. Med. Res.* **2024**, *52*, 3000605231224161. [CrossRef] [PubMed]
57. Simão, B.S., Jr.; Costa, D.D.; Cangussu, M.C.T.; Sotto-Maior, B.S.; Devita, R.L.; de Carvalho, J.J.; da Silva Brum, I. Observational Study on the Success Rate of Osseointegration: A Prospective Analysis of 15,483 Implants in a Public Health Setting. *BioMed* **2022**, *2*, 422–430. [CrossRef]
58. Parithimarkalaignan, S.; Padmanabhan, T.V. Osseointegration: An update. *J. Indian Prosthodont. Soc.* **2013**, *13*, 2–6. [CrossRef] [PubMed]
59. Branemark, P.I.; Zarb, G.A.; Albrekson, T. *Tissue-Integrated Prostheses: Osseointegration in Clinical Dentistry*; Quintessence Publishing Company: Batavia, IL, USA, 1985.
60. Gill, T.; Kühl, S.; Rawlinson, S.; Pippenger, B.; Bellon, B.; Shahdad, S. Primary stability and osseointegration comparing a novel tapered design tissue-level implant with a parallel design tissue-level implant. An experimental in vivo study. *Clin. Oral Implants Res.* **2024**, *35*, 1114–1127. [CrossRef] [PubMed]
61. Lioubavina-Hack, N.; Lang, N.P.; Karring, T. Significance of primary stability for osseointegration of dental implants. *Clin. Oral Implants Res.* **2006**, *17*, 244–250. [CrossRef]
62. Javed, F.; Ahmed, H.B.; Crespi, R.; Romanos, G.E. Role of primary stability for successful osseointegration of dental implants: Factors of influence and evaluation. *Interv. Med. Appl. Sci.* **2013**, *5*, 162–167. [CrossRef]
63. Xu, L.; Jacobs, R.; Cao, Y.; Sun, X.; Qin, X. Tissue-engineered bone construct promotes early osseointegration of implants with low primary stability in oversized osteotomy. *BMC Oral Health* **2024**, *24*, 69. [CrossRef]
64. Lee, J.; Lim, Y.-J.; Ahn, J.-S.; Kim, B.; Baek, Y.-W.; Lim, B.-S. Correlation of two different devices for the evaluation of primary implant stability depending on dental implant length and bone density: An in vitro study. *PLoS ONE* **2024**, *19*, e0290595. [CrossRef]
65. Barikani, H.; Rashtak, S.; Akbari, S.; Badri, S.; Daneshparvar, N.; Rokn, A. The effect of implant length and diameter on the primary stability in different bone types. *J. Dent.* **2013**, *10*, 449–455.
66. Stoilov, M.; Shafaghi, R.; Stark, H.; Marder, M.; Kraus, D.; Enkling, N. Influence of Implant Macro-Design, -Length, and -Diameter on Primary Implant Stability Depending on Different Bone Qualities Using Standard Drilling Protocols-An In Vitro Analysis. *J. Funct. Biomater.* **2023**, *14*, 469. [CrossRef]
67. Gómez-Polo, M.; Ortega, R.; Gómez-Polo, C.; Martín, C.; Celemín, A.; del Río, J. Does Length, Diameter, or Bone Quality Affect Primary and Secondary Stability in Self-Tapping Dental Implants? *J. Oral Maxillofac. Surgery* **2016**, *74*, 1344–1353. [CrossRef]
68. Cucinelli, C.; Pereira, M.S.; Borges, T.; Figueiredo, R.; Leitão-Almeida, B. The Effect of Increasing Thread Depth on the Initial Stability of Dental Implants: An In Vitro Study. *Surgeries* **2024**, *5*, 817–825. [CrossRef]
69. Hiranmayi, V.K. Factors influencing implant stability. *J. Dent. Implants* **2018**, *8*, 69–76. [CrossRef]
70. Huang, S.; Murphy, L.; Xu, W. Genes and functions from breast cancer signatures. *BMC Cancer* **2018**, *18*, 473. [CrossRef]
71. El-Anwar, M.I.; El-Zawahry, M.M.; El-Mofty, M. Load Transfer on Dental Implants and Surrounding Bones. *Aust. J. Basic Appl. Sci.* **2012**, *6*, 551–560.
72. Hansson, S.; Norton, M. The relation between surface roughness and interfacial shear strength for bone-anchored implants. A mathematical model. *J. Biomech.* **1999**, *32*, 829–836. [CrossRef]
73. Bianchi, A.E.; Dolci, G., Jr.; Sberna, M.T.; Sanfilippo, F. Factors affecting bone response around loaded titanium dental implants: A literature review. *J. Appl. Biomater. Biomech.* **2005**, *3*, 135–140.
74. Stanford, C.M. Surface modifications of dental implants. *Aust. Dent. J.* **2008**, *53*, S26–S33.
75. Skalak, R.; Zhao, Y. Interaction of force-fitting and surface roughness of implants. *Clin. Implant Dent. Relat. Res.* **2000**, *2*, 219–224. [CrossRef]
76. Barfeie, A.; Wilson, J.; Rees, J. Implant surface characteristics and their effect on osseointegration. *Br. Dent. J.* **2015**, *218*, E9. [CrossRef]
77. Romero, M.; Herrero-Climent, M.; Ríos-Carrasco, B.; Brizuela, A.; Romero, M.M.; Gil, J. Investigation of the Influence of Roughness and Dental Implant Design on Primary Stability via Analysis of Insertion Torque and Implant Stability Quotient: An In Vitro Study. *J. Clin. Med.* **2023**, *12*, 4190. [CrossRef] [PubMed]

78. Le Guéhennec, L.; Soueidan, A.; Layeolle, P.; Amouriq, Y. Surface treatments of titanium dental implants for rapid osseointegration. *Dent. Mater.* **2007**, *23*, 844–854. [CrossRef] [PubMed]
79. Hosseini, S.H.; Kazemian, M.; Ghorbanzadeh, S. A brief overview of cellular and molecular mechanisms of osseointegration. *Int. J. Contemp. Dent. Med. Rev.* **2015**, *12*, 13.
80. Yu, M.; Yang, H.; Li, B.; Wang, R.; Han, Y. Molecular mechanisms of interrod spacing-mediated osseointegration via modulating inflammatory response and osteogenic differentiation. *Chem. Eng. J.* **2023**, *454*, 140141. [CrossRef]
81. Nishimura, I. Genetic Networks in Osseointegration. *J. Dent. Res.* **2013**, *92* (Suppl. 12), 109S–118S. [CrossRef]
82. Ma, Y.; Wang, S.; Wang, H.; Chen, X.; Shuai, Y.; Wang, H.; Mao, Y.; He, F. Mesenchymal stem cells and dental implant osseointegration during aging: From mechanisms to therapy. *Stem. Cell Res. Ther.* **2023**, *14*, 382. [CrossRef]
83. Trindade, R.; Albrektsson, T.; Galli, S.; Prgomet, Z.; Tengvall, P.; Wennerberg, A. Osseointegration and foreign body reaction: Titanium implants activate the immune system and suppress bone resorption during the first 4 weeks after implantation. *Clin. Implant Dent. Relat. Res.* **2018**, *20*, 82–91. [CrossRef]
84. Yin, X.; Yang, C.; Wang, Z.; Zhang, Y.; Li, Y.; Weng, J.; Feng, B. Alginate/chitosan modified immunomodulatory titanium implants for promoting osteogenesis in vitro and in vivo. *Mater. Sci. Eng. C Mater. Biol. Appl.* **2021**, *124*, 112087. [CrossRef]
85. Wu, J.; Chen, M.; Xiao, Y.; Yang, H.; Wang, G.; Zhang, X.; Dai, L.; Yuan, Z. The Bioactive Interface of Titanium Implant with Both Anti-Oxidative Stress and Immunomodulatory Properties for Enhancing Osseointegration under Diabetic Condition. *Adv. Healthc. Mater.* **2024**, *13*, e2401974. [CrossRef]
86. Chen, L.; Wang, D.; Qiu, J.; Zhang, X.; Liu, X.; Qiao, Y.; Liu, X. Synergistic effects of immunoregulation and osteoinduction of ds-block elements on titanium surface. *Bioact. Mater.* **2020**, *6*, 191–207. [CrossRef]
87. Sun, H.; Yang, Y.; Yu, L.; Liu, K.; Fei, Y.; Guo, C.; Zhou, Y.; Hu, J.; Shi, L.; Ji, H. Inhibition of Inflammatory Response and Promotion of Osteogenic Activity of Zinc-Doped Micro-Arc Titanium Oxide Coatings. *ACS Omega* **2022**, *7*, 14920–14932. [CrossRef] [PubMed]
88. Dong, J.; Wang, W.; Zhou, W.; Zhang, S.; Li, M.; Li, N.; Pan, G.; Zhang, X.; Bai, J.; Zhu, C. Immunomodulatory biomaterials for implant-associated infections: From conventional to advanced therapeutic strategies. *Biomater. Res.* **2022**, *26*, 72. [CrossRef] [PubMed]
89. Lee, U.L.; Yun, S.; Lee, H.; Cao, H.L.; Woo, S.H.; Jeong, Y.H.; Jung, T.G.; Kim, C.M.; Choung, P.H. Osseointegration of 3D-printed titanium implants with surface and structure modifications. *Dent. Mater.* **2022**, *38*, 1648–1660. [CrossRef] [PubMed]
90. Kurup, A.; Dhatrak, P.; Khasnis, N. Surface modification techniques of titanium and titanium alloys for biomedical dental applications: A review. *Mater. Today Proc.* **2021**, *39*, 84–90. [CrossRef]
91. Abrahamsson, I.; Zitzmann, N.U.; Berglundh, T.; Wennerberg, A.; Lindhe, J. Bone and soft tissue integration to titanium implants with different surface topography: An experimental study in the dog. *Int. J. Oral Maxillofac. Implant.* **2001**, *16*, 323–332.
92. Stich, T.; Alagboso, F.; Křenek, T.; Kovářík, T.; Alt, V.; Docheva, D. Implant-bone-interface: Reviewing the impact of titanium surface modifications on osteogenic processes in vitro and in vivo. *Bioeng. Transl. Med.* **2022**, *7*, e10239. [CrossRef]
93. Ikeda, E.; Tsuji, T. Growing bioengineered teeth from single cells: Potential for dental regenerative medicine. *Expert Opin. Biol. Ther.* **2008**, *8*, 735–744. [CrossRef]
94. Biguetti, C.C.; Cavalla, F.; Silveira, E.M.; Fonseca, A.C.; Vieira, A.E.; Tabanez, A.P.; Rodrigues, D.C.; Trombone, A.P.F.; Garlet, G.P. Oral implant osseointegration model in C57Bl/6 mice: Microtomographic, histological, histomorphometric and molecular characterization. *J. Appl. Oral Sci.* **2017**, *1*, e20170601. [CrossRef]
95. Perlman, R.L. Mouse models of human disease: An evolutionary perspective. *Evol. Med. Public Health* **2016**, *1*, 170–176. [CrossRef]
96. Setiawati, R.; Rahardjo, P. *Bone Development and Growth*; IntechOpen: London, UK, 2019. [CrossRef]
97. Scarano, A.; Khater, A.G.A.; Gehrke, S.A.; Inchingolo, F.; Tari, S.R. Animal Models for Investigating Osseointegration: An Overview of Implant Research over the Last Three Decades. *J. Funct. Biomater.* **2024**, *15*, 83. [CrossRef]
98. Pazzaglia, U.E. Periosteal and endosteal reaction to reaming and nailing: The possible role of revascularization on the endosteal anchorage of cementless stems. *Biomaterials* **1996**, *17*, 1009–1014. [CrossRef] [PubMed]
99. Quarterman, J.C.; Phruttiwanichakun, P.; Fredericks, D.C.; Salem, A.K. Zoledronic Acid Implant Coating Results in Local Medullary Bone Growth. *Mol. Pharm.* **2022**, *19*, 4654–4664. [CrossRef] [PubMed]
100. Niehaus, A.J.; Anderson, D.E.; Samii, V.F.; Weisbrode, S.E.; Johnson, J.K.; Noon, M.S.; Tomasko, D.L.; Lannutti, J.L. Effects of orthopedic implants with a polycaprolactone polymer coating containing bone morphogenetic protein-2 on osseointegration in bones of sheep. *Am. J. Vet. Res.* **2009**, *70*, 1416–1425. [CrossRef] [PubMed]
101. Gupta, S. Bone Healing in the Presence of Orthopaedic Implants. In *Handbook of Orthopaedic Trauma Implantology*; Banerjee, A., Biberthaler, P., Shanmugasundaram, S., Eds.; Springer: Singapore, 2023; pp. 869–904. [CrossRef]
102. Maruyama, M.; Rhee, C.; Utsunomiya, T.; Zhang, N.; Ueno, M.; Yao, Z.; Goodman, S.B. Modulation of the Inflammatory Response and Bone Healing. *Front. Endocrinol.* **2020**, *11*, 386. [CrossRef]
103. Loi, F.; Córdova, L.A.; Pajarinen, J.; Lin, T.H.; Yao, Z.; Goodman, S.B. Inflammation, fracture and bone repair. *Bone* **2016**, *86*, 119–130. [CrossRef]

104. Gupta, S. Bone Healing in the Presence of Orthopedic Implants. In *Handbook of Orthopaedic Trauma Implantology*; Banerjee, A., Biberthaler, P., Shanmugasundaram, S., Eds.; Springer: Singapore, 2022; pp. 1–36. [CrossRef]
105. Yang, Y.; Xiao, Y. Biomaterials Regulating Bone Hematoma for Osteogenesis. *Adv. Healthc. Mater.* **2020**, *9*, e2000726. [CrossRef]
106. Shiu, H.T.; Leung, P.C.; Ko, C.H. The roles of cellular and molecular components of a hematoma at early stage of bone healing. *J. Tissue Eng. Regen. Med.* **2018**, *12*, e1911–e1925. [CrossRef]
107. Milillo, L.; Cinone, F.; Lo Presti, F.; Lauritano, D.; Petruzzi, M. The Role of Blood Clot in Guided Bone Regeneration: Biological Considerations and Clinical Applications with Titanium Foil. *Materials* **2021**, *14*, 6642. [CrossRef]
108. Shiu, H.T.; Goss, B.; Lutton, C.; Crawford, R.; Xiao, Y. Formation of blood clot on biomaterial implants influences bone healing. *Tissue Eng. Part B Rev.* **2014**, *20*, 697–712. [CrossRef]
109. Duda, G.N.; Geissler, S.; Checa, S.; Tsitsilonis, S.; Petersen, A.; Schmidt-Bleek, K. The decisive early phase of bone regeneration. *Nat. Rev. Rheumatol.* **2023**, *19*, 78–95. [CrossRef]
110. Pathak, U.S.; Balasubramanian, A.; Beilan, J.A.; Butaney, M.; Tatem, A.J.; Thirumavalavan, N.; Lipshultz, L.I. Vasoepididymostomy: An insight into current practice patterns. *Transl. Androl. Urol.* **2019**, *8*, 728–735. [CrossRef] [PubMed]
111. Han, S.; Proctor, A.R.; Ren, J.; Benoit, D.S.W.; Choe, R. Temporal blood flow changes measured by diffuse correlation tomography predict murine femoral graft healing. *PLoS ONE* **2018**, *13*, e0197031. [CrossRef] [PubMed]
112. Kurian, M.; Stevens, R.; McGrath, K.M. Towards the Development of Artificial Bone Grafts: Combining Synthetic Biomineralisation with 3D Printing. *J. Funct. Biomater.* **2019**, *10*, 12. [CrossRef] [PubMed]
113. Naito, H.; Iba, T.; Takakura, N. Mechanisms of new blood-vessel formation and proliferative heterogeneity of endothelial cells. *Int. Immunol.* **2020**, *32*, 295–305. [CrossRef]
114. Rodrigues, M.; Kosaric, N.; Bonham, C.A.; Gurtner, G.C. Wound Healing: A Cellular Perspective. *Physiol. Rev.* **2019**, *99*, 665–706. [CrossRef]
115. Scridon, A. Platelets and Their Role in Hemostasis and Thrombosis-From Physiology to Pathophysiology and Therapeutic Implications. *Int. J. Mol. Sci.* **2022**, *23*, 12772. [CrossRef]
116. de Sousa Gomes, P.; Daugela, P.; Poskevicius, L.; Mariano, L.; Fernandes, M.H. Molecular and Cellular Aspects of Socket Healing in the Absence and Presence of Graft Materials and Autologous Platelet Concentrates: A Focused Review. *J. Oral Maxillofac. Res.* **2019**, *10*, e2. [CrossRef]
117. Kenkre, J.S.; Bassett, J.H.D. The bone remodelling cycle. *Ann. Clin. Biochem.* **2018**, *55*, 308–327. [CrossRef]
118. Feng, X.; McDonald, J.M. Disorders of bone remodeling. *Annu. Rev. Pathol.* **2011**, *6*, 121–145. [CrossRef]
119. Xiao, W.; Wang, Y.; Pacios, S.; Li, S.; Graves, D.T. Cellular and Molecular Aspects of Bone Remodeling. *Front. Oral Biol.* **2016**, *18*, 9–16. [CrossRef]
120. Udagawa, N.; Koide, M.; Nakamura, M.; Nakamichi, Y.; Yamashita, T.; Uehara, S.; Kobayashi, Y.; Furuya, Y.; Yasuda, H.; Fukuda, C.; et al. Osteoclast differentiation by RANKL and OPG signaling pathways. *J. Bone Miner. Metab.* **2021**, *39*, 19–26. [CrossRef] [PubMed]
121. Boyce, B.F.; Xing, L. Biology of RANK, RANKL, and osteoprotegerin. *Arthritis. Res. Ther.* **2007**, *9* (Suppl. 1), S1. [CrossRef] [PubMed]
122. Boyce, B.F.; Xing, L. Functions of RANKL/RANK/OPG in bone modeling and remodeling. *Arch. Biochem. Biophys.* **2008**, *473*, 139–146. [CrossRef] [PubMed]
123. Tobeiha, M.; Moghadasian, M.H.; Amin, N.; Jafarnejad, S. RANKL/RANK/OPG Pathway: A Mechanism Involved in Exercise-Induced Bone Remodeling. *BioMed Res. Int.* **2020**, *2020*, 6910312. [CrossRef]
124. Yun, J.H.; Han, S.H.; Choi, S.H.; Lee, M.H.; Lee, S.J.; Song, S.U.; Oh, N. Effects of bone marrow-derived mesenchymal stem cells and platelet-rich plasma on bone regeneration for osseointegration of dental implants: Preliminary study in canine three-wall intrabony defects. *J. Biomed. Mater. Res. B Appl. Biomater.* **2014**, *102*, 1021–1030. [CrossRef]
125. Lin, H.; Tang, Y.; Lozito, T.P.; Oyster, N.; Wang, B.; Tuan, R.S. Efficient in vivo bone formation by *BMP-2* engineered human mesenchymal stem cells encapsulated in a projection stereolithographically fabricated hydrogel scaffold. *Stem Cell Res. Ther.* **2019**, *10*, 254. [CrossRef]
126. Fernandes, G.; Yang, S. Application of platelet-rich plasma with stem cells in bone and periodontal tissue engineering. *Bone Res.* **2016**, *4*, 16036. [CrossRef]
127. Aniołek, K.; Łosiewicz, B.; Kubisztal, J.; Osak, P.; Stróż, A.; Barylski, A.; Kaptacz, S. Mechanical Properties, Corrosion Resistance and Bioactivity of Oxide Layers Formed by Isothermal Oxidation of Ti-6Al-7Nb Alloy. *Coatings* **2021**, *11*, 505. [CrossRef]
128. Alla, R.K.; Ginjupalli, K.; Upadhya, N.; Mohammed, S.; Sekar, R.; Ravi, R. Surface Roughness of Implants: A Review. *Trends Biomater. Artif. Organs* **2011**, *25*, 112–118.
129. Łosiewicz, B.; Osak, P.; Maszybrocka, J.; Kubisztal, J.; Stach, S. Effect of Autoclaving Time on Corrosion Resistance of Sandblasted Ti G4 in Artificial Saliva. *Materials* **2020**, *13*, 4154. [CrossRef]
130. Stróż, A.; Maszybrocka, J.; Goryczka, T.; Dudek, K.; Osak, P.; Łosiewicz, B. Influence of Anodizing Conditions on Biotribological and Micromechanical Properties of Ti–13Zr–13Nb Alloy. *Materials* **2023**, *16*, 1237. [CrossRef] [PubMed]

131. Osak, P.; Maszybrocka, J.; Kubisztal, J.; Łosiewicz, B. Effect of amorphous calcium phosphate coatings on tribological properties of titanium grade 4 in protein-free artificial saliva. *Biotribology* **2022**, *32*, 100219. [CrossRef]
132. Łosiewicz, B.; Osak, P.; Maszybrocka, J.; Kubisztal, J.; Bogunia, S.; Ratajczak, P.; Aniołek, K. Effect of Temperature on Electrochemically Assisted Deposition and Bioactivity of CaP Coatings on CpTi Grade 4. *Materials* **2021**, *14*, 5081. [CrossRef] [PubMed]
133. Cylindrical Dental Implant MACHINED WINSIX®. Available online: https://www.medicalexpo.com/prod/biosaf/product-124601-1082647.html (accessed on 28 October 2024).
134. Ballo, A.M.; Omar, O.; Xia, W.; Palmquist, A. *Dental Implant Surfaces—Physicochemical Properties, Biological Performance, and Trends*; IntechOpen: London, UK, 2011. [CrossRef]
135. Cochran, D.L. A comparison of endosseous dental implant surfaces. *J. Periodontol.* **1999**, *70*, 1523–1539. [CrossRef]
136. Hong, D.G.K.; Oh, J.H. Recent advances in dental implants. *Maxillofac. Plast. Reconstr. Surg.* **2017**, *39*, 33. [CrossRef]
137. Matos, G.R.M. Surface Roughness of Dental Implant and Osseointegration. *J. Maxillofac. Oral Surg.* **2021**, *20*, 1–4. [CrossRef]
138. Rupp, F.; Liang, L.; Geis-Gerstorfer, J.; Scheideler, L.; Hüttig, F. Surface characteristics of dental implants: A review. *Dent. Mater.* **2018**, *34*, 40–57. [CrossRef]
139. Pham, M.H.; Haugen, H.J.; Reseland, J.E. Fluoride Modification of Titanium Surfaces Enhance Complement Activation. *Materials* **2020**, *13*, 684. [CrossRef]
140. Addy, L. An introduction to dental implants. *Br. Dent. J.* **2024**, *236*, 753–757. [CrossRef]
141. Vörös, J.; Wieland, M.; Ruiz-Taylor, L.; Textor, M.; Brunette, D.M. Characterization of Titanium Surfaces. In *Titanium in Medicine. Engineering Materials*; Brunette, D.M., Tengvall, P., Textor, M., Thomson, P., Eds.; Springer: Berlin/Heidelberg, Germany, 2001; p. 114. [CrossRef]
142. Machined Surface Coronal (MSc) Dental Implants. Available online: https://southernimplants.com/solutions/innovative-products/msc-implants/ (accessed on 28 October 2024).
143. Brånemark System® Mk III. Available online: https://store.nobelbiocare.com/us/en/media/eifu/IFU1014_EN_US_00.pdf (accessed on 28 October 2024).
144. Bredent Medical. Available online: https://www.medicalexpo.com.cn/prod/bredent-medical/product-71642-1088938.html (accessed on 28 October 2024).
145. Fousová, M.; Vojtech, D.; Jablonska, E.; Fojt, J.; Lipov, J. Novel Approach in the Use of Plasma Spray: Preparation of Bulk Titanium for Bone Augmentations. *Materials* **2017**, *10*, 987. [CrossRef]
146. Jiang, P.; Zhang, Y.; Hu, R.; Shi, B.; Zhang, L.; Huang, Q.; Yang, Y.; Tang, P.; Lin, C. Advanced surface engineering of titanium materials for biomedical applications: From static modification to dynamic responsive regulation. *Bioact. Mater.* **2023**, *27*, 15–57. [CrossRef] [PubMed]
147. Cizek, J.; Matejicek, J. Medicine Meets Thermal Spray Technology: A Review of Patents. *J. Therm. Spray Tech* **2018**, *27*, 1251–1279. [CrossRef]
148. IMZ Original. Available online: https://www.spotimplant.com/en/dental-implants/imz/imz-original (accessed on 28 October 2024).
149. Seidling, R.; Lehmann, L.J.; Lingner, M.; Mauermann, E.; Obertacke, U.; Schwarz, M.L.R. Analysis of the osseointegrative force of a hyperhydrophilic and nanostructured surface refinement for TPS surfaces in a gap healing model with the Göttingen minipig. *J. Orthop. Surg. Res.* **2016**, *11*, 119. [CrossRef] [PubMed]
150. Simmons, C.A.; Valiquette, N.; Pillar, R.M. Osseointegration of Sintered Porous-surfaced and Plasma Spray–Coated Implants: An Animal Model Study of Early Postimplantation Healing Response and Mechanical Stability. *J. Biomed. Mater. Res.* **1999**, *47*, 127–138. [CrossRef]
151. Hadzik, J.; Jurczyszyn, K.; Gębarowski, T.; Trytek, A.; Gedrange, T.; Kozakiewicz, M.; Dominiak, M.; Kubasiewicz-Ross, P.; Trzcionka-Szajna, A.; Szajna, E.; et al. An Experimental Anodized and Low-Pressure Oxygen Plasma-Treated Titanium Dental Implant Surface-Preliminary Report. *Int. J. Mol. Sci.* **2023**, *24*, 3603. [CrossRef]
152. Rodriguez y Baena, R.; Rizzo, S.; Manzo, L.; Lupi, S.M. Nanofeatured Titanium Surfaces for Dental Implantology: Biological Effects, Biocompatibility, and Safety. *J. Nanomater.* **2017**, *2017*, 6092895. [CrossRef]
153. Vasilev, O.; Hayles, A.; Campbell, D.; Jaarsma, R.; Johnson, L.; Vasilev, K. Nanoscale antibacterial coatings incorporating silver nanoparticles derived by plasma techniques—A state-of-the-art perspective. *Mater. Today Chem.* **2024**, *41*, 102341. [CrossRef]
154. Ten Good Reasons for IMZ®-TwinPlus—DENTSPLY Friadent. Available online: https://www.yumpu.com/en/document/view/9007132/ten-good-reasons-for-imzr-twinplus-dentsply-friadent (accessed on 28 October 2024).
155. Bruggenkate, C.M.; Sutter, F.; Schroeder, A.; Oosterbeek, H.S. Explantation procedure in the F-type and Bonefit ITI implant system. *Int. J. Oral Maxillofac. Surg.* **1991**, *20*, 155–158. [CrossRef]
156. Lifecore Dental in the Restore TPS System. In Brief: Lifecore. Available online: https://insights.citeline.com/MT003637/In-Brief-Lifecore/ (accessed on 28 October 2024).
157. Steri-Oss®. Available online: https://www.spotimplant.com/en/dental-implants/steri-oss (accessed on 28 October 2024).

158. Fintová, S.; Kuběna, I.; Palán, J.; Mertová, K.; Duchek, M.; Hutař, P.; Pastorek, F.; Kunz, L. Influence of Sandblasting and Acid Etching on Fatigue Properties of Ultra-Fine Grained Ti Grade 4 for Dental Implants. *J. Mech. Behav. Biomed. Mater.* **2020**, *111*, 104016. [CrossRef]
159. Kim, H.-K.; Ahn, B. Effect of Al_2O_3 Sandblasting Particle Size on the Surface Topography and Residual Compressive Stresses of Three Different Dental Zirconia Grades. *Materials* **2021**, *14*, 610. [CrossRef]
160. Gil, F.; Pérez, R.; Olmos, J.; Herraez-Galindo, C.; Gutierrez-Pérez, J.; Torres-Lagares, D. The Effect of Using Al_2O_3 and TiO_2 in Sandblasting of Titanium Dental Implants. *J. Mater. Res.* **2022**, *37*, 2604–2613. [CrossRef]
161. Guo, C.Y.; Matinlinna, J.P.; Tang, A.T. Effects of surface charges on dental implants: Past, present, and future. *Int. J. Biomater.* **2012**, *2012*, 381535. [CrossRef] [PubMed]
162. Smeets, R.; Stadlinger, B.; Schwarz, F.; Beck-Broichsitter, B.; Jung, O.; Precht, C.; Kloss, F.; Gröbe, A.; Heiland, M.; Ebker, T. Impact of Dental Implant Surface Modifications on Osseointegration. *BioMed Res. Int.* **2016**, *2016*, 6285620. [CrossRef] [PubMed]
163. Lukaszewska-Kuska, M.; Leda, B.; Gajdus, P.; Hedzelek, W. Evaluation of modified titanium surfaces physical and chemical characteristics. *Nucl. Instrum. Methods Phys. Res. B* **2017**, *411*, 94–99. [CrossRef]
164. Kasemo, B.; Lausmaa, J. Surface science aspects on inorganic biomaterials. *Crit. Rev. Biocompat.* **1986**, 2, 335–380.
165. Prima Plus 4.1 (RD). Available online: https://osseosource.com/prima-plus-4-1-rd-/p-3167.html (accessed on 28 October 2024).
166. Osteoplant Hex. Available online: https://www.spotimplant.com/en/dental-implants/osteoplant/osteoplant-hex (accessed on 28 October 2024).
167. Collaert, B.; De Bruyn, H. Immediate functional loading of TiOblast dental implants in full-arch edentulous mandibles: A 3-year prospective study. *Clin. Oral Implants Res.* **2008**, *19*, 1254–1260. [CrossRef]
168. Al-Nawas, B.; Kämmerer, P.W.; Morbach, T.; Ladwein, C.; Wegener, J.; Wagner, W. Ten-Year Retrospective Follow-Up Study of the TiOblast™ Dental Implant. *Clin. Implant Dent. Rel. Res.* **2012**, *14*, 127–134. [CrossRef]
169. Ferguson, R. Renova Internal Hex Implant System: Surgical and restorative versatility. *Dent. Implantol. Update* **2005**, *16*, 49–54.
170. Che Isa, N.N.; Mohd, Y.; Yury, N. Electrochemical Deposition and Characterization of Hydroxyapatite (HAp) on Titanium Substrate. *APCBEE Procedia* **2012**, *3*, 46–52. [CrossRef]
171. Usinskas, P.; Stankeviciute, Z.; Beganskiene, A.; Kareiva, A. Sol-Gel Derived Porous and Hydrophilic Calcium Hydroxyapatite Coating on Modified Titanium Substrate. *Surf. Coat. Technol.* **2016**, *307 Pt A*, 935–940. [CrossRef]
172. Jaafar, A.; Schimpf, C.; Mandel, M.; Hecker, C.; Rafaja, D.; Krüger, L.; Arki, P.; Joseph, Y. Sol–gel derived hydroxyapatite coating on titanium implants: Optimization of sol–gel process and engineering the interface. *J. Mater. Res.* **2022**, *37*, 2558–2570. [CrossRef]
173. Łukaszewska-Kuska, M.; Krawczyk, P.; Martyla, A.; Hędzelek, W.; Dorocka-Bobkowska, B. Hydroxyapatite coating on titanium endosseous implants for improved osseointegration: Physical and chemical considerations. *Adv. Clin. Exp. Med.* **2018**, *27*, 1055–1059. [CrossRef] [PubMed]
174. Świeczko-Żurek, B.; Bartmański, M. Investigations of Titanium Implants Covered with Hydroxyapatite Layer. *Adv. Mater. Sci.* **2016**, *16*, 78–86. [CrossRef]
175. Kuroda, K.; Okido, M. Hydroxyapatite coating of titanium implants using hydroprocessing and evaluation of their osteoconductivity. *Bioinorg. Chem. Appl.* **2012**, *2012*, 730693. [CrossRef]
176. Baltatu, M.S.; Sandu, A.V.; Nabialek, M.; Vizureanu, P.; Ciobanu, G. Biomimetic Deposition of Hydroxyapatite Layer on Titanium Alloys. *Micromachines* **2021**, *12*, 1447. [CrossRef]
177. Park, Y.S.; Yi, K.Y.; Lee, I.S.; Han, C.H.; Jung, Y.C. The Effects of Ion Beam– Assisted Deposition of Hydroxyapatite on the Grit-Blasted Surface of Endosseous Implants in Rabbit Tibiae. *Int. J. Oral Maxillofac. Implants* **2005**, *20*, 31–38.
178. 3i T3 Implant. Available online: https://www.dentalproductshopper.com/implants-edentulous-solutions/implants/3i-t3-implant (accessed on 28 October 2024).
179. Mautsch, C.; Wolfart, S.; Mautsch, W.; Rittich, A.B. Long-term outcome of the IMZ implant system: A retrospective clinical study with a follow-up between 23 and 34 years. *Int. J. Implant Dent.* **2022**, *8*, 54. [CrossRef]
180. Kallus, T.; Bessing, C.; Homsi, G.; Eklund, I. Five-year evaluation of Lifecore Restore implants: A retrospective comparison with Nobel Biocare MK II implants. *Clin. Implant Dent. Relat. Res.* **2009**, *11*, 167–177. [CrossRef]
181. Nobel Replace External Hex (Steri-Oss). Available online: https://www.spotimplant.com/en/dental-implants/nobel-biocare/nobel-replace-external-hex-steri-oss (accessed on 28 October 2024).
182. Petrini, M.; Giuliani, A.; Di Campli, E.; Di Lodovico, S.; Iezzi, G.; Piattelli, A.; D'Ercole, S. The Bacterial Anti-Adhesive Activity of Double-Etched Titanium (DAE) as a Dental Implant Surface. *Int. J. Mol. Sci.* **2020**, *21*, 8315. [CrossRef]
183. Xie, Y.; Zuo, J.; Zhou, B.; Ma, L.; Yu, Z.M.; Wei, Q.; Tang, Z.G. Sandblast-free double-etched titanium for dental implants application. *Mater. Lett.* **2016**, *176*, 74–77. [CrossRef]
184. Giner, L.; Mercadé, M.; Torrent, S.; Punset, M.; Pérez, R.A.; Delgado, L.M.; Gil, F.J. Double acid etching treatment of dental implants for enhanced biological properties. *J. Appl. Biomater. Funct. Mater.* **2018**, *16*, 83–89. [CrossRef] [PubMed]

185. Santos Marino, J.; Cortés-Bretón Brinkmann, J.; García-Gil, I.; Martínez-Rodríguez, N.; Fraile, J.F.; Barona Dorado, C.; Martínez-González, J.M. Clinical Evaluation of Dental Implants with a Double Acid-Etched Surface Treatment: A Cohort Observational Study with Up to 10-Year Follow-Up. *Materials* **2021**, *14*, 6483. [CrossRef] [PubMed]
186. Dhaliwal, J.S.; David, S.R.N.; Zulhilmi, N.R.; Dhaliwal, S.K.S.; Knights, J.; Junior, R.F.d.A. Contamination of titanium dental implants: A narrative review. *SN. Appl. Sci.* **2020**, *2*, 1011. [CrossRef]
187. The Osseotite® Dental Implant System. Available online: https://www.zimvie.com/en/dental/dental-implant-systems/3i-osseotite-implant.html (accessed on 28 October 2024).
188. Osseotite® Implant Reference List. Available online: https://www.biomax.it/wp-content/uploads/2020/10/ZB0120_OsseotiteImplantReference_EN.pdf (accessed on 28 October 2024).
189. The Osseotite® Implant. Available online: https://www.biomet3i.cz/userFiles/pdf/zb0067_rev_a_osseotite_implant_brochure_final_secured.pdf (accessed on 28 October 2024).
190. del Olmo, R.; Czerwiński, M.; Santos-Coquillat, A.; Dubey, V.; Dhoble, S.J.; Michalska-Domańska, M. Nano-scale Surface Modification of Dental Implants: Fabrication. In *Surface Modification of Titanium Dental Implants*; Gulati, K., Ed.; Springer: Cham, Switzerland, 2023. [CrossRef]
191. Rungcharassaeng, K.; Kan, J.Y.K. Fabricating a stable record base for completely edentulous patients treated with osseointegrated implants using healing abutments. *J. Prosthet. Dent.* **1999**, *81*, 224–227. [CrossRef] [PubMed]
192. Hasegawa, M.; Saruta, J.; Hirota, M.; Taniyama, T.; Sugita, Y.; Kubo, K.; Ishijima, M.; Ikeda, T.; Maeda, H.; Ogawa, T. A Newly Created Meso-, Micro-, and Nano-Scale Rough Titanium Surface Promotes Bone-Implant Integration. *Int. J. Mol. Sci.* **2020**, *21*, 783. [CrossRef]
193. Lee, J.H.; Kwon, Y.H.; Herr, Y.; Shin, S.; Chung, J.H. Effect of Erbium-Doped: Yttrium, Aluminium and Garnet Laser Irradiation on the Surface Microstructure and Roughness of Sand-Blasted, Large Grit, Acid-Etched Implants. *J. Periodontal Implant. Sci.* **2011**, *41*, 135–142. [CrossRef]
194. Velasco-Ortega, E.; Ortiz-Garcia, I.; Jiménez-Guerra, A.; Núñez-Márquez, E.; Moreno-Muñoz, J.; Rondón-Romero, J.L.; Cabanillas-Balsera, D.; Gil, J.; Muñoz-Guzón, F.; Monsalve-Guil, L. Osseointegration of Sandblasted and Acid-Etched Implant Surfaces. A Histological and Histomorphometric Study in the Rabbit. *Int. J. Mol. Sci.* **2021**, *22*, 8507. [CrossRef]
195. Muhammed, H.A.; Mahmoud, E.M.; Fahmy, A.E.; Nasr, D.M. The effect of sandblasting versus acid etching on the surface roughness and biaxial flexural strength of CAD/CAM resin-matrix ceramics (In vitro study). *BMC Oral Health* **2023**, *23*, 169. [CrossRef]
196. Patcas, R.; Zinelis, S.; Eliades, G.; Eliades, T. Surface and interfacial analysis of sandblasted and acid-etched enamel for bonding orthodontic adhesives. *Am. J. Orthod. Dentofacial Orthop.* **2015**, *147* (Suppl. 4), S64–S75. [CrossRef]
197. Bok, W.M.; Kim, S.Y.; Lee, S.J.; Shin, G.-S.; Park, J.-M.; Lee, M.-H. Surface characteristics and bioactivation of sandblasted and acid-etched (SLA) Ti-10Nb-10Ta alloy for dental implant. *Int. J. Precis. Eng. Manuf.* **2015**, *16*, 2185–2192. [CrossRef]
198. Dental News®. Implant Standard. Available online: https://dentalnews.pl/produkt/implant-standard/ (accessed on 28 October 2024).
199. Stafford, G.L. Review Found Little Difference between Sandblasted and Acid-etched (SLA) Dental Implants and Modified Surface (SLActive) Implants. *Evid. Based Dent.* **2014**, *15*, 87–88. [CrossRef] [PubMed]
200. Schupbach, P.; Glauser, R.; Bauer, S. Al_2O_3 Particles on Titanium Dental Implant Systems following Sandblasting and Acid-Etching Process. *Int. J. Biomater.* **2019**, *1*, 6318429. [CrossRef] [PubMed]
201. Spiral (SPI) By Alpha Bio Tec®. Available online: https://www.spotimplant.com/en/dental-implants/alpha-bio-tec/spi (accessed on 28 October 2024).
202. About the DFI Implant. Available online: https://info.alpha-bio.net/dfi-implant (accessed on 28 October 2024).
203. Rocci, M.; Rocci, A.; Martignoni, M.; Albrektsson, T.; Barlattani, A.; Gargari, M. Comparing the TiOblast and Osseospeed surfaces. Histomorphometric and histological analysis in humans. *Oral Implantol.* **2008**, *1*, 34–42.
204. Straumann Group. Available online: https://www.straumann.com/group/en/home/about/our-history.html (accessed on 22 December 2024).
205. Wennerberg, A.; Galli, S.; Albrektsson, T. Current knowledge about the hydrophilic and nanostructured SLActive surface. *Clin. Cosmet. Investig. Dent.* **2011**, *3*, 59–67. [CrossRef]
206. Zinelis, S.; Silikas, N.; Thomas, A.; Syres, K.; Eliades, G. Surface characterization of SLActive dental implants. *Eur. J. Esthet. Dent.* **2012**, *7*, 72–92.
207. SLActive. Available online: http://www.schmidt-dental.pl/wp-content/uploads/2015/11/Straumann_SLActive_Studies.pdf (accessed on 28 October 2024).
208. Romero-Ruiz, M.M.; Gil-Mur, F.J.; Ríos-Santos, J.V.; Lázaro-Calvo, P.; Ríos-Carrasco, B.; Herrero-Climent, M. Influence of a Novel Surface of Bioactive Implants on Osseointegration: A Comparative and Histomorfometric Correlation and Implant Stability Study in Minipigs. *Int. J. Mol. Sci.* **2019**, *20*, E2307. [CrossRef]

209. Straumann® SLActive®. Beyond Hydrophilicity—The Science of High Performance. Available online: https://www.straumann.com/en/discover/slactive.html (accessed on 28 October 2024).
210. Distinct Nano-Structures Present on the SLActive® Surface27,28. Available online: https://www.straumann.com/en/discover/slactive.html (accessed on 28 October 2024).
211. Advanced In-Vitro Research Shows Nano-Structure Support Early Osseointegration23,24. Available online: https://www.straumann.com/en/discover/slactive.html (accessed on 28 October 2024).
212. Baier, R.E.; Meyer, A.E. Future directions in surface preparation of dental implants. *J. Dent. Educ.* **1988**, *52*, 788–791. [CrossRef]
213. Şener, I.; Yamaner, G.; Sertgoz, A. Clinical Outcomes of Patients Treated with SLA and SLActive Implants. In Proceedings of the IADR/PER General Session 2010, Barcelona, Spain, 14–17 July 2010.
214. Birch, J.; Burleigh, T. Oxides Formed on Titanium by Polishing, Etching, Anodizing, or Thermal Oxidizing. *Corrosion* **2000**, *56*, 1233–1241. [CrossRef]
215. Huang, Y.H.; Xiropaidis, A.; Sorensen, R.; Hall, J.; Wikesjö, U. Bone Formation at Titanium Porous Oxide (TiUnite (TM)) Oral Implants in Type IV Bone. *Clin. Oral Implants Res.* **2005**, *16*, 105–111. [CrossRef]
216. Badekas, H.; Panagopoulos, C. Titanium anodization under constant voltage conditions. *Surf. Coat. Technol.* **1987**, *31*, 381–388. [CrossRef]
217. Nowińska, D.; Osak, P.; Maszybrocka, J.; Łosiewicz, B. Anodic Production and Characterization of Biomimetic Oxide Layers on Grade 4 Titanium for Medical Applications. *J. Funct. Biomater.* **2024**, *15*, 180. [CrossRef] [PubMed]
218. Chen, J.; Zhang, Z.; Ouyang, J.; Chen, X.; Xu, Z.; Sun, X. Bioactivity and Osteogenic Cell Response of TiO_2 Nanotubes Coupled with Nanoscale Calcium Phosphate via Ultrasonification-Assisted Electrochemical Deposition. *Appl. Surf. Sci.* **2014**, *305*, 24–32. [CrossRef]
219. Kahar, S.; Singh, A.; Patel, V.; Kanetkar, U. Anodizing of Ti and Ti Alloys for Different Applications: A Review. *Int. J. Sci. Res. Dev.* **2020**, *8*, 272–276.
220. Replace Select Tapered TiUnite RP 4.3 × 13 mm. Available online: https://store.nobelbiocare.com/us/en/replace-select-tapered-tiunite-rp-4-3-x-13-mm (accessed on 28 October 2024).
221. McCracken, M. Dental implant materials: Commercially pure titanium and titanium alloys. *J. Prosthodont.* **1999**, *8*, 40–43. [CrossRef]
222. Product Catalog 2017/18 Complete Assortment. Available online: https://www.nobelbiocare.com/sites/g/files/wdvifx201/files/81206_ProdCatalog2017-18_GB.pdf (accessed on 28 October 2024).
223. Maló, P.; de Araújo Nobre, M.; Gonçalves, Y.; Lopes, A.; Ferro, A. Immediate Function of Anodically Oxidized Surface Implants (TiUnite™) for Fixed Prosthetic Rehabilitation: Retrospective Study with 10 Years of Follow-Up. *BioMed Res. Int.* **2016**, *2016*, 2061237. [CrossRef]
224. Traini, T.; Murmura, G.; Sinjari, B.; Perfetti, G.; Scarano, A.; D'Arcangelo, C.; Caputi, S. The Surface Anodization of Titanium Dental Implants Improves Blood Clot Formation Followed by Osseointegration. *Coatings* **2018**, *8*, 252. [CrossRef]
225. Karl, M.; Albrektsson, T. Clinical performance of dental implants with a moderately rough (TiUnite) surface: A meta-analysis of prospective clinical studies. *Int. J. Oral Maxillofac. Implants.* **2017**, *32*, 717–734. [CrossRef]
226. Jungner, M.; Lundqvist, P.; Lundgren, S. Oxidized titanium implants (Nobel Biocare TiUnite) compared with turned titanium implants (Nobel Biocare mark III) with respect to implant failure in a group of consecutive patients treated with early functional loading and two-stage protocol. *Clin. Oral Implants Res.* **2005**, *16*, 308–312. [CrossRef]
227. Li, G.; Ma, F.; Liu, P.; Qi, S.; Li, W.; Zhang, K.; Chen, X. Review of micro-arc oxidation of titanium alloys: Mechanism, properties and applications. *J. Alloys Compd.* **2023**, *948*, 169773. [CrossRef]
228. Ming, X.; Wu, Y.; Zhang, Z.; Li, Y. Micro-Arc Oxidation in Titanium and Its Alloys: Development and Potential of Implants. *Coatings* **2023**, *13*, 2064. [CrossRef]
229. Wen, X.; Liu, Y.; Xi, F.; Zhang, X.; Kang, Y. Micro-arc oxidation (MAO) and its potential for improving the performance of titanium implants in biomedical applications. *Front. Bioeng. Biotechnol.* **2023**, *11*, 1282590. [CrossRef] [PubMed]
230. Friedemann, A.E.R.; Thiel, K.; Haßlinger, U.; Ritter, M.; Gesing, T.M.; Plagemann, P. Investigations into the Structure of PEO-Layers for Understanding of Layer Formation. *Appl. Surf. Sci.* **2018**, *443*, 467–474. [CrossRef]
231. Sikdar, S.; Menezes, P.V.; Maccione, R.; Jacob, T.; Menezes, P.L. Plasma Electrolytic Oxidation (PEO) Process—Processing, Properties, and Applications. *Nanomaterials* **2021**, *11*, 1375. [CrossRef] [PubMed]
232. Biomimetic Implants. Available online: https://implantsystem.avinent.com/wp-content/uploads/sites/4/2019/06/biomimetic-implants-avinent-eng.pdf (accessed on 28 October 2024).
233. Łosiewicz, B.; Stróż, A.; Osak, P.; Maszybrocka, J.; Gerle, A.; Dudek, K.; Balin, K.; Łukowiec, D.; Gawlikowski, M.; Bogunia, S. Production, Characterization and Application of Oxide Nanotubes on Ti–6Al–7Nb Alloy as a Potential Drug Carrier. *Materials* **2021**, *14*, 6142. [CrossRef]
234. Osak, P.; Skwarek, S.; Łukowiec, D.; Przeliorz, G.; Łosiewicz, B. Preparation and Characterization of Oxide Nanotubes on Titanium Surface for Use in Controlled Drug Release Systems. *Materials* **2024**, *17*, 3753. [CrossRef]

235. Stróż, A.; Gawlikowski, M.; Balin, K.; Osak, P.; Kubisztal, J.; Zubko, M.; Maszybrocka, J.; Dudek, K.; Łosiewicz, B. Biological Activity and Thrombogenic Properties of Oxide Nanotubes on the Ti-13Nb-13Zr Biomedical Alloy. *J. Funct. Biomater.* **2023**, *14*, 375. [CrossRef]
236. Tran, C.; Walsh, L.J. *Novel Models to Manage Biofilms on Microtextured Dental Implant Surfaces*; IntechOpen: London, UK, 2016. [CrossRef]
237. Lutz, R.; Srour, S.; Nonhoff, J.; Weisel, T.; Damien, C.J.; Schlegel, K.A. Biofunctionalization of Titanium Implants with a Biomimetic Active Peptide (P-15) Promotes Early Osseointegration. *Clin. Oral Implants Res.* **2010**, *21*, 726–734. [CrossRef]
238. Fu, L.; Omi, M.; Sun, M.; Cheng, B.; Mao, G.; Liu, T.; Mendonça, G.; Averick, S.E.; Mishina, Y.; Matyjaszewski, K. Covalent Attachment of P15 Peptide to Ti Alloy Surface Modified with Polymer to Enhance Osseointegration of Implants. *ACS Appl. Mater. Interfaces* **2019**, *11*, 38531–38536. [CrossRef]
239. Chang, Y.-C.; Ho, K.-N.; Feng, S.-W.; Huang, H.-M.; Chang, C.-H.; Lin, C.-T.; Teng, N.-C.; Pan, Y.-H.; Chang, W.-J. Fibronectin-Grafted Titanium Dental Implants: An In Vivo Study. *BioMed Res. Int.* **2016**, *2016*, 2414809. [CrossRef]
240. Lo, V.; I-Chun Lai, J.; Sunde, M. Fungal Hydrophobins and Their Self-Assembly into Functional Nanomaterials. *Adv. Exp. Med. Biol.* **2019**, *1174*, 161–185. [CrossRef]
241. Pawar, V.; Bulbake, U.; Khan, W.; Srivastava, R. Chitosan Sponges as a Sustained Release Carrier System for the Prophylaxis of Orthopedic Implant-Associated Infections. *Int. J. Biol. Macromol.* **2019**, *134*, 100–112. [CrossRef]
242. Łosiewicz, B.; Osak, P.; Kubisztal, J. The effect of a composite chitosan/copper(II) ion coating on the corrosion resistance of grade 4 titanium in saline: Preliminary results. *Prog. Chem. Appl. Chitin Deriv.* **2023**, *28*, 89–102. [CrossRef]
243. Szklarska, M.; Łosiewicz, B.; Dercz, G.; Maszybrocka, J.; Rams-Baron, M.; Stach, S. Electrophoretic deposition of chitosan coatings on the Ti15Mo biomedical alloy from a citric acid solution. *RSC Adv.* **2020**, *10*, 13386–13393. [CrossRef] [PubMed]
244. OsseoSpeed TX S (Astra Tech). Available online: https://stg.spotimplant.com/en/dental-implants/dentsply-implants/osseospeed-tx-s (accessed on 28 October 2024).
245. Homa, K.; Zakrzewski, W.; Dobrzyński, W.; Piszko, P.J.; Piszko, A.; Matys, J.; Wiglusz, R.J.; Dobrzyński, M. Surface Functionalization of Titanium-Based Implants with a Nanohydroxyapatite Layer and Its Impact on Osteoblasts: A Systematic Review. *J. Funct. Biomater.* **2024**, *15*, 45. [CrossRef] [PubMed]
246. Yu, C.; Yu, Y.; Lu, Y.; Quan, K.; Mao, Z.; Zheng, Y.; Qin, L.; Xia, D. UiO-66/AgNPs Coating for Dental Implants in Preventing Bacterial Infections. *J. Dent. Res.* **2024**, *103*, 516–525. [CrossRef]
247. Zhang, Y.; Cheng, Z.; Liu, Z.; Shen, X.; Cai, C.; Li, M.; Luo, Z. Functionally Tailored Metal-Organic Framework Coatings for Mediating Ti Implant Osseointegration. *Adv. Sci.* **2023**, *10*, e2303958. [CrossRef]
248. Wu, J.; Jiang, S.; Xie, W.; Xue, Y.; Qiao, M.; Yang, X.; Zhang, X.; Wan, Q.; Wang, J.; Chen, J.; et al. Surface modification of the Ti surface with nanoscale bio-MOF-1 for improving biocompatibility and osteointegration *in vitro* and *in vivo*. *Mater. Chem. B* **2022**, *10*, 8535–8548. [CrossRef]
249. Sabzehmeidani, M.M.; Kazemzad, M. Recent advances in surface-mounted metal-organic framework thin film coatings for biomaterials and medical applications: A review. *Biomater. Res.* **2023**, *27*, 115. [CrossRef]
250. Kaur, G. Apatites: A Mark of BioactivityBioactivity. In *Bioactive Glasses: Potential Biomaterials for Future Therapy*; Kaur, G., Ed.; Series in BioEngineering; Springer International Publishing: Cham, Switzerland, 2017; pp. 145–172, ISBN 978-3-319-45716-1.
251. Tang, W.; Fischer, N.G.; Kong, X.; Sang, T.; Ye, Z. Hybrid coatings on dental and orthopedic titanium implants: Current advances and challenges. *BMEMat* **2024**, *2*, e12105. [CrossRef]
252. Bravo, E.; Serrano, B.; Ribeiro-Vidal, H.; Virto, L.; Sánchez, I.S.; Herrera, D.; Sanz, M. Biofilm formation on dental implants with a hybrid surface microtopography: An in vitro study in a validated multispecies dynamic biofilm model. *Clin. Oral Implant. Res.* **2023**, *34*, 475–485. [CrossRef]
253. Serrano, B.; Sanz-Sánchez, I.; Serrano, K.; Montero, E.; Sanz, M. One-year outcomes of dental implants with a hybrid surface macro-design placed in patients with history of periodontitis: A randomized clinical trial. *J. Clin. Periodontol.* **2022**, *49*, 90–100. [CrossRef] [PubMed]
254. Jemat, A.; Ghazali, M.J.; Razali, M.; Otsuka, Y. Surface Modifications and Their Effects on Titanium Dental Implants. *BioMed Res. Int.* **2015**, *2015*, 791725. [CrossRef] [PubMed]
255. Sun, X.D.; Liu, T.T.; Wang, Q.Q.; Zhang, J.; Cao, M.S. Surface Modification and Functionalities for Titanium Dental Implants. *ACS Biomater. Sci. Eng.* **2023**, *9*, 4442–4461. [CrossRef]
256. IMAX NHSI. Available online: https://www.spotimplant.com/en/dental-implants/ires/imax-nhsi-internal-hex-c (accessed on 28 October 2024).
257. Morra, M.; Cassinelli, C.; Torre, E.; Iviglia, G. Permanent wettability of a novel, nanoengineered, clinically available, hyaluronan-coated dental implant. *Clin. Exp. Dent. Res.* **2018**, *4*, 196–205. [CrossRef] [PubMed]
258. Tarnow, D.P. Dental implants in periodontal care. *Curr. Opin. Periodontol.* **1993**, *157*, 157–162.
259. Leesungbok, R.; Hong, S.O.; Lee, S.W.; Htay, P.E.E.; Choi, J.J.; Park, J.J. An eight-year retrospective study on the clinical outcomes of laser surface-treated implants. *Int. J. Implant. Dent.* **2024**, *10*, 38. [CrossRef]

260. Lackington, W.A.; Schweizer, P.; Khokhlova, M.; Cancellieri, C.; Guimond, S.; Chopard-Lallier, A.-L.; Hofstetter, J.; Schmutz, P.; Maeder, X.; Rottmar, M. Femtosecond Laser-Texturing the Surface of Ti-Based Implants to Improve Their Osseointegration Capacity. *Adv. Mater. Interfaces* **2022**, *9*, 2201164. [CrossRef]
261. Ionescu, A.C.; Brambilla, E.; Azzola, F.; Ottobelli, M.; Pellegrini, G.; Francetti, L.A. Laser microtextured titanium implant surfaces reduce in vitro and in situ oral biofilm formation. *PLoS ONE* **2018**, *13*, e0202262. [CrossRef]
262. Veiko, V.; Karlagina, Y.; Zernitckaia, E.; Egorova, E.; Radaev, M.; Yaremenko, A.; Chernenko, G.; Romanov, V.; Shchedrina, N.; Ivanova, E.; et al. Laser-Induced μ-Rooms for Osteocytes on Implant Surface: An In Vivo Study. *Nanomaterials* **2022**, *12*, 4229. [CrossRef]
263. Khalil, M.I.; Sakr, H. Implant Surface Topography Following Different Laser Treatments: An In Vitro Study. *Cureus* **2023**, *15*, e38731. [CrossRef]
264. Shapoff, C.A.; Lahey, B.; Wasserlauf, P.; Kim, D. Radiographic Analysis of Crestal Bone Levels on Laser-Lok® Collar Dental Implants. *Int. J. Periodontics Restor. Dent.* **2010**, *30*, 129–137.
265. Laser Etching: Everything You Need to Know. Available online: https://www.laserax.com/blog/laser-etching (accessed on 28 October 2024).
266. Tapered Internal Plus (4.5). Available online: https://osseosource.com/tapered-internal-plus-4-5-/p-2495.html (accessed on 18 October 2024).
267. Laser-Lok Microchannels. Clinical Overview. Available online: https://www.laser-lok.com (accessed on 18 October 2024).

coatings

MDPI

Review

Effect of 3D-Printed Porous Titanium Alloy Pore Structure on Bone Regeneration: A Review

Si He †, Jiang Zhu †, Yiwan Jing, Shuai Long, Lu Tang, Lijia Cheng * and Zheng Shi *

Clinical Medical College & Affiliated Hospital, School of Basic Medical Sciences, Mechanical Engineering College, Chengdu University, Chengdu 610106, China; hesisi0420@163.com (S.H.); 13378119012@163.com (J.Z.); m18380728911@163.com (Y.J.); 11319288735@gmail.com (S.L.); tl20120613@163.com (L.T.)

* Correspondence: chenglijia@cdu.edu.cn (L.C.); drshiz1002@hotmail.com (Z.S.)

† These authors contributed equally to this work.

Abstract: As a biomedical material, porous titanium alloy has gained widespread recognition and application within the field of orthopedics. Its remarkable biocompatibility, bioactivity, and mechanical properties establish it as a promising material for facilitating bone regeneration. A well-designed porous structure can lower the material's modulus while retaining ample strength, rendering it more akin to natural bone tissue. The progression of additive manufacturing (AM) technology has significantly propelled the advancement of porous implants, simplifying the production of such structures. AM allows for the customization of porous implants with various shapes and sizes tailored to individual patients. Additionally, it enables the design of microscopic-scale porous structures to closely mimic natural bone, thus opening up avenues for the development of porous titanium alloy bone implants that can better stimulate bone regeneration. This article reviews the research progress on the structural design and preparation methods of porous titanium alloy bone implants, analyzes the porous structure design parameters that affect the performance of the implant, and discusses the application of porous medical titanium alloys. By comparing the effects of the parameters of different porosity, pore shape, and pore size on implant performance, it was concluded that pore diameters in the range of 500~800 μm and porosity in the range of 70%–90% have better bone-regeneration effects. At the same time, when the pore structure is a diamond, rhombohedral, or cube structure, it has better mechanical properties and bone-regeneration effects, providing a reference range for the application of clinical porous implants.

Keywords: 3D printing; titanium alloys; bone implant; porous structure

1. Introduction

The repair and replacement of a wide range of bone defects caused by diseases, trauma, and aging has been an important subject for centuries. So far, the regeneration of bone defects caused by war, infection, car accidents, tumors, and genetics is still a clinical challenge [1]. Bone tissue possesses intrinsic regenerative capabilities, but it has a critical size limit. Within this critical size limit, bone defects can be repaired through the self-healing capacity of the bone tissue itself. However, when bone defects exceed this critical size limit, such as those resulting from trauma, disease, tumor resection, or osteomyelitis, spontaneous healing becomes challenging in the absence of external intervention. A segmental bone defect is defined as a defect whose length exceeds 1.5 times its diameter. At present, the most common treatment methods for bone defects are autologous bone transplantation and allograft bone transplantation. Among them, autologous bone transplantation is considered to be the "gold standard" in clinical practice; that is, it is the most effective bone-regeneration method [2,3]. However, these two treatments have obvious limitations, such as limited autogenous bone, more pain for patients [4], and immune rejection [5]. In view of this, orthopedic alternatives have rapidly developed in response to clinical needs, including skull prostheses, dental implants, interbody fusion devices, femur prostheses, and tibial prostheses.

Currently, the biomaterials used for implants include the following four categories: metals and their alloys, polymers, ceramics, and natural materials. The properties of several commonly used materials are compared with those of natural human bone, as shown in Table 1. Metal materials exhibit superior mechanical strength and wear resistance in comparison to polymer and ceramic materials, so metal materials are often used as the preferred materials for load-bearing implants [6]. Among all metal materials, titanium alloy (Ti6Al4V) is widely used in the manufacture of orthopedic and dental implants because of its high strength, low density, corrosion resistance, low elastic modulus, and good biocompatibility [7,8]. The elastic modulus of human cortical bone and trabecular bone ranges from 3 GPa to 30 GPa and 0.01 GPa to 3 GPa, respectively [9–11], while the elastic modulus of a traditional solid titanium alloy can reach 110 GPa, which is much higher than the elastic modulus of human bone, so the stress-shielding problem may be faced after implantation [12]. The high contrast between the mechanical properties of bone tissue and implant biomaterials may lead to bone resorption, a phenomenon defined as stress shielding [13], which limits the normal growth of bone and ultimately leads to implantation failure. Secondly, the dense titanium alloy was connected to the host-only interface, and the intensity after implantation was also prone to loosening [14]. To address these issues, titanium alloys with a porous structure have been introduced, and this porous structure is regarded as an effective method for mitigating the mismatch in the elastic modulus [15–17]. Porous titanium alloys combine the advantages of titanium alloys and porous structures and can be used to manufacture implants with a more perfect structure and performance to achieve the effect of reducing the elastic modulus and regulating mechanical properties.

The ideal orthopedic implant requires its structure to conform to the anatomy of natural bone tissue and have connected pores that can satisfy the inward growth and vascularization of cells. It has certain mechanical properties to provide mechanical support and avoid the fatigue fracture of the materials; has a surface suitable for cell adhesion, growth, and reproduction; and should have good biocompatibility and a goodbone-tissue-integration ability [18–20]. The design of porous structures involves the porosity, pore size, and shape. The porosity, pore size, and shape of porous scaffolds can play an important role in the growth of cells and the mechanics of the scaffold and will affect cell nutrients, the flow of oxygen, and the biological responses of the cells (such as their proliferation, differentiation, and signaling). The pore structure creates conditions for processes such as cell adhesion, growth, and reproduction [21], allowing for cell migration, the influx of oxygen and nutrients, and the circulation of body fluids [22,23]. Since the introduction of porous Ti6Al4V bone implants, researchers have been diligently striving to discover a porous implant design that is better suited for promoting bone tissue regeneration [4].

Table 1. Comparison of properties of various biomaterials.

Material	Yield Strength (MPa)	Strength of Extension (MPa)	Elasticity Modulus (GPa)	Cite
Human bone (Cortical bone)	30–70	70–150	4–30	[24]
CpTi	320	465	110	[25]
Ti6Al4V	585–1060	690–1100	55–110	[26]
Stainless steel	190–690	490–1350	200–210	[27,28]
Co-based alloys	310–1586	655–1793	210–253	[29,30]
Hydroxyapatite	—	40–300	80–120	[31]
Bioglass45S5	—	42	35	[32]

2. Preparation of Porous Titanium Alloys

Traditional processing methods such as metal foaming [33], freeze casting [34], and powder metallurgy [35] can be used to manufacture porous structures. However, for implants with complex structures, the traditional methods cannot obtain an accurate porosity, and it is difficult to control their porous structures [36], which limits the application of porous structures in bone implants. In recent years, more and more studies have adopted additive manufacturing technology (also known as 3D printing technology) to manufacture implants with porous structures [37]. Three-dimensional printing technology is a bottom-up processing method. It operates by creating solid objects layer by layer, following computer-generated 3D models. This technology excels in meeting the demands of personalized customization to the fullest extent [38]. Three-dimensional printing technology can further change its elastic modulus and mechanical properties by adjusting parameters such as the pore size, porosity, pore shape, and surface topography. Today, 3D printing technology has been widely used in various fields, including agriculture, healthcare, the automotive industry, and the aerospace industry for mass customization. Figure 1 illustrates the utilization of 3D printing technology across various fields. The advantage of 3D printing in the medical field is that it can be tailored for patients by computer, the size and shape of the defect can be determined by a CT scan, and then the matching 3D model can be built by reverse modeling [39], as shown in Figure 1a. Three-dimensional printing technology can further modify its mechanical and bone-inducing properties through parameter adjustments, such as the pore size, shape, porosity, and surface topography.

Metal additive manufacturing technology can be divided into several different forming processes, including Selective Laser Sintering (SLS), Direct Metal Laser Sintering (DMLS), Selective Hot Sintering (SHS), Selective Laser Melting (SLM), and Electron Beam Selective Melting (SEBM), among others. SLM can make porous titanium structures well, and the Young's modulus of the structure is well matched with that of the skeleton [40]. Therefore, it is often used in the medical field for the molding of porous titanium alloys. Cui et al. [41] prepared porous titanium alloys by using SLM technology. Compared to solid metals, the elastic modulus of porous titanium has been reduced to 0.74 GPa while achieving a compressive yield strength of 201.91 MPa. This meets the requirements for compatibility with human bone tissue, highlighting the advantages of 3D printing technology in fabricating porous structures.

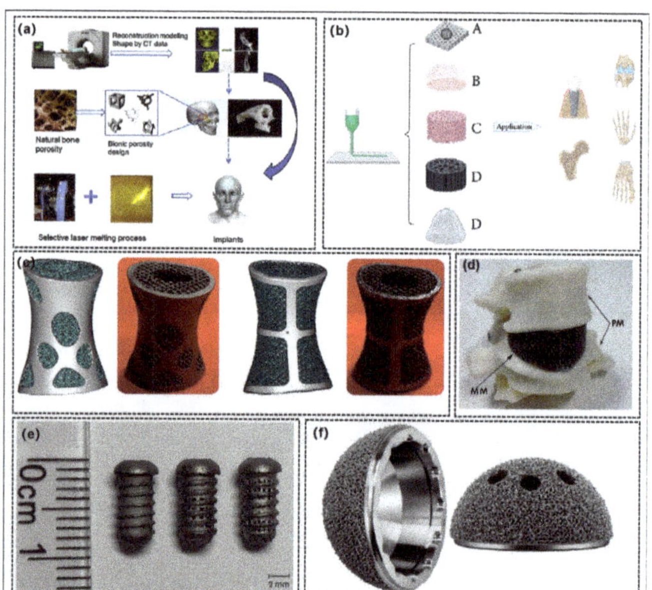

Figure 1. (**a**) Customization process of 3D-printed implants [15]. (**b**) Application of 3D printing in orthopedics [42]. (**c**) Three-dimensionally printed artificial vertebral body model and actual pictures [43].

(**d**) SLM-3D-printed personalized Ti6Al4V spinal cage (mesh) implant [44]. (**e**) Three-dimensionally printed Ti6Al4V threaded implant specimen [45]. (**f**) Trabecular titanium acetabular cup produced by 3D ACT EBM [46]. (Reprinted with permission from Ref. [15]. Copyright 2019 Elsevier; Ref. [43]. Copyright 2021 Elsevier. Reprinted from Ref. [42]; Ref. [44]; Ref. [45]; Ref. [46]).

3. Effect of 3D Printing Porous Titanium Alloy Pore Structure on Bone Regeneration

Although a large number of porous metal implants have been designed, the design and properties of porous structures are still subject to extensive research, especially the effects of porosity, pore size, and shape. At present, there is no unified conclusion on the optimal values of the pore size, porosity, and other parameters. In order to find suitable porous structures, the effects of the above three structures on the physical properties and endogenetic bone growth of titanium implantation are discussed below. In order to find suitable porous structures, the effects of the above three structures on the physical properties and endogenetic bone growth of titanium implantation are discussed below.

3.1. Porosity

Porosity is the ratio of the porous portion to the solid portion of the scaffold. Porosity characteristics are usually obtained indirectly through physical measurements (for example, using the density principle, immersing the sample in water, placing it in a graduated cylinder, replacing the volume of water with the actual volume of the scaffold, and obtaining porosity data based on the difference in volume) or by using digital image processing and analysis, as well as computed tomography techniques that can provide a more direct way to obtain porosity data [47]. The porosity is calculated as follows:

$$\eta = \frac{Vp}{Vz} \times 100~\%$$

Here, Vp represents the pore volume of the scaffold, while Vz denotes the original volume of the scaffold.

The elastic modulus of implants is mainly regulated by porosity and can be changed by adjusting the porosity. At the same time, porosity plays an important role in establishing early bone integration and forming strong interface bonding between porous implants and surrounding tissues [48]. Increasing porosity provides more room for cells to grow, and the transport of oxygen and nutrients is correspondingly increased [49]. However, adding too high a porosity with the same structural design can lead to a significant decrease in mechanical properties, so finding the optimal porosity range is crucial for the successful application of the implant. The right porosity can provide cells with room to reproduce while mimicking the pore structure and mechanical strength of natural bone tissue [15]. When the porosity of the implant matches that of the human bone, the optimal bone growth environment can be obtained. The structural variations in porous scaffolds with different porosities are illustrated in Figure 2. Natural bone tissue comprises two distinct structures: cancellous bone and cortical bone. The interior of cancellous bone features a spongy structure with a porosity ranging from 50% to 90%. The internal structure of cortical bone is dense, the bone density is much lower than that of cancellous bone, and the porosity is only 5% to 10%. Scaffolds with a porosity comparable to that of human bone trabeculae (70% to 90%) have been shown to enhance cell viability and inward bone growth [50,51]. Moreover, studies have shown that when the porosity is greater than 70%, the porosity of the porous structure can have a beneficial effect on bone tissue [52]. A scaffold structure with a 600–900 μm pore size and 60%–90% porosity is recommended as the best structure [53]. Zhang et al. [54] differentiated the pore rate of a 3D-printed preparation (40%, 70%, and 90%), and the pore diameter was 700 μm multipores. Micro-CT results showed that the bone integration effect of the implant with a porosity of 40% (P40) was inferior to that of the implant with a porosity of 70% (P70) and 90% (P90). In addition, it is suggested that the change in pore size has a more significant effect on osteogenesis when the porosity is in the range of 70%–90%. In recent years, more and more researchers have prepared porous titanium alloys with a gradient structure. Hindy et al. [40] follow the biomimetic approach and apply the porosity gradient visible in natural bone to the fabrication of orthopedic and dental implants. The replication of this functional gradient ensures the correct distribution of the compression stiffness in different regions.

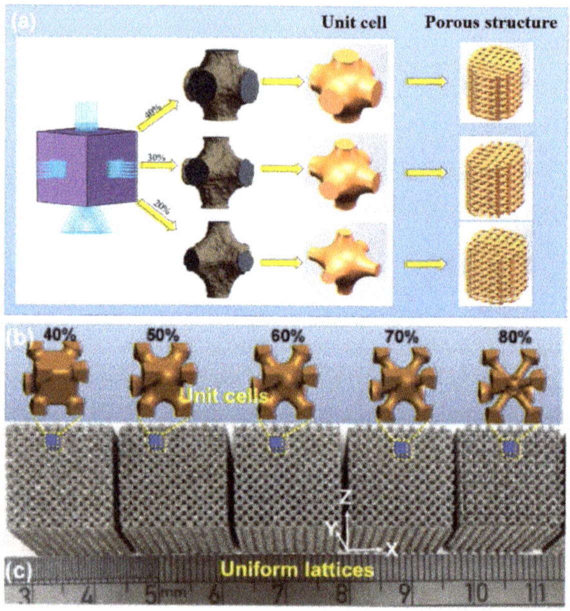

Figure 2. (**a**) Structural change model of porous scaffolds with different porosity [55]. (**b**) Unit cells with different structural porosity [56]. (**c**) The LPBF-processed uniform lattice samples [56]. (Reprinted with permission from Ref. [55]. Copyright 2023 Elsevier; Ref. [56]. Copyright 2021 Elsevier).

3.2. Pore Configuration

Most studies only explored the porosity and pore size suitable for endogenic bone growth under the condition of a fixed pore shape and ignored the fact that the pore shape of a porous structure would also affect the effective spatial distribution of the cells inside the scaffold and affect the mechanical properties of the scaffold. The pore shape was originally designed to mimic the shape of micropores inside the natural human bone, which is a complex tissue with a precise porous structure. There are different opinions about the micropore structure of bone tissue. Some people think that the microholes inside the human bones are round, some people observe that they are square holes, and some people think that they are hexagonal honeycomb holes. The geometry of the holes in bone implants can be square, rectangular, spherical, trabecular, or hexagonal, and more complex shapes can be made by using solid free-form fabrication techniques, such as cubes [57], diamonds [58], rhombohedrons [41], and variations of these structures. With different pore shapes, the mechanical properties and osteogenic properties of the scaffolds are also different. For instance, the diamond structure has two additional angles compared to the cubic structure, thereby offering a larger adhesive surface area for cells. The internal topologies of the porous materials designed by computer-aided methods can be roughly divided into (1) spatially arranged units composed of pillars, (2) three-period minimum surfaces (TPMSs), and (3) irregular bioinspired or Voronoi Mosaic structures. In recent years, TPMSs have also been widely applied to the field of bone tissue engineering based on naturally occurring nanoscale spiral structures found on butterfly wings that have an average curvature value of zero while the average curvature value of human trabecular bones is also close to zero [17]. A TPMS is an infinite and periodic surface, and the TPMS is often described by the following types of equations:

$$\cos \alpha x + \cos \beta y + \cos \gamma z = c$$

This equation satisfies the equation φ (x, y, z) = c, and this function φ (x, y, z) is the isosurface evaluated by the isosurface c.

TPMS structures, including gyro, primitive, and diamond structures, are generated by using mathematical formulas to tune their mechanical properties by changing various parameters such as the periodicity and relative density. Kelly et al. [17] evaluated the performance of TPMS titanium scaffolds produced by AM to repair femoral defects in rats and confirmed that TPMS scaffolds can repair segmental bone defects. Table 2 summarizes studies of the biological properties of AM implants with different structures. Jahir-Hussain et al. [59] conducted a comparative analysis of the mechanical properties of 3D-printed polylactide (PLA) scaffolds with four distinct pore structures, including round, square, hexagonal, and triangular, by utilizing finite element analysis (FEA). Their findings revealed that scaffolds featuring hexagonal pore shapes exhibited mechanical properties consistent with those of human bones. Van Bael et al. [60] discovered that, in comparison to hexagonal holes, triangular holes were more favorable for cell growth and differentiation, whereas rectangular holes were more prone to causing cell blockage. By examining local curvature and pore shapes, it was determined that obtuse angles were more likely to result in cell blockage compared to acute angles. However, Xu et al. [61] reported that the osteogenic ability of hexagonal prism scaffolds was higher than that of triangular prism scaffolds through in vivo and in vitro studies. Zhao et al. [62] reported the influence of tetrahedral and octahedral cell scaffolds on cell affinity and found that octahedral cells exhibit better static mechanical properties and a longer fatigue life than tetrahedral cells. At the same time, cells spread better on the scaffold on the octahedron than on the tetrahedron.

Kovács et al. [63] studied the mechanical properties and bone inward growth effect of titanium alloy scaffolds with six lattice shapes, including a gyro type, cube, cylinder, tetrahedron, diagonal cone, and Tyson polygon. The efficiency of the bone inward growth of several lattice shapes was compared, and the results showed that the bone growth degree of the gyro, conical, and cubic lattices was the best. Lim et al. [64] also came to the same conclusion by implanting titanium scaffolds of three different structures (octadense, gyroid, and dode) into the femur of rabbits, and no differences in bone formation in the titanium scaffolds were observed between the three types of pore structures. Farazin et al. [65] compared the biocompatibility of the cube, pyramid, and diagonal pore structures and found that the pyramid structures had the highest cell viability and migration ability. Deng et al. [66] conducted a study to investigate the effect of 3D-printed scaffolds with four different pore structures (i.e., diamond, tetrahedral cells, round pores, and cubes) on the osteogenic properties. The results showed that the diamond structure produced the best bone growth, possibly because the structure's strut angles are similar to the angles between the trabeculae of cancellous bone in humans. At the same time, fluid dynamics (CFD) studies also show that the diamond structure has the smallest fluid velocity difference and the longest fluid flow path. This property is very beneficial for promoting blood vessel development, promoting nutrient transport, and enhancing bone formation. Therefore, the diamond structure is more conducive to bone growth. Compared with diamond structures, rhombohedral dodecahedrons have been shown to have better mechanical strength and moderate biological properties and can be applied to body parts with relatively high mechanical properties requirements [67]. Zhao et al. [15] conducted a study on the mechanical properties of supports featuring various pore structure elements. The findings revealed that supports with diamond-shaped pore elements exhibited the lowest compressive strength, measuring only approximately 38.2 MPa. Supports featuring cyclopore elements displayed a lower compressive strength, around 57.0 MPa, while those with cube-shaped pore elements demonstrated a higher compressive strength, approximately 142.8 MPa. In summary, the diamond structure, rhombohedral dodecahedron structure, and cube structure show great potential in promoting bone regeneration.

Table 2. Common porous scaffold structures and their characteristics.

Pore Size (μm)	Porosity (%)	Pore Structure	Cell Structure	Conclusions	Ref.
500	80 75 70 65 60	Diamond Gyroid Orthogonal Cube Truss	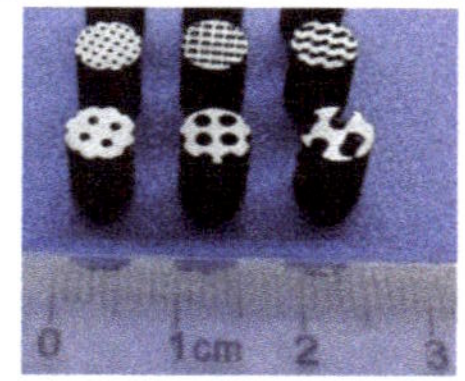	Truss and cube structures have higher compressive strength. Diamond and gyroid structures have lower compressive strength, which may be due to the complex porosity and small vertical solid-bearing surface of these two structures.	[15]
100	—	Circular Square Hexagonal Triangular	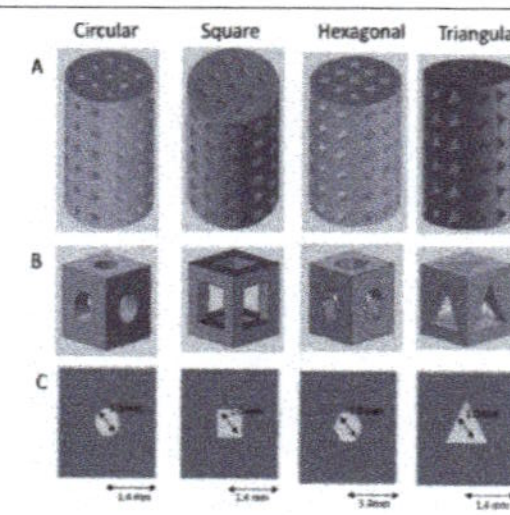	The results of the finite element analysis (FEA) indicate that scaffolds with a hexagonal pore shape exhibit greater similarity in performance to human bones.	[59]
500	—	Triangle Hexagon Rectangle	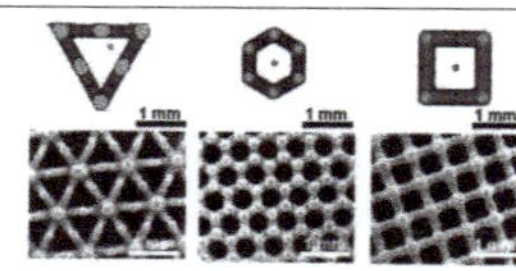	Rectangular pore is easy to cause cell blockage. Compared with hexagonal, triangular pore structure is more conducive to cell growth and differentiation.	[60]
500 1000	67 62.87 84 77	Tetrahedron Octahedron	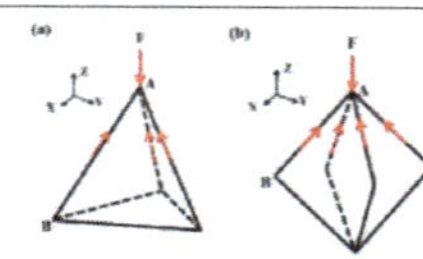 F represents compressive stress, and the red arrow represents tensile stress.	The adhesion of scaffolds with 1000 apertures was superior, but their compressive and fatigue properties were inferior to those of scaffolds with 500 apertures. Octahedral scaffolds exhibited better compression performance and fatigue life compared to tetrahedral scaffolds, and they also displayed a greater capacity for cell proliferation.	[62]

Table 2. *Cont.*

Pore Size (μm)	Porosity (%)	Pore Structure	Cell Structure	Conclusions	Ref.
1076 739	70	Gyroid	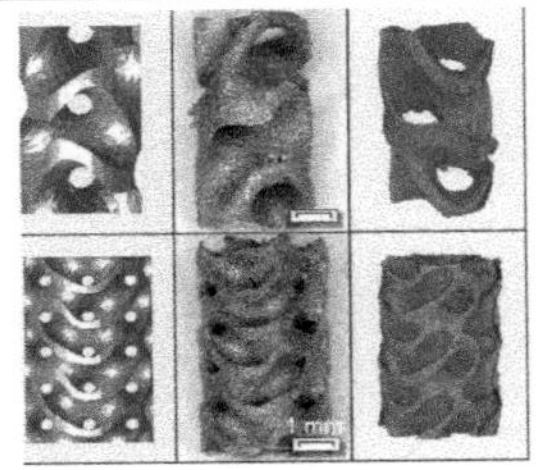	Titanium gyroid-sheet scaffolds can be used to repair segmental defects, and small-hole gyroscaffolds exhibit considerable inward-facing growth compared to large-hole gyroscaffolds. There was no significant difference in torsional strength and stiffness of the small pore implant compared with intact femur.	[17]
600	—	Gyroscope Cube cylinder tetrahedron Double pyramid Voronoi	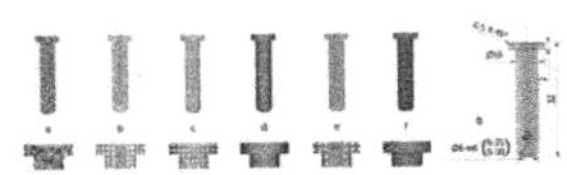	Six lattice shapes, gyroscope, cube, cylinder, tetrahedron, double pyramid, and Voronoi, were ranked for bone growth efficiency, and gyro, double pyramid, and cube lattice implants had the highest bone tissue growth per unit time.	[63]
1070 300 760	75	Octedens Gyroid Dode	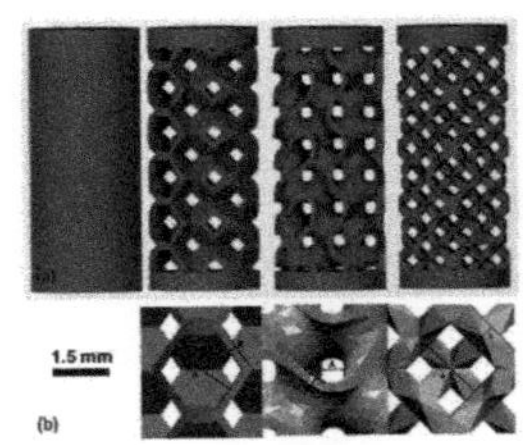 A: pore size, B: strut thickness.	No differences in bone formation in titanium scaffolds were observed between the three types of pore structures.	[64]
650	65	Diamond Tetrahedro Cell Circular Cube	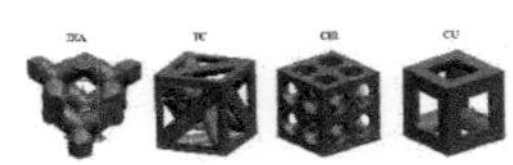	Diamond structure has the best bone growth, and fluid dynamic analysis also shows that this structure is conducive to blood vessel growth and bone formation.	[66]

3.3. Pore Size

The porous structure not only efficiently lowers the elastic modulus, encouraging the formation of a mineralized layer on the implant's surface and promoting protein adsorption, but also offers a conducive environment for cell adhesion, thereby facilitating the proliferation and differentiation of bone cells. Additionally, it serves as a channel for the transmission of metabolism and nutrition [68]. The pore size is very important for porous implants, and it affects the expression of osteogenic genes and the differentiation of osteoblasts. Wang et al. [69] confirmed that pore size and structure also play a certain role in regulating the expression of genes related to angiogenesis. The aperture of the scaffold should ensure that bone cells, nerve fibers, and blood vessels can grow into the scaffold. When the aperture is too large, it increases air permeability, which prevents cells from adhering to the surface [70]. When the aperture is too small, cells cannot enter the scaffold, resulting in cell accumulation, reducing cell migration in the scaffold and even affecting the circulation of nutrients and metabolic waste, which is not conducive to the growth of bone tissue [52]. At present, there is no precise definition of the most suitable pore size for bone growth. Some studies generally believe that the pore size of 100 μm to 400 μm can promote angiogenesis and bone growth, and below this range will limit bone cell growth [71]. Through a comprehensive analysis of the pore size required for the internal growth space of bone tissue and the formation of blood vessels, it is recommended that the optimal pore size is 300–600 μm [72,73]. The comprehensive impact of pore size on implants is summarized and drawn Figure 3. After staining with Toluidine Blue, it was observed that new bone tissue had developed within nearly all of the surface micropores of the 600 μm implants. The results indicate that, in comparison to scaffolds with 200 μm and 1000 μm apertures, the scaffolds featuring 600 μm apertures were more favorable for the growth of new bone tissue. Zhao et al. [62] reported tetrahedral cell titanium alloy scaffolds with pore sizes of 500 μm and 1000 μm. Figure 4a shows that cells on the scaffold with a pore size of 1000 μm exhibit better spread and more filamentous pseudopods. Hara et al. [74] conducted an experiment in which they implanted four types of cylindrical porous titanium alloys with varying pore sizes (500 μm, 640 μm, 800 μm, and 1000 μm) into the distal end of rabbit femurs. The findings revealed that porous titanium alloys with pore sizes smaller than 800 μm offered bioactive surfaces and maintained mechanical stability for bone fixation through implants. Meanwhile, Zhou et al. [75] proposed that the pore size of ideal bone tissue engineering scaffolds should be 300–900 μm. Zhang et al. [54] prepared titanium alloy scaffolds with a constant porosity of 70% and different pore sizes (400 μm, 700 μm, and 900 μm) and porous titanium alloy scaffolds with constant pore sizes of 700 μm and different porosities (40%, 70%, and 90%). The effect of the pore size and porosity on osteogenesis was discussed. The micro-CT results showed that a scaffold with a pore size of 700 μm can better induce cell ingrowth and new bone formation. An interesting phenomenon was discovered through fluorescence images; that is, cells are more likely to grow at the edges and then spread toward the center. Within the recommended porosity range (70%–90%), changes in pore size have a more significant impact on osteogenesis. Similarly, the same phenomenon was found in a study by Ran et al. [76]. The actual pore sizes of P500/P700/P900 implants prepared by SLM were 401 ± 26 μm, 607 ± 24 μm, and 801 ± 33 μm, respectively. Through observation, the morphology of osteoblasts on different implants found that the larger the pore size, the higher the cell density, as shown in Figure 4b,c. Finally, by implanting the implant into rabbits, it was concluded that the biological performance of the P700 group with an actual pore diameter of approximately 600 μm was better than that of the other two groups. Wang et al. [77] created consistently sized cubic pores measuring 300 μm, 400 μm, 500 μm, 600 μm, 700 μm, 800 μm, 900 μm, and 1000 μm through a combination of in vivo and in vitro experiments. The structural modifications and experimental outcomes are depicted in Figure 4d–f. The cell adhesion, proliferation, and differentiation of the 500 μm, 600 μm, and 700 μm porous scaffolds were superior to those of the other groups. Subsequent in vivo experiments showed that the 600 μm porous scaffolds had a better ability to induce new bone formation. Ouyang [78] compared the correlation between the mechanical properties and bone regeneration of scaffolds with a 400 μm, 650 μm, 850 μm, and 1100 μm aperture prepared by SLM and finally showed that the scaffolds with a 650 μm aperture showed the best bone inward growth.

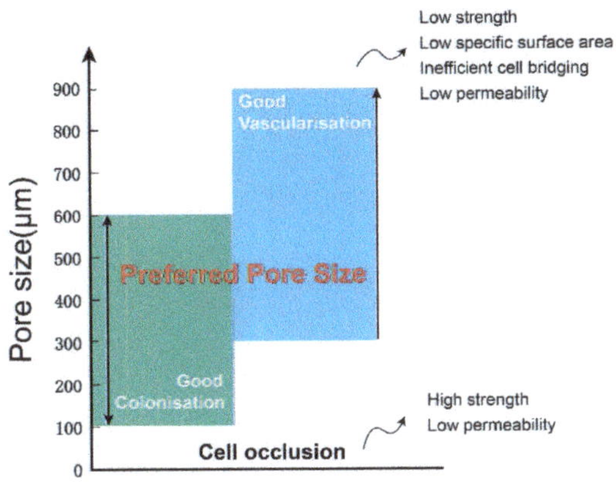

Figure 3. Comprehensive effect of pore size on implants.

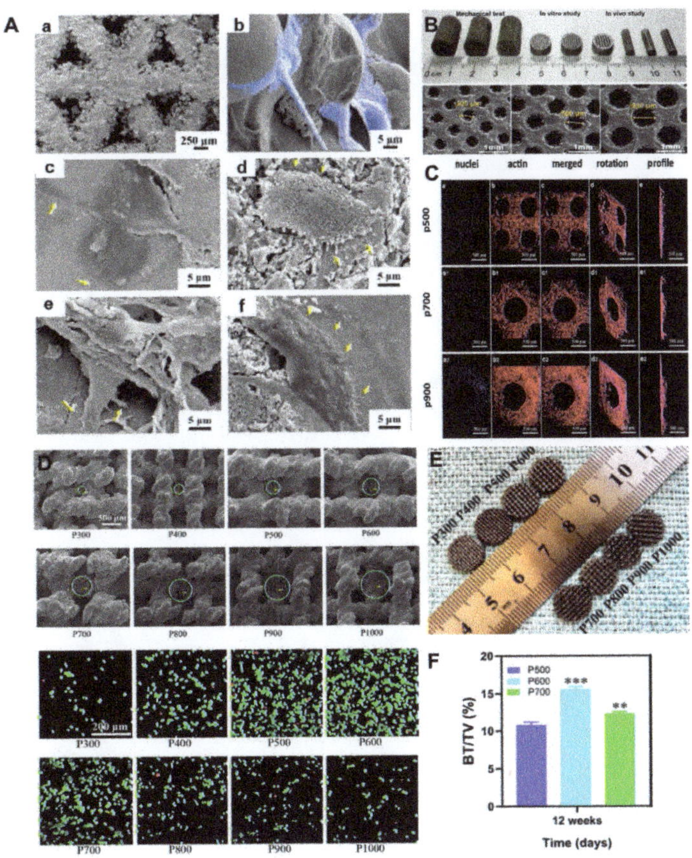

Figure 4. Effect of porous titanium scaffolds with different pore size and porosity on osteogenesis and cells. **(A)**: SEM micrographs of cells after 3 days of culture with scaffolds: a on the surface, b into

pores. Pseudopodia are in blue. SEM morphology of cells after 3 days of culture on c T500, d T1000, e O500, f O1000. Lamellipodia and filopodia are indicated by yellow arrows in c–f [62]. (**B**): Macrophotographs of 3D-printed porous Ti6Al4V samples used for mechanical testing and in vitro and in vivo studies, and SEM images of p500, p700, and p900 [76]. (**C**): Representative fluorescence images of osteoblast adhesion to porous Ti6Al4V implants after 14 days of culture. Cells are stained with actin filaments (red) and nuclei (blue) [76]. (**D**): Pore size analysis of Ti6Al4V and cell-staining results(live: green; dead: red), the green circle represents the pore size measured by Image-J software (Version 1.54) [77]. (**E**): Optical pictures of porous titanium alloys with different pore sizes [77]. (**F**): Quantitative analysis of BV/TV [77] (* $p < 0.05$, ** $p < 0.01$, and *** $p < 0.001$, when compared with P500). (Reprinted with permission from Ref. [62]. Copyright 2023 Elsevier. Reprinted from Ref. [76]; Ref. [77]).

4. Application of 3D-Printed Titanium Alloy Bone-Repair Scaffolds

The utilization of 3D-printed porous titanium alloys addresses the issue of a mismatched elastic modulus between implants and human bone, enhancing compatibility with host bone tissue. Consequently, this technology finds frequent application in orthopedic bone defect repair and dental implant procedures. This section is dedicated to discussing the use of titanium-alloy-based orthopedic and dental implants. A single porous titanium alloy alone may not suffice to address the common challenges encountered in clinical practice. In practical research and application, it is often combined with other functional materials to achieve enhanced performance and functionality [62].

Biomimetic modifications can optimize the biocompatibility of the surface of 3D-printed porous structures, imparting various desirable properties. The judicious selection of biomaterial coatings represents a straightforward and effective approach to augment the biological activity of 3D-printed implants. For example, Wei et al. [79] used 3D printing technology to prepare a porous titanium alloy scaffold with a porosity of 68% and a pore size of 710 μm and then used multiarc ion plating technology to prepare a magnesium coating on the scaffold. Magnesium and its alloy also have good biocompatibility, and the most important thing is that it is a degradable material. The results showed that magnesium ions would be released after implantation, and the appropriate concentration of magnesium ions could inhibit the proliferation of bone tumor cells. Porous titanium is combined with antibacterial hydrogel to fill the micropores of porous titanium, which can be used to treat infected bone defects to induce bone repair and bone integration. Qiao et al. [80] 3D printed a titanium scaffold with a porosity of 70% and a pore size of 600 μm and found that the regenerated bone tissue around the bare titanium scaffold was very limited while the composite implant showed an antibacterial ability and the ability to promote the bone formation differentiation of bone marrow mesenchymal stem cells (BMSCs) in both in vivo and in vitro experiments. More applications are shown in Table 3.

Table 3. Application case of porous titanium alloy implants.

Material	Method	Feature	Outcome	Ref.
Ti-6Al-4V	EBM	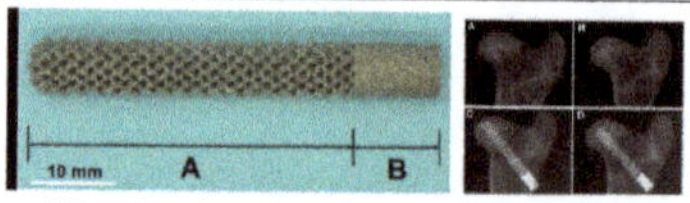 3D printed porous titanium is a porous titanium alloy rod with a diamond lattice prepared using electron beam melting (EBM) technology. A: the body of the rod; B: the end of the rod.	Following Ti-Rod implantation, the femoral head showed good osseointegration, with tight integration between the peripheral bone and the rod, and the new bone grew along the metal trabecula without the intervention of fibrous tissue.	[81]

Table 3. *Cont.*

Material	Method	Feature	Outcome	Ref.
Ti-6Al-4V	SLM	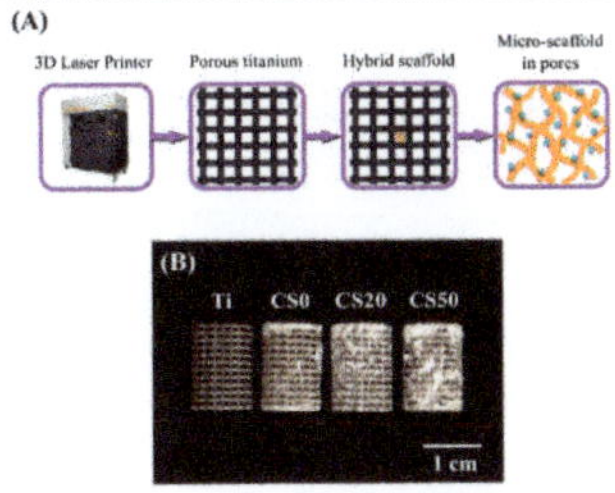 The aperture of the 3D-printed Ti-6Al-4V stent is about 350 μm, and the maximum compressive strength is 49.3 ± 0.9 MPa. (A) is the preparation and coating scheme of 3D printed Ti6Al4V scaffold. (B) is a photographic image of Mg-CS/CH coated Ti6-Al-4V scaffold.	The Mg-CS/CH-coated Ti-6Al-4V scaffold enhanced cell adhesion, proliferation, and differentiation, thereby enhancing downstream osteogenesis and mineralization.	[6]
Ti-6Al-4V	EBM	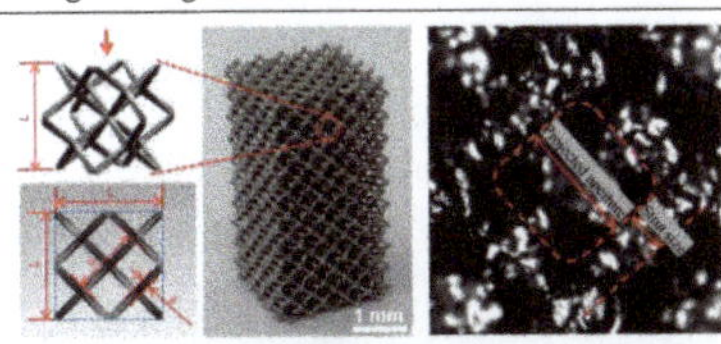 Titanium alloy cells (L, a and s are unit, projected aperture and strut size, respectively) were designed to simulate trabecular structure.	Compared to the solid titanium alloy structure, the elastic modulus of the 3D-printed titanium alloy scaffold with a trabecular structure ranges from 0.39 to 0.618 GPa, which closely approximates that of natural bone. This characteristic helps mitigate the occurrence of stress-shielding phenomena.	[82]
Ti-6Al-4V	EBM	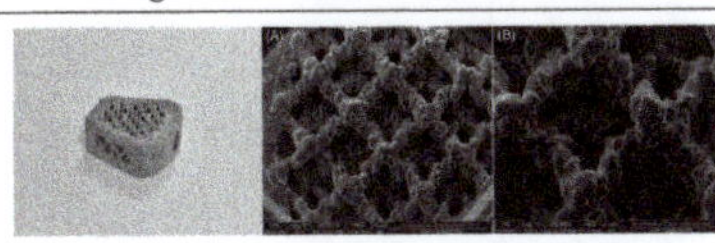 (a) SEM image of the new 3D cage at 50 times; (b) SEM image of the new 3D cage at 100 times. The interior of the 3D-printed cage has an octahedral porous structure with uniform pore size and interconnectivity.	New bone grows inside the cage through pores on the surface of the newly 3D-printed cage. This 3D-printed porous titanium cage exhibits excellent biocompatibility and osseointegration capabilities, making it a potential candidate for clinical applications.	[83]
Ti-6Al-4V	EBM	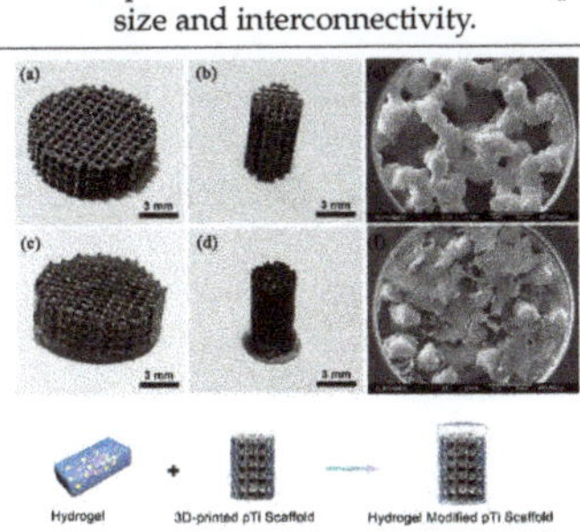 Visual images of (a) disk-shaped and (b) columnar-shaped pTi scaffolds. Visual images of supramolecular hydrogel modified (c) disk-shaped and (d) columnar-shaped pTi scaffolds. SEM microphotographs of (e) pTi scaffolds and (f) supramolecular hydrogel modified pTi scaffolds. Hydrogels composed of sodium tetraborate, polyvinyl alcohol, silver nanoparticles, and tetraethyl orthosilicate were combined with titanium alloy scaffolds for the treatment of infected bone defects.	Hydrogels composed of sodium tetraborate, polyvinyl alcohol, silver nanoparticles, and tetraethyl orthosilicate were combined with titanium alloy scaffolds for the treatment of infected bone defects. In vivo experiments verified that these implants can promote bone regeneration while effectively exhibiting antibacterial properties.	[80]

Table 3. *Cont.*

Material	Method	Feature	Outcome	Ref.
Ti-6Al-4V	—	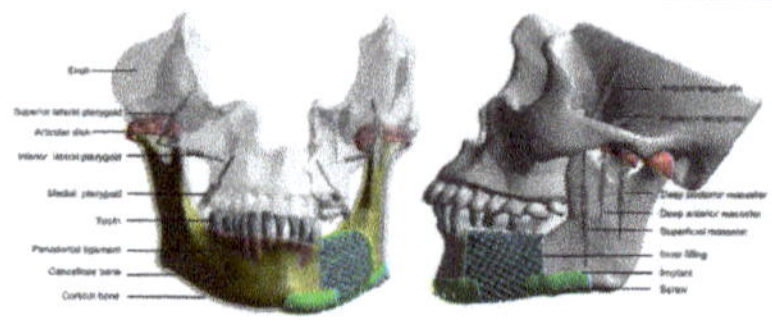 A 3D-printed titanium mesh used to repair mandibular defects.	Through finite element analysis, the optimized implant can provide an excellent mechanical environment for bone regeneration, so as to achieve long-term stability and occlusion reconstruction of the implant.	[84]

5. Conclusions

The 3D-printed porous titanium alloy bone-repair scaffold can effectively solve the stress-shielding problem between the implant and human bone, solve the bone mismatch problem, shorten the operation time, and reduce the surgical failure rate, and it is expected to solve the clinical problem of large bone defects repair, which has been widely studied and applied in recent years. In this study, we conducted a comprehensive review of the influence of structural design in 3D-printed orthopedic titanium alloy implants on bone regeneration. We examined the critical factors such as pore size, porosity, and pore shape that can effectively enhance bone regeneration. This analysis offers valuable insights and establishes a reference framework for future research endeavors focused on the structural aspects of porous implants. Based on the discussion of the above literature, the pore size range of 500–800 μm and the porosity range of 60%–90% can achieve a better bone growth effect. The cell structures with a better bone regeneration effect were diamond, rhombohedral dodecahedron, and cube.

6. Future Direction

While 3D printing porous titanium alloy offers numerous advantages, it still encounters certain practical challenges. There is a need for further refinement in both the design of porous structures and the selection of titanium alloy materials for 3D printing porous titanium alloy brackets. First of all, the porous structure of a human skeleton is not uniformly distributed on the whole but presents a gradient porous structure with a dense exterior and loose interior. In the future, bionic porous scaffolds that can balance biological and mechanical properties should be further studied to achieve the perfect combination of implants and human bones. Secondly, efforts should be made to research the composition of titanium alloys to improve the performance of titanium alloys, thereby preparing stents with better performance. Finally, in order to verify the long-term safety and effectiveness of porous titanium alloys, more clinical experiments and research are needed to combine porous titanium implants with other functional materials to achieve antibacterial, osteogenic, and other effects and obtain more clinical data and experience. With the deepening of research, 3D printing will be able to exert greater value in combination with artificial intelligence and big data in the future.

Author Contributions: Writing—original draft preparation, S.H. and L.C.; writing—review and editing, J.Z. and Z.S.; formal analysis, Y.J. and S.L.; data curation, L.T. All authors have read and agreed to the published version of the manuscript.

Funding: This work was supported by the Natural Science Foundation of Sichuan Province, China (2022NSFSC1510), the Medical Scientific Research Project of Chengdu City, China (2021043), the Sichuan Provincial Science and Technology Foundation (22NZZH0031), and the higher education talent training quality and teaching reform project of the education department of Sichuan Province, China (JG2021-1102).

Institutional Review Board Statement: Not applicable.

Informed Consent Statement: Not applicable.

Data Availability Statement: Data are contained within the article.

Conflicts of Interest: The authors declare no conflicts of interest.

References

1. Ruales-Carrera, E.; Engler, M.L.P.D.; Vaz, P.; Özcan, M.; Volpato, C.A.M. Esthetic and functional rehabilitation of bilateral congenital absence of maxillary lateral incisors: Minimally invasive surgical and prosthetic approach. *J. Esthet. Restor. Dent.* **2019**, *31*, 5–12. [CrossRef]
2. García-Gareta, E.; Coathup, M.J.; Blunn, G.W. Osteoinduction of bone grafting materials for bone repair and regeneration. *Bone* **2015**, *81*, 112–121. [CrossRef]
3. Sun, Q.; Li, Z.; Liu, B.; Yuan, X.; Guo, S.; Helms, J.A. Improving intraoperative storage conditions for autologous bone grafts: An experimental investigation in mice. *J. Tissue Eng. Regener. Med.* **2019**, *13*, 2169–2180. [CrossRef]
4. Li, J.; Zhong, H.; Cao, B.; Ran, Z.; Tan, J.; Deng, L.; Hao, Y.; Yan, J. Comparative Study of 3D-Printed Porous Titanium Alloy with Rod Designs of Three Different Geometric Structures for Orthopaedic Implantation. *Acta Metall. Sin. Engl.* **2023**, *37*, 54–66. [CrossRef]
5. Donkiewicz, P.; Benz, K.; Kloss-Brandstätter, A.; Jackowski, J. Survival rates of dental implants in autogenous and allogeneic bone blocks: A systematic review. *Medicina* **2021**, *57*, 1388. [CrossRef]
6. Tsai, C.H.; Hung, C.H.; Kuo, C.N.; Chen, C.Y.; Peng, Y.N.; Shie, M.Y. Improved bioactivity of 3D printed porous titanium alloy scaffold with chitosan/magnesium-calcium silicate composite for orthopaedic applications. *Materials* **2019**, *12*, 203. [CrossRef] [PubMed]
7. Zhang, B.; Pei, X.; Zhou, C.; Fan, Y.; Jiang, Q.; Ronca, A.; D'Amora, U.; Chen, Y.; Li, H.; Sun, Y. The biomimetic design and 3D printing of customized mechanical properties porous Ti6Al4V scaffold for load-bearing bone reconstruction. *Mater. Des.* **2018**, *152*, 30–39. [CrossRef]
8. Chowdhury, S.; Arunachalam, N. Surface functionalization of additively manufactured titanium alloy for orthopaedic implant applications. *J. Manuf. Process.* **2023**, *102*, 387–405. [CrossRef]
9. Lu, X.; Zhang, D.; Xu, W.; Yu, A.; Zhang, J.; Tamaddon, M.; Zhang, J.; Qu, X.; Liu, C.; Su, B. The effect of Cu content on corrosion, wear and tribocorrosion resistance of Ti-Mo-Cu alloy for load-bearing bone implants. *Corros. Sci.* **2020**, *177*, 109007. [CrossRef]
10. Gautam, D.; Rao, V.K. Nondestructive evaluation of mechanical properties of femur bone. *J. Nondestr. Eval.* **2021**, *40*, 22. [CrossRef]
11. Katz, J.L. Anisotropy of Young's modulus of bone. *Nature* **1980**, *283*, 106–107. [CrossRef]
12. Zhang, Y.; Sun, B.; Zhao, L.; Yang, G. Design and manufacturing of a novel trabecular tibial implant. *Materials* **2023**, *16*, 4720. [CrossRef]
13. Raffa, M.L.; Nguyen, V.H.; Hernigou, P.; Flouzat-Lachaniette, C.H.; Haiat, G. Stress shielding at the bone-implant interface: Influence of surface roughness and of the bone-implant contact ratio. *J. Orthop. Res.* **2021**, *39*, 1174–1183. [CrossRef]
14. Wang, S.; Liu, L.; Li, K.; Zhu, L.; Chen, J.; Hao, Y. Pore functionally graded Ti6Al4V scaffolds for bone tissue engineering application. *Mater. Des.* **2019**, *168*, 107643. [CrossRef]
15. Zhao, L.; Pei, X.; Jiang, L.; Hu, C.; Sun, J.; Xing, F.; Zhou, C.; Fan, Y.; Zhang, X. Bionic design and 3D printing of porous titanium alloy scaffolds for bone tissue repair. *Compos. B Eng.* **2019**, *162*, 154–161. [CrossRef]
16. Zhang, L.C.; Liu, Y.; Li, S.; Hao, Y. Additive manufacturing of titanium alloys by electron beam melting: A review. *Adv. Eng. Mater.* **2018**, *20*, 1700842. [CrossRef]
17. Kelly, C.N.; Lin, A.S.; Leguineche, K.E.; Shekhar, S.; Gall, K. Functional repair of critically sized femoral defects treated with bioinspired titanium gyroid-sheet scaffolds. *J. Mech. Behav. Biomed. Mater.* **2021**, *116*, 104380. [CrossRef]
18. Vetrik, M.; Parizek, M.; Hadraba, D.; Kukackova, O.; Brus, J.; Hlidkova, H.; Komankova, L.; Hodan, J.; Sedlacek, O.; Slouf, M. Porous heat-treated polyacrylonitrile scaffolds for bone tissue engineering. *ACS Appl. Mater. Interfaces* **2018**, *10*, 8496–8506. [CrossRef]
19. Yao, Q.; Liu, Y.; Selvaratnam, B.; Koodali, R.T.; Sun, H. Mesoporous silicate nanoparticles/3D nanofibrous scaffold-mediated dual-drug delivery for bone tissue engineering. *J. Control Release* **2018**, *279*, 69–78. [CrossRef]
20. Sharma, A.; Molla, M.S.; Katti, K.S.; Katti, D.R. Multiscale models of degradation and healing of bone tissue engineering nanocomposite scaffolds. *J. Nanomech. Micromech.* **2017**, *7*, 04017015. [CrossRef]
21. Liu, C.; Wang, C.-y.; Liu, H.; Wang, Z.-h.; Lin, G.-y. Mechanical properties and biocompatibility of 3D printing Ti6Al4V titanium alloy scaffolds. *Zhongguo Youse Jinshu Xue Chin. J. Nonferrous Met.* **2018**, *28*, 758–765.
22. Oladapo, B.I.; Zahedi, S.A.; Ismail, S.O. Mechanical performances of hip implant design and fabrication with PEEK composite. *Polymer* **2021**, *227*, 123865. [CrossRef]
23. Xu, C.; Zhang, H.; Yu, S.; Wu, W.; Zhang, L.; Liu, Q.; Ren, L. Direct ink writing of porous Fe scaffolds for bone implants: Pore size evolution and effect on degradation and mechanical properties. *J. Mater. Res. Technol.* **2023**, *25*, 4901–4912. [CrossRef]
24. Nguyen, L.H.; Annabi, N.; Nikkhah, M.; Bae, H.; Binan, L.; Park, S.; Kang, Y.; Yang, Y.; Khademhosseini, A. Vascularized bone tissue engineering: Approaches for potential improvement. *Tissue Eng. Part B Rev.* **2012**, *18*, 363–382. [CrossRef]
25. Hacisalihoglu, I.; Samancioglu, A.; Yildiz, F.; Purcek, G.; Alsaran, A. Tribocorrosion properties of different type titanium alloys in simulated body fluid. *Wear* **2015**, *332*, 679–686. [CrossRef]

26. Yu, G.; Li, Z.; Li, S.; Zhang, Q.; Hua, Y.; Liu, H.; Zhao, X.; Dhaidhai, D.T.; Li, W.; Wang, X. The select of internal architecture for porous Ti alloy scaffold: A compromise between mechanical properties and permeability. *Mater. Des.* **2020**, *192*, 108754. [CrossRef]
27. Gerlich, D.; Hart, S. Pressure dependence of the elastic moduli of three austenitic stainless steels. *J. Appl. Phys.* **1984**, *55*, 880–884. [CrossRef]
28. Heary, R.F.; Naresh, P.; Sujitha, S.; Nitin, A. Elastic modulus in the selection of interbody implants. *J. Spine Surg.* **2017**, *3*, 163. [CrossRef]
29. Davis, J.R. *Handbook of Materials for Medical Devices*; ASM International: Materials Park, OH, USA, 2003.
30. Park, J.; Lakes, R.S. *Biomaterials: An Introduction*; Springer: Berlin/Heidelberg, Germany, 2007.
31. Osuchukwu, O.A.; Salihi, A.; Abdullahi, I.; Abdulkareem, B.; Nwannenna, C.S. Synthesis techniques, characterization and mechanical properties of natural derived hydroxyapatite scaffolds for bone implants: A review. *SN Appl. Sci.* **2021**, *3*, 1–23. [CrossRef]
32. Aufa, A.; Hassan, M.Z.; Ismail, Z. Recent advances in Ti-6Al-4V additively manufactured by selective laser melting for biomedical implants: Prospect development. *J. Alloys Compd.* **2022**, *896*, 163072. [CrossRef]
33. Parthasarathy, J.; Starly, B.; Raman, S.; Christensen, A. Mechanical evaluation of porous titanium (Ti6Al4V) structures with electron beam melting (EBM). *J. Mech. Behav. Biomed. Mater.* **2010**, *3*, 249–259. [CrossRef] [PubMed]
34. Yook, S.W.; Kim, H.E.; Koh, Y.H. Fabrication of porous titanium scaffolds with high compressive strength using camphene-based freeze casting. *Mater. Lett.* **2009**, *63*, 1502–1504. [CrossRef]
35. Oh, I.-H.; Nomura, N.; Masahashi, N.; Hanada, S. Mechanical properties of porous titanium compacts prepared by powder sintering. *Scr. Mater.* **2003**, *49*, 1197–1202. [CrossRef]
36. Fiorilli, S.; Baino, F.; Cauda, V.; Crepaldi, M.; Vitale-Brovarone, C.; Demarchi, D.; Onida, B. Electrophoretic deposition of mesoporous bioactive glass on glass–ceramic foam scaffolds for bone tissue engineering. *J. Mater. Sci. Mater. Med.* **2015**, *26*, 21. [CrossRef]
37. Du, X.; Yu, B.; Pei, P.; Ding, H.; Yu, B.; Zhu, Y. 3D printing of pearl/CaSO 4 composite scaffolds for bone regeneration. *J. Mater. Chem. B* **2018**, *6*, 499–509. [CrossRef]
38. Jariwala, S.H.; Lewis, G.S.; Bushman, Z.J.; Adair, J.H.; Donahue, H.J. 3D printing of personalized artificial bone scaffolds. *3D Print. Addit. Manuf.* **2015**, *2*, 56–64. [CrossRef]
39. Liu, H.; Li, W.; Liu, C.; Tan, J.; Wang, H.; Hai, B.; Cai, H.; Leng, H.-J.; Liu, Z.-J.; Song, C.-L. Incorporating simvastatin/poloxamer 407 hydrogel into 3D-printed porous Ti6Al4V scaffolds for the promotion of angiogenesis, osseointegration and bone ingrowth. *Biofabrication* **2016**, *8*, 045012. [CrossRef]
40. Hindy, A.; Farahmand, F.; Pourdanesh, F.; Torshabi, M.; Al Janabi, A.H.; Rasoulianboroujeni, M.; Tayebi, L.; Tabatabaei, F.S. Synthesis and characterization of 3D-printed functionally graded porous titanium alloy. *J. Mater. Sci.* **2020**, *55*, 9082–9094. [CrossRef]
41. Cui, J.; Yi, Y.; Zhang, J.; Chai, L.; Jin, H. Preparation and mechanical properties analysis of porous structure for bone tissue engineering. *Biomed. Mater. Eng.* **2022**, *33*, 465–476. [CrossRef]
42. Li, B.; Zhang, M.; Lu, Q.; Zhang, B.; Miao, Z.; Li, L.; Zheng, T.; Liu, P. Application and development of modern 3D printing technology in the field of orthopedics. *Biomed Res. Int.* **2022**, *2022*, 8759060. [CrossRef]
43. Kang, J.; Dong, E.; Li, X.; Guo, Z.; Wang, L. Topological design and biomechanical evaluation for 3D printed multi-segment artificial vertebral implants. *Mater. Sci. Eng. C* **2021**, *127*, 112250. [CrossRef]
44. Murr, L.E. Metallurgy principles applied to powder bed fusion 3D printing/additive manufacturing of personalized and optimized metal and alloy biomedical implants: An overview. *Prog. Artif. Intell.* **2020**, *9*, 1087–1103. [CrossRef]
45. Zhang, T.; Zhang, X.; Mao, M.; Li, J.; Sun, H. Chitosan/hydroxyapatite composite coatings on porous Ti6Al4V titanium implants: In vitro and in vivo studies. *J. Periodontal Implant Sci.* **2020**, *50*, 392. [CrossRef] [PubMed]
46. Xiao, G.; Yang, L.; Feng, L.; Xinguang, W.; Ke, Z.; Zhongjun, L.; Hua, T. A new 3D printing porous trabecular titanium metal acetabular cup for primary total hip arthroplasty: A minimum 2-year follow-up of 92 consecutive patients. *J. Orthop. Surg. Res.* **2020**, *15*, 383.
47. Sarna-Bo, K.; Skic, K.; Sobieszczański, J.; Boguta, P.; Chaas, R. Contemporary Approach to the Porosity of Dental Materials and Methods of Its Measurement. *Int. J. Mol. Sci.* **2021**, *22*, 8903. [CrossRef]
48. Bandyopadhyay, A.; Shivaram, A.; Tarafder, S.; Sahasrabudhe, H.; Banerjee, D.; Bose, S. In vivo response of laser processed porous titanium implants for load-bearing implants. *Ann. Biomed. Eng.* **2017**, *45*, 249–260. [CrossRef]
49. Takahashi, Y.; Tabata, Y. Effect of the fiber diameter and porosity of non-woven PET fabrics on the osteogenic differentiation of mesenchymal stem cells. *J. Biomater. Sci. Polym. Ed.* **2004**, *15*, 41–57. [CrossRef]
50. Wang, Z.; Wang, C.; Li, C.; Qin, Y.; Zhong, L.; Chen, B.; Li, Z.; Liu, H.; Chang, F.; Wang, J. Analysis of factors influencing bone ingrowth into three-dimensional printed porous metal scaffolds: A review. *J. Alloys Compd.* **2017**, *717*, 271–285. [CrossRef]
51. Arabi, N.; Zamanian, A.; Rashvand, S.N.; Ghorbani, F. The tunable porous structure of gelatin–bioglass nanocomposite scaffolds for bone tissue engineering applications: Physicochemical, mechanical, and in vitro properties. *Macromol. Mater. Eng.* **2018**, *303*, 1700539. [CrossRef]
52. Murphy, C.M.; Haugh, M.G.; O'Brien, F.J. The effect of mean pore size on cell attachment, proliferation and mig ration in collagen-glycosaminoglycan scaffolds for bone tissue enginee ring. *Biomaterials* **2010**, *31*, 461–466. [CrossRef]

53. Pei, X.; Wu, L.; Zhou, C.; Fan, H.; Gou, M.; Li, Z.; Zhang, B.; Lei, H.; Sun, H.; Liang, J. 3D printed titanium scaffolds with homogeneous diamond-like structures mimicking that of the osteocyte microenvironment and its bone regeneration study. *Biofabrication* **2020**, *13*, 015008. [CrossRef] [PubMed]
54. Zhang, Y.; Sun, N.; Zhu, M.; Qiu, Q.; Zhao, P.; Zheng, C.; Bai, Q.; Zeng, Q.; Lu, T. The contribution of pore size and porosity of 3D printed porous titanium scaffolds to osteogenesis. *Biomater. Adv.* **2022**, *133*, 112651. [CrossRef] [PubMed]
55. Li, H.; Yao, B.; Li, Z.; Peng, Y.; Fan, H. Compressive properties and deformation mechanism of selective laser melting of Ti6Al4V porous femoral implants based on topological optimization. *Compos. Struct.* **2023**, *321*, 117326. [CrossRef]
56. Tan, C.; Zou, J.; Li, S.; Jamshidi, P.; Abena, A.; Forsey, A.; Moat, R.J.; Essa, K.; Wang, M.; Zhou, K. Additive manufacturing of bio-inspired multi-scale hierarchically strengthened lattice structures. *Int. J. Mach. Tools Manuf.* **2021**, *167*, 103764. [CrossRef]
57. Ahmadi, S.M.; Amin Yavari, S.; Wauthle, R.; Pouran, B.; Schrooten, J.; Weinans, H.; Zadpoor, A.A. Additively manufactured open-cell porous biomaterials made from six different space-filling unit cells: The mechanical and morphological properties. *Materials* **2015**, *8*, 1871–1896. [CrossRef] [PubMed]
58. Herrera, A.; Yánez, A.; Martel, O.; Afonso, H.; Monopoli, D. Computational study and experimental validation of porous structures fabricated by electron beam melting: A challenge to avoid stress shielding. *Mater. Sci. Eng. C* **2014**, *45*, 89–93. [CrossRef] [PubMed]
59. Jahir-Hussain, M.J.; Maaruf, N.A.; Esa, N.E.F.; Jusoh, N. The effect of pore geometry on the mechanical properties of 3D-printed bone scaffold due to compressive loading. *IOP Conf. Ser. Mater. Sci. Eng.* **2021**, *1051*, 012016. [CrossRef]
60. Van Bael, S.; Chai, Y.C.; Truscello, S.; Moesen, M.; Kerckhofs, G.; Van Oosterwyck, H.; Kruth, J.-P.; Schrooten, J. The effect of pore geometry on the in vitro biological behavior of human periosteum-derived cells seeded on selective laser-melted Ti6Al4V bone scaffolds. *Acta Biomater.* **2012**, *8*, 2824–2834. [CrossRef]
61. Xu, J.; Weng, X.-J.; Wang, X.; Huang, J.-Z.; Zhang, C.; Muhammad, H.; Ma, X.; Liao, Q.-D. Potential use of porous titanium–niobium alloy in orthopedic implants: Preparation and experimental study of its biocompatibility in vitro. *PLoS ONE* **2013**, *8*, e79289. [CrossRef]
62. Danlei, Z.; Yutian, H.; Yong, A.; Changjun, H.; Qian, W.; Yan, L.; Jie, L.; Qingsong, W.; Zhen, Z. Effect of pore geometry on the fatigue properties and cell affinity of porous titanium scaffolds fabricated by selective laser melting. *J. Mech. Behav. Biomed. Mater.* **2018**, *88*, 478–487.
63. Kovács, Á.É.; Csernátony, Z.; Csámer, L.; Méhes, G.; Szabó, D.; Veres, M.; Braun, M.; Harangi, B.; Serbán, N.; Zhang, L. Comparative analysis of bone ingrowth in 3D-printed titanium lattice structures with different patterns. *Materials* **2023**, *16*, 3861. [CrossRef]
64. Lim, H.-K.; Ryu, M.; Woo, S.-H.; Song, I.-S.; Choi, Y.-J.; Lee, U.-L. Bone conduction capacity of highly porous 3D-printed titanium scaffolds based on different pore designs. *Materials* **2021**, *14*, 3892. [CrossRef]
65. Farazin, A.; Zhang, C.; Gheisizadeh, A.; Shahbazi, A. 3D bio-printing for use as bone replacement tissues: A review of biomedical application. *Biomed. Eng. Adv.* **2023**, *5*, 100075. [CrossRef]
66. Deng, F.; Liu, L.; Li, Z.; Liu, J. 3D printed Ti6Al4V bone scaffolds with different pore structure effects on bone ingrowth. *J. Biol. Eng.* **2021**, *15*, 4. [CrossRef]
67. Huang, G.; Pan, S.-T.; Qiu, J.-X. The osteogenic effects of porous Tantalum and Titanium alloy scaffolds with different unit cell structure. *Colloids Surf. B* **2022**, *210*, 112229. [CrossRef]
68. Yan, Y.; Kang, Y.; Li, D.; Yu, K.; Zhang, Y. Microstructure, mechanical properties and corrosion behavior of porous Mg-6wt.% Zn scaffolds for bone tissue engineering. *J. Mater. Eng. Perform.* **2018**, *27*, 970–984. [CrossRef]
69. Wang, C.; Xu, D.; Lin, L.; Li, S.; Hou, W.; He, Y.; Sheng, L.; Yi, C.; Zhang, X.; Li, H. Large-pore-size Ti6Al4V scaffolds with different pore structures for vascularized bone regeneration. *Mater. Sci. Eng. C* **2021**, *131*, 112499. [CrossRef] [PubMed]
70. Song, C.; Liu, L.; Deng, Z.; Lei, H.; Yuan, F.; Yang, Y.; Li, Y.; Yu, J. Research progress on the design and performance of porous titanium alloy bone implants. *J. Mater. Res. Technol.* **2023**, *23*, 2626–2641. [CrossRef]
71. Karageorgiou, V.; Kaplan, D. Porosity of 3D biomaterial scaffolds and osteogenesis. *Biomaterials* **2005**, *26*, 5474–5491. [CrossRef] [PubMed]
72. Taniguchi, N.; Fujibayashi, S.; Takemoto, M.; Sasaki, K.; Otsuki, B.; Nakamura, T.; Matsushita, T.; Kokubo, T.; Matsuda, S. Effect of pore size on bone ingrowth into porous titanium implants fabricated by additive manufacturing: An in vivo experiment. *Mater. Sci. Eng. C* **2016**, *59*, 690–701. [CrossRef] [PubMed]
73. Guo, A.X.; Cheng, L.; Zhan, S.; Zhang, S.; Xiong, W.; Wang, Z.; Wang, G.; Cao, S.C. Biomedical applications of the powder-based 3D printed titanium alloys: A review. *J. Mater. Sci. Technol.* **2022**, *125*, 252–264. [CrossRef]
74. Hara, D.; Nakashima, Y.; Sato, T.; Hirata, M.; Kanazawa, M.; Kohno, Y.; Yoshimoto, K.; Yoshihara, Y.; Nakamura, A.; Nakao, Y. Bone bonding strength of diamond-structured porous titanium-alloy implants manufactured using the electron beam-melting technique. *Mater. Sci. Eng. C* **2016**, *59*, 1047–1052. [CrossRef] [PubMed]
75. Zhou, C.; Ye, X.; Fan, Y.; Ma, L.; Tan, Y.; Qing, F.; Zhang, X. Biomimetic fabrication of a three-level hierarchical calcium phosphate/collagen/hydroxyapatite scaffold for bone tissue engineering. *Biofabrication* **2014**, *6*, 035013. [CrossRef] [PubMed]
76. Ran, Q.; Yang, W.; Hu, Y.; Shen, X.; Yu, Y.; Xiang, Y.; Cai, K. Osteogenesis of 3D printed porous Ti6Al4V implants with different pore sizes. *J. Mech. Behav. Biomed. Mater.* **2018**, *84*, 1–11. [CrossRef] [PubMed]
77. Wang, C.; Wu, J.; Liu, L.; Xu, D.; Liu, Y.; Li, S.; Hou, W.; Wang, J.; Chen, X.; Sheng, L. Improving osteoinduction and osteogenesis of Ti6Al4V alloy porous scaffold by regulating the pore structure. *Front. Chem.* **2023**, *11*, 1190630. [CrossRef] [PubMed]

78. Ouyang, P.; Dong, H.; He, X.; Cai, X.; Wang, Y.; Li, J.; Li, H.; Jin, Z. Hydromechanical mechanism behind the effect of pore size of porous titanium scaffolds on osteoblast response and bone ingrowth. *Mater. Des.* **2019**, *183*, 108151. [CrossRef]
79. Wei, X.; Tang, Z.; Wu, H.; Zuo, X.; Dong, H.; Tan, L.; Wang, W.; Liu, Y.; Wu, Z.; Shi, L. Biofunctional magnesium-coated Ti6Al4V scaffolds promote autophagy-dependent apoptosis in osteosarcoma by activating the AMPK/mTOR/ULK1 signaling pathway. *Mater. Today Bio* **2021**, *12*, 100147. [CrossRef] [PubMed]
80. Qiao, S.; Wu, D.; Li, Z.; Zhu, Y.; Zhan, F.; Lai, H.; Gu, Y. The combination of multi-functional ingredients-loaded hydrogels and three-dimensional printed porous titanium alloys for infective bone defect treatment. *J. Tissue Eng.* **2020**, *11*, 2041731420965797. [CrossRef]
81. Wang, C.; Liu, D.; Xie, Q.; Liu, J.; Deng, S.; Gong, K.; Huang, C.; Yin, L.; Xie, M.; Guo, Z. A 3D printed porous titanium alloy rod with diamond crystal lattice for treatment of the early-stage femoral head osteonecrosis in sheep. *Int. J. Med. Sci.* **2019**, *16*, 486. [CrossRef]
82. Zhang, C.; Zhang, L.; Liu, L.; Lv, L.; Gao, L.; Liu, N.; Wang, X.; Ye, J. Mechanical behavior of a titanium alloy scaffold mimicking trabecular structure. *J. Orthop. Surg. Res.* **2020**, *15*, 40. [CrossRef]
83. Li, P.; Jiang, W.; Yan, J.; Hu, K.; Han, Z.; Wang, B.; Zhao, Y.; Cui, G.; Wang, Z.; Mao, K. A novel 3D printed cage with microporous structure and in vivo fusion function. *J. Biomed. Mater. Res. Part A* **2019**, *107*, 1386–1392. [CrossRef] [PubMed]
84. Gao, H.; Li, X.; Wang, C.; Ji, P.; Wang, C. Mechanobiologically optimization of a 3D titanium-mesh implant for mandibular large defect: A simulated study. *Mater. Sci. Eng. C* **2019**, *104*, 109934. [CrossRef] [PubMed]

coatings

Review

Biomaterials in Orthopedic Devices: Current Issues and Future Perspectives

Nina Filip [1,†], Iulian Radu [2,*], Bogdan Veliceasa [3,†], Cristiana Filip [1,*], Mihaela Pertea [4,†], Andreea Clim [1], Alin Constantin Pinzariu [1,†], Ilie Cristian Drochioi [5], Remus Lucian Hilitanu [6] and Ionela Lacramioara Serban [1]

1 Department of Morpho-Functional Sciences (II), Faculty of Medicine, "Grigore T. Popa" University of Medicine and Pharmacy, 700020 Iasi, Romania
2 Department of Surgery, Regional Institute of Oncology, I-st Surgical Oncology, Faculty of Medicine, "Grigore T. Popa" University of Medicine and Pharmacy, 700020 Iasi, Romania
3 Department of Orthopedics and Traumatology, Surgical Science (II), Faculty of Medicine, "Grigore T. Popa" University of Medicine and Pharmacy, 700115 Iasi, Romania
4 Department Plastic Surgery and Reconstructive, Faculty of Medicine, "Grigore T. Popa" University of Medicine and Pharmacy, 700020 Iasi, Romania
5 Department of Oral and Maxillofacial Surgery and Reconstructive, Faculty of Dental Medicine, "Grigore T. Popa" University of Medicine and Pharmacy, 700020 Iasi, Romania
6 Department of Preventive Medicine and Interdisciplinarity, Faculty of Medicine, "Grigore T. Popa" University of Medicine and Pharmacy, 700020 Iasi, Romania
* Correspondence: iulian.radu@umfiasi.ro (I.R.); cristiana.filip@umfiasi.ro (C.F.)
† These authors contributed equally to this work.

Abstract: In orthopedics, bone fixation imposes the use of implants in almost all cases. Over time, the materials used for the implant have evolved from inert materials to those that mimic the morphology of the bone. Therefore, bioabsorbable, biocompatible, and bioactive materials have emerged. Our study aimed to review the main types of implant materials used in orthopedics and present their advantages and drawbacks. We have searched for the pros and cons of the various types of material in the literature from over the last twenty years. The studied data show that consecrated metal alloys, still widely used, can be successfully replaced by new types of polymers. The data from the literature show that, by manipulating their composition, the polymeric compounds can simulate the structure of the different layers of human bone, while preserving its mechanical characteristics. In addition, manipulation of the polymer composition can provide the initiation of desired cellular responses. Among the implanting materials, polyurethane is distinguished as the most versatile polymeric material for use both as orthopedic implants and as material for biomechanical testing of various bone reduction and fixation techniques.

Keywords: orthopedic devices; implants materials; biomechanical tests; polyurethane; titanium-based alloys; biocompatibility; bioceramics

1. Introduction

Over time, for medical purposes, different materials foreign to the human body have been used, such as hair or cellulose used as a suture material, and wood or horn proved to be effective as a material for fixing bone fractures [1,2]. Modern times have brought metals and alloys to the fore as the first choice for the composition of implants, due to their properties: superior strength, reduced risk of rejection, and, above all, biological inertness [3]. Since the Second World War, a new type of compounds, namely synthetically obtained polymers, began to be increasingly used in medical applications, including orthopedics [3,4]. Over time, synthetic materials have evolved from biocompatible and biodegradable materials to bioactive materials today. Data from the literature show that, by manipulating the composition, polymeric compounds can simulate the structure of different tissues while maintaining their mechanical characteristics [3,4]. In addition, by choosing the appropriate type of composition, polymers can initiate targeted cellular responses that favor the healing/repair process. Among the synthetic compounds, polyurethane stands out as one of the most versatile materials, being able to be used as orthopedic implants, covering materials, drainage devices, but also as a material for biomechanical testing of different fracture reduction and fixation techniques [5,6]. The diversity of uses those polyurethanes allow is generated by the fact that, through chemical manipulation, the resulting polyurethanes can mimic many of the structures of human tissues [7]. Knowledge of the mechanical properties of materials and their biocompatibility are the key points for their use in orthopedic surgery.

2. Requirements for Materials Used to Fix Bones

From ancient times, mankind has used different foreign materials for medical purposes, such as employing hair or cellulose as suture material and using wood or horn for bones fracture fixation [1,8–11]. In modern times, metals and alloys have become the first choice in implants utilization, due to their superior strength, lower risk of rejection, and inertness. Since the Second World War, a new class of compounds, the synthetic polymers, have rapidly developed and been increasingly used [12–14]. Regardless of the type of material used, the implant devices must meet general and specific requirements, such as mechanical and biological characteristics.

2.1. General Requirements of Materials for Orthopedic Implant

The basic requirements for implant materials are the clinical and manufacturing properties. As clinical criteria, the materials used for implants must neither be rejected by the body nor generate harmful products [15]. As manufacturing criteria, the materials must allow for the fabrication of the optimum configuration at an affordable price.

2.2. Specific Requirements of Materials for Orthopedic Implant

2.2.1. Mechanical Requirements

The mechanical requirements for an implant are dictated by its use; e.g., strength for loading, elasticity for shear stress. In orthopedics, the implant materials must work in repeated upload/download cycles under different types of forces such as bending, twisting, and shearing stress. Furthermore, implantation devices are subjected to corrosive conditions over long periods of time, which may affect their properties. Therefore, mechanical properties must be accurately assessed in order to maintain fracture reduction. The mechanical properties of the material are evaluated in terms of deformation (strain) produced by an applied force (stress) [16]. The applied stress may be produced by loading, bending, torsion, and compression, or by shear actions. The mechanical properties of an implant material can be assessed using the general stress-strain diagram for the deformation of a material under an imposed external force [17]. The diagram indicates two separate regions: the elastic region, where no deformation occurs, and the plastic region, where deformation is permanent. In the elastic region, the slope of the curve expresses intrinsic stiffness or rigidity (also known as Young's modulus). Rigidity defines the property of a solid body to resist deformation. As stress increases, micro–fractures accumulate, plastic deformations appear, and finally the material breaks up [18]. In the specific case of bones, rigidity is not constant, as is the case for implanted materials. It is said that bone tissue is an anisotropic material. Anisotropy means that a material exhibits different mechanical properties depending on the direction of their determination. Thus, the mechanical properties of the bones depend on the direction of the force application. For example, bone tissue is more rigid when an external force acts in the longitudinal direction of the bone than when it acts on its surface [19].

In the particular case of a pelvis fracture, the main stress is the loading stress. Here, the general load-curve can be converted into a load deformation curve [20]. The converted diagram allows the calculation of the implantation device's rigidity, in order to compare it with the rigidity of a healthy bone. It is also useful in the case of a simulation, for the assessment of the stiffness of artificial bones that mimic certain bone pathologies; e.g., the osteoporotic bone. In the plastic region, it is possible to determine the final stress that causes the breakage of material through fracture. Thus, parameters that are widely used for characterization of a mechanical implant material are rigidity and failure-load [18,20]. As a conclusion, using load-deformation diagrams, the mechanical properties of the implant can be calculated and the implant resistance estimated.

2.2.2. Biological Requirements

The most important property for an orthopedic device is biological inertness, meaning the lack of reactivity with the biological surroundings. In body conditions, a certain degree of reactivity takes place; thus, limited implant degradation may occur. This is considered acceptable if the process does not impair the mechanical strength, or generate harmful byproducts [21]. Contrary to this initial concept, the most recent approaches consider that a certain controlled reactivity between the implant material and the biological surroundings can be used to accelerate the healing process [22]. The introduction of biomaterials has opened up the possibilities of designing materials with particular properties that facilitate the adaptation of the implant to the mechanical and morphological characteristics of the receiving structure. Dolcimascolo et al. [23] consider that biomaterials used in the last 60 years can be grouped into three generations: bio-inert materials (first generation), bioactive or biodegradable materials (second generation), and the current third generation. The design of the third generation intends that the implanted material triggers specific molecular responses that will accelerate the healing process. In recent studies, a new class of biomaterials has been presented: the fourth generation [24–27]. Categories of biomaterials can be seen in Figure 1.

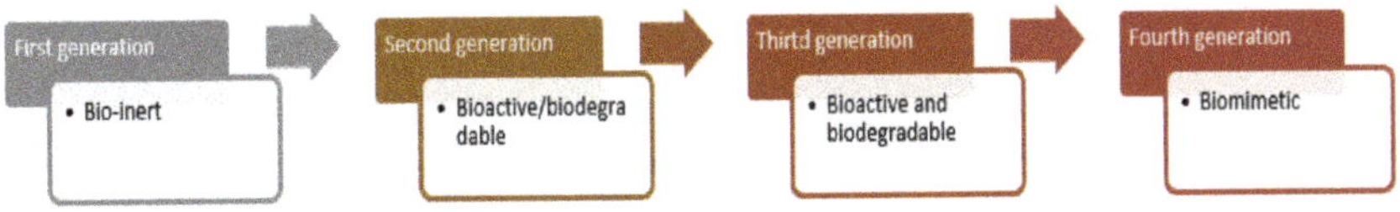

Figure 1. Categories of biomaterials.

This classification outlines the conceptual evolution of the requirements for implant materials. The major requirements pertaining to the first generation had been stiffness and the biological inertness of the materials. Implantation triggers the absorption of non-specific proteins on the surface of the material; a fibrous tissue capsule is formed, and the implant is encapsulated inside, which is a disadvantage for the implant [28]. This led to the design of the second generation of materials. For the second generation, which occurred between 1980 and 2000, the major requirements were the development of bio-absorbable and bioactive materials, while maintaining the mechanical properties of the implant [29,30]. A bio-absorbable material is able to degrade in a progressive manner, allowing the healing and regeneration of tissues. The bioactive material is designed to generate a layer of hydroxyapatite (HA), a natural component of the bone, on the surface of the material instead of the fibrous capsule. It is considered that in vivo stimulation of the production of a hydroxyapatite layer on the implant surface improves the process of mineralization, fixation, and bone regeneration [31,32]. The chemical attachment of various reactive groups on the surface of the polymer causes it to become bio-functional. Thus, this improvement by the second generation materials was possible by modifying the implant surface to promote specific cellular responses, instead of nonspecific responses, as was the case with the first generation [29]. The third generation of biomaterials orientate around the ability to trigger signals that stimulate specific cellular responses at the molecular level. These biomaterials are related to tissue engineering, regenerative medicine, tissue transplantation, and grafting. These biomaterials are designed to be temporary three-dimensional porous structures that are capable of stimulating tissue regeneration, nutrient supply, and possibly, angiogenesis [33,34]. The fourth generation of biomaterials elicits a personalized interaction with cellular processes and microenvironments, which includes four requirements: inertia, activity, receptivity, and autonomy [35].

3. Specific Orthopedic Implant Materials

3.1. Metal Materials

The first materials used for implant devices were metals and alloys, due to their superior strength and a certain biological inertness. Metals selected for the implant include: iron, cobalt, nickel, titanium, and zirconium. The metal combinations aim to obtain specific properties in the final mixture, such as: elasticity, strength, ductility (to reduce the risk of crack failure), and corrosion resistance [36]. Current alloys used in orthopedic metal-based implants include: stainless steels, cobalt-based alloys, and titanium-based alloys. Table 1 summarizes both the advantages and disadvantages of the most important metal materials used in orthopedics.

Table 1. Advantages and disadvantages of metal materials used in orthopedic devices.

Metal Materials	Advantages	Disadvantages	References
Stainless steel (316L)	high resistance, less expensive, easy fabrication	allergic reaction, stress shielding effect	[37–39]
Cobalt-chromium-based alloys (Co-Cr-Mo, Cr-Ni-Cr-Mo)	high corrosion resistance, wear resistance	early implant loosening rate, difficult fabrication	[36,40–42]
Titanium –based alloys (Ti-4Al-4V, Ti-6Al-7Nb, Ti-13Nb-13Zr)	biocompatibility, Young modulus close to bone, excellent corrosion resistance, good osteointegration	expensive, intoxication, bone resorption, allergy	[43–46]
Mg based alloys	biodegradability in vivo, biocompatibility	low mechanical strength, fast degradation	[36,47,48]

3.1.1. Stainless Steel

Stainless steel 18-8 (18% chromium, 8% nickel) is the most common alloy. It has superior corrosion resistance obtained through compositional modifications by using additional metals, especially Cr [37]. The presence of Cr allows Cr_2O_3 to form a strong and adherent layer that favors healing. Stainless steel is widely used in removable orthopedic devices (fracture plates, hip screws) due to its low cost [49,50]. Currently, the new stainless steel-based alloys contain Co-Cr, Ni, Mn, and a high nitrogen content [50]. These types of alloys can be used in combination with polyethylene (PE) for disc prosthesis [51].

3.1.2. Cobalt-Based Alloys

Cobalt-based alloys are superior to stainless steel in terms of strength [42,52]. They have the advantage of better biocompatibility, being more resistant to corrosion than stainless steel, but more expensive to manufacture. Some variants of the alloy composed of cobalt-chromium-molybdenum are especially used for implants in hip prosthesis [22,53]. This alloy version is reserved for metal-to-metal devices due to its high abrasion resistance [54,55].

3.1.3. Titanium-Based Alloys

For orthopedic devices, titanium may be used alone or in alloys with other metals. The use of pure titanium has the following advantages: low weight, very good corrosion resistance, especially in saline solution (due to the formation of an adhesive layer of TiO_2), and the ability to become tightly integrated into the bone [56–59]. This last property greatly improves the long-term behavior of the implant, as well as reduces the risk of loosening and failure of the device. Though titanium-based alloys have proven to be highly corrosion-resistant and biocompatible, there are concerns regarding long-term implantation because of the release of potentially toxic alloying elements and the risk of stress shielding, as their elastic modulus values are still relatively high compared to the elastic modulus of bone. Being an expensive material, titanium-based alloys are currently only used for patients with hypersensitivity reactions to steel or cobalt-chromium alloys [45,46,60].

In conclusion, even though metallic implants belong to the concepts of the first-generation, they are still in use and they continuously evolve. Proper features of the implant surface, such as roughness, wettability, and electrostatic charges that finally dictate the quality of implant anchorage in bone, can be obtained through a wide variety of surface treatments applied to the metal alloys.

3.2. Non-Metal Materials

3.2.1. Polymeric Materials

Today, polymeric materials are good replacements for many types of materials in all medical fields. The increasing use of polymers is dictated by low production costs and high versatility. Full data on all types of synthetic polymers used in clinical medicine can be found in Mainz's review [61]. In the case of orthopedics, the use of polymers is steadily increasing due to the unlimited possibilities of manipulating their bio-mechanical properties [6,62]. The polymeric biomaterials belonging to the first generation include: polyethylene (PE), polymethylmethacrylate (PMMA), and polyurethanes (PU). These materials are still used today in both the first- and third-generation versions.

Both the advantages and disadvantages of the most important non-metal materials used in orthopedic devices are presented in Table 2.

Table 2. Advantages and disadvantages of some non-metal materials used in orthopedics.

Non-Metal Materials	Advantages	Disadvantages	References
Polyethylene (PE)/ultrahigh molecular weight polyethylene (UHMWPE)	low resistance to friction, resistance to wear, biocompatibility	debris generation along time as a result of wear, osteolysis	[63–70]
Polymethylmethacrylate (PMMA)	good tensile properties and tensile strength	long-term usage can produce cement fragmentation	[71–80]

Polyethylene

Polyethylene is used in total hip and knee arthroplasty, insertion of the tibia, or as a spacer in disc replacement [64,67]. The major benefits of PE use are: low resistance to friction, resistance to abrasion or to impact, and good biocompatibility. A particular type of polyethylene polymer, named ultrahigh molecular weight polyethylene (UHMWPE), associated with metal, is highly used in orthopedic surgery or joint prostheses [70] due to its high load bearing capacity. The main drawback is the possibility of debris generation over time as a result of wear.

Polymethylmethacrylate

Polymethylmethacrylate is also known as "acrylic cement". It is a versatile compound used in ophthalmology, dentistry, as well as orthopedics. In orthopedics, PMMA is used for hip arthroplasty, spinal fracture fixation, internal fracture-fixation plates (so-called "luting"), and as a permanent bone substitute in the treatment of pathologic fractures [81–84]. The most important feature of PMMA is that it can be molded into particular shapes dictated by implant requirements, or be polymerized in situ during the time of surgery. The polymerization process lasts from 6 to 7 min. Used as cement that anchors the prostheses to the bone, PMMA must ensure bone adhesion. For this purpose, the polymer is filled with hydroxyapatite particles; thus, providing a homogeneous load transfer from the implant to the bone [85]. Other advantages of PMMA include good tensile properties, tensile strength, as well as good flexural rigidity. Although it has many advantages, the use of PMMA does have some disadvantages, such as the release of heat and methyl methacrylate monomer (MMA) in the in situ polymerization process. Released MMA causes hemodynamic effects (hypotension and hypoxemia). Another disadvantage of PMMA utilization is its brittleness compared to metal materials. In the particular case of fragility fractures, such as vertebral or sacral osteoporotic fractures, the injection of PMMA has the advantage of improving stability and reducing pain [86]. The main disadvantage of the procedure is the extravasation of the PMMA in the pre-sacral space, spinal canal, sacral foramen, which may affect sacral nerve root, or sacral spinal canal [86].

Polyurethanes

Polyurethane is a very versatile and inexpensive material and is therefore used for many medical purposes. It can be produced in various types that offer specific properties depending on the purpose of the implanted device. These include stiffness, flexibility, mechanical strength, or elasticity. However, the most important property is that it can mimic certain biological structures of the body, especially bone structure. Furthermore, polyurethanes can be made entirely biocompatible (for permanent implantation) and also biodegradable (absorbable scaffolds for tissue regeneration) [87]. The medical use of polyurethane includes vascular catheters [88,89], transparent semipermeable films for wound dressings [90], heart valves [91], ureteral [92], stents, or orthopedic implants [93]. Third-generation PU are used for tissue reinforcement scaffolds (as biodegradable material in combination with urea) [94] or for regeneration of peripheral nerves [95,96]. Polyurethanes are by far the most used material in mechanical testing of orthopedic fixation devices. Their use is justified by the fact that, through chemical modification, polyurethanes can faithfully simulate both the compact structure of the cortical bone and the trabecular structure of the spongy bone. When tested with an externally applied force, the stress-strain curve of the polyurethane foam exhibited a similar behavior to the spongy bone. Thompson et al. [97] tested different types of rigid polyurethane foams and found that the polymer and the spongy bone exhibit similar behavior in the elastic area. Due to the similarity between the human bone and the polymer, a standardization of the polyurethane foams was established based on mechanical behavior for different bone density [98]. As a conclusion, PU foams can be used [99] as human bone replacements to measure important functional parameters (resistance, stability, and rigidity) of orthopedic implants in biomechanical tests.

3.2.2. Ceramics

Due to the increase in the life expectancy of the population and the number of surgical interventions, it is expected that the implants will be very reliable and resistant to breakage in vivo, so that they can provide a service life of more than 30 years. Because they present biocompatibility, high hardness, and high resistance to wear, ceramic materials are generally suitable for bone replacement bearings [36]. The development of ceramics in joint replacement bearings aims to reduce clinical wear, and consequently, reduce the risk of debris-induced osteolysis [100]. Bioceramics are classified into three types: bioinert, that do not interact with living tissues and are non-toxic (zirconia, alumina); biodegradables, that are absorbed and dissolved inside the body (calcium phosphates and hydroxyapatite); and bioactive, that are able to form (bioactive glass). Alumina-zirconia ceramic composites show remarkable stability and mechanical properties, but they have high production costs [101]. Calcium phosphate coatings have been used in orthopedics due to their similarity to the mineral, the bone phase, and presents the advantage of significant biocompatibility and osseointegration with the host tissue. Plasma sprayed calcium phosphate coatings are not uniform, and there is little control over thickness and surface topography, which can lead to implant inflammation when particles are released from them [102].

4. Future Perspectives for the Materials Used in Orthopedics

Currently, the materials widely used in orthopedics are metal, bioceramics, and polymers [103–110]. Recent data suggest that a special interest is being given to biomedical nanotechnology. Ceria nanoparticles or nano-ceria (CeO_2-NPs) were presented with an increased potential for applicability in orthopedics [111–114]. Luo et al. [115] investigated the effect and mechanism of cerium oxide nanoparticles (CeO_2 NPs) in MC3T3-E1 mouse osteoblast precursors, and reported that they improved matrix mineralization and increased osteogenic gene expression. Castiglioni et al. [116] proposed the use of silver nanoparticles to fight infections in orthopedic implants, because those that show antimicrobial activity are smaller than those that exert toxic effects on bone-forming cells in vitro. Samanta et al. [117] highlighted that gold nanoparticles (Au NPs) have antibacterial action. These results demonstrate the positive impact that nanomedicine can have on improving the effectiveness of materials used in orthopedics.

Due to biocompatibility and appropriate mechanical properties, magnesium alloys have recently become the focus of research. At this moment, Mg-based metal alloys represent the new generation of biodegradable metal materials, with a good osseointegration property [105]. In a case study, Holweg et al. [118] focused on intraoperative clinical sites of human bone stabilized with magnesium screws and reported homogeneous degradation with good bone-implant interface. Future research must be concentrated on the direction of alloys with a low degradation rate and an improved mechanical strength, in order to solve load-bearing zone fractures [119].

Current research focuses on both the development of new materials and surface modification strategies. The development of new techniques and strategies on composite coatings to better mimic the structure of human bone would lead to a new generation of orthopedic implants with improved implant integration and bone healing.

The recent development and use of 3D printing technology is rapidly becoming more valuable to the field of orthopedics, but the field of orthopedic implants has not been sufficiently explored. Polymers are one of the most common materials used in 3D-printed bone replacements because of their potential use as filaments for fused deposition modeling, solutions for stereolithography apparatus, and gels for direct ink writing [120]. Feltz et al. [121] evaluated the feasibility of using desktop 3D printers to reproduce surgical implant models using biocompatible materials and reported reduced manufacturing costs, but did not achieve mechanical properties similar to standard stainless-steel implants. In orthopedics, 3D printed materials can be made into implants, prostheses, and create life-size anatomical models [122].

Materials used in orthopedics will continue to evolve in order to reduce implant costs, maintain patient safety, optimize surgical techniques, and reduce the risk of infection.

5. Conclusions

The challenges encountered in orthopedics are: fractures, broken joints, and already diseased bone traumas. In order to restore normal bone function or joints, the use of prosthesis or implantation devices is inevitable. Certain properties of the implant material are mandatory to ensure a total recovery. Current data show that consecrated metal alloys, still widely used, can be successfully replaced by new types of polymers. Synthetic materials based on biocompatible polymers belonging to the latest generation are able to provide multiple possibilities for reproducing the anatomical structure, reabsorbing over a period of time, or generating a specific response from the biological environment. Literature indicates that polyurethanes are the most suitable material for the mechanical testing of orthopedic devices. We also believe that nanotechnology is very important for the success of orthopedic implants. Finally, synthetic materials allow simulations to evaluate the behavior of an implant device; thus, increasing the chances of a successful surgical implantation procedure.

Author Contributions: Conceptualization, N.F. and I.R.; methodology, C.F.; formal analysis, B.V. and M.P.; investigation, N.F., I.R., A.C.P. and C.F; resources, I.C.D.; data curation, A.C.; writing—original draft preparation, A.C.P., A.C. and I.C.D.; writing—review and editing, R.L.H.; visualization, B.V. and M.P.; supervision, I.L.S.; roject administration, N.F. and I.L.S.; All authors have equal contributions. All authors have read and agreed to the published version of the manuscript.

Funding: This research received no external funding.

Institutional Review Board Statement: Not applicable.

Informed Consent Statement: Not applicable.

Data Availability Statement: Not applicable.

Conflicts of Interest: The authors declare no conflict of interest.

References

1. Marin, E.; Boschetto, F.; Pezzotti, G. Biomaterials and biocompatibility: An historical overview. *J. Biomed. Mater. Res. Part A* **2020**, *108*, 1617–1633. [CrossRef] [PubMed]
2. Chaudhari, A.A.; Vig, K.; Baganizi, D.R.; Sahu, R.; Dixit, S.; Dennis, V.; Singh, S.R.; Pillai, S.R. Future prospects for scaffolding methods and biomaterials in skin tissue engineering: A review. *Int. J. Mol. Sci.* **2016**, *17*, 1974. [CrossRef] [PubMed]
3. Fan, J.; Abedi-Dorcheh, K.; Sadat Vaziri, A.; Kazemi-Aghdam, F.; Rafieyan, S.; Sohrabinejad, M.; Ghorbani, M.; Rastegar Adib, F.; Ghasemi, Z.; Klavins, K.; et al. A Review of Recent Advances in Natural Polymer-Based Scaffolds for Musculoskeletal Tissue Engineering. *Polymers* **2022**, *14*, 2097. [CrossRef] [PubMed]
4. Tsegay, F.; Elsherif, M.; Butt, H. Smart 3D Printed Hydrogel Skin Wound Bandages: A Review. *Polymers* **2022**, *14*, 1012. [CrossRef] [PubMed]
5. Ramakrishna, S.; Mayer, J.; Wintermantel, E.; Leong, K.W. Biomedical applications of polymer-composite materials: A review. *Compos. Sci. Technol.* **2001**, *61*, 1189–1224. [CrossRef]
6. Francis, A. Biological evaluation of preceramic organosilicon polymers for various healthcare and biomedical engineering applications: A review. *J. Biomed. Mater. Res. Part B Appl. Biomater.* **2021**, *109*, 744–764. [CrossRef] [PubMed]
7. Oladapo, B.I.; Zahedi, S.A.; Ismail, S.O.; Olawade, D.B. Recent advances in biopolymeric composite materials: Future sustainability of bone-implant. *Renew. Sustain. Energy Rev.* **2021**, *150*, 111505. [CrossRef]
8. Ghasemi-Mobarakeh, L.; Kolahreez, D.; Ramakrishna, S.; Williams, D. Key terminology in biomaterials and biocompatibility. *Curr. Opin. Biomed. Eng.* **2019**, *10*, 45–50. [CrossRef]
9. Heimann, R.B. (Ed.) *Materials for Medical Application*; Walter de Gruyter GmbH & Co KG: Görlitz, Germany, 2020.
10. Love, B. *Biomaterials: A Systems Approach to Engineering Concepts*; Academic Press: Cambridge, MA, USA, 2017.
11. Hacker, M.C.; Krieghoff, J.; Mikos, A.G. Synthetic polymers. In *Principles of Regenerative Medicine*; Academic Press: Cambridge, MA, USA, 2019; pp. 559–590.
12. Xu, X.; Song, J. Segmental long bone regeneration guided by degradable synthetic polymeric scaffolds. *Biomater. Transl.* **2020**, *1*, 33.
13. Mir, M.; Ali, M.N.; Barakullah, A.; Gulzar, A.; Arshad, M.; Fatima, S.; Asad, M. Synthetic polymeric biomaterials for wound healing: A review. *Prog. Biomater.* **2018**, *7*, 1–21. [CrossRef]
14. Donnaloja, F.; Jacchetti, E.; Soncini, M.; Raimondi, M.T. Natural and Synthetic Polymers for Bone Scaffolds Optimization. *Polymers* **2020**, *12*, 905. [CrossRef]
15. Tian, L.; Tang, N.; Ngai, T.; Wu, C.; Ruan, Y.; Huang, L.; Qin, L. Hybrid fracture fixation systems developed for orthopaedic applications: A general review. *J. Orthop. Transl.* **2019**, *16*, 1–13. [CrossRef]
16. Pal, S. Mechanical properties of biological materials. In *Design of Artificial Human Joints & Organs*; Springer: Boston, MA, USA, 2014; pp. 23–40.
17. Silver, F.H.; Shah, R. Measurement of mechanical properties of natural and engineered implants. *Adv. Tissue Eng. Regen. Med. Open Access* **2016**, *1*, 20–25. [CrossRef]
18. Li, J.; Zhao, Z.; Yin, P.; Zhang, L.; Tang, P. Comparison of three different internal fixation implants in treatment of femoral neck fracture—A finite element analysis. *J. Orthop. Surg. Res.* **2019**, *14*, 76. [CrossRef]
19. Bankoff, A.D.P. Biomechanical characteristics of the bone. *Hum. Musculoskelet. Biomech.* **2012**, *61*, 86.
20. Basso, T. Internal fixation of fragility fractures of the femoral neck: Ex vivo biomechanical studies. *Acta Orthop.* **2015**, *86* (Suppl. S361), S1–S36. [CrossRef]
21. Todros, S.; Todesco, M.; Bagno, A. Biomaterials and Their Biomedical Applications: From Replacement to Regeneration. *Processes* **2021**, *9*, 1949. [CrossRef]
22. Thanigaivel, S.; Priya, A.K.; Balakrishnan, D.; Dutta, K.; Rajendran, S.; Soto-Moscoso, M. Insight on recent development in metallic biomaterials: Strategies involving synthesis, types and surface modification for advanced therapeutic and biomedical applications. *Biochem. Eng. J.* **2022**, *187*, 108522. [CrossRef]
23. Dolcimascolo, A.; Calabrese, G.; Conoci, S.; Parenti, R. Innovative biomaterials for tissue engineering. In *Biomaterial-Supported Tissue Reconstruction or Regeneration*; IntechOpen: London, UK, 2019. [CrossRef]
24. Festas, A.J.; Ramos, A.; Davim, J.P. Medical devices biomaterials—A review. *Proc. Inst. Mech. Eng. Part L J. Mater. Des. Appl.* **2020**, *234*, 218–228. [CrossRef]
25. Bhaskar, B.; Nagarjuna, V. Biomaterials, Tissue Engineering, and Regenerative Medicine: A Brief Outline. In *Biomaterials in Tissue Engineering and Regenerative Medicine*; Springer: Singapore, 2021; pp. 3–17.
26. Basu, B.; Gowtham, N.H.; Xiao, Y.; Kalidindi, S.R.; Leong, K.W. Biomaterialomics: Data science-driven pathways to develop fourth-generation biomaterials. *Acta Biomater.* **2022**, *143*, 1–25. [CrossRef]

27. Allo, B.A.; Costa, D.O.; Dixon, S.J.; Mequanint, K.; Rizkalla, A.S. Bioactive and Biodegradable Nanocomposites and Hybrid Biomaterials for Bone Regeneration. *J. Funct. Biomater.* **2012**, *3*, 432–463. [CrossRef] [PubMed]
28. Gautam, G.; Kumar, S.; Kumar, K. Processing of biomaterials for bone tissue engineering: State of the art. *Mater. Today Proc.* **2022**, *50*, 2206–2217. [CrossRef]
29. Saad, M.; Akhtar, S.; Srivastava, S. Composite polymer in orthopedic implants: A review. *Mater. Today Proc.* **2018**, *5*, 20224–20231. [CrossRef]
30. Sheikh, Z.; Najeeb, S.; Khurshid, Z.; Verma, V.; Rashid, H.; Glogauer, M. Biodegradable Materials for Bone Repair and Tissue Engineering Applications. *Materials* **2015**, *8*, 5744–5794. [CrossRef] [PubMed]
31. Arifin, A.; Sulong, A.B.; Muhamad, N.; Syarif, J.; Ramli, M.I. Material processing of hydroxyapatite and titanium alloy (HA/Ti) composite as implant materials using powder metallurgy: A review. *Mater. Des.* **2014**, *55*, 165–175. [CrossRef]
32. Zakaria, S.M.; Sharif Zein, S.H.; Othman, M.R.; Yang, F.; Jansen, J.A. Nanophase hydroxyapatite as a biomaterial in advanced hard tissue engineering: A review. *Tissue Eng. Part B Rev.* **2013**, *19*, 431–441. [CrossRef] [PubMed]
33. Ning, C.; Zhou, L.; Tan, G. Fourth-generation biomedical materials. *Mater. Today* **2016**, *19*, 2–3. [CrossRef]
34. Serrano, M.C.; Ameer, G.A. Recent insights into the biomedical applications of shape-memory polymers. *Macromol. Biosci.* **2012**, *12*, 1156–1171. [CrossRef]
35. Montoya, C.; Du, Y.; Gianforcaro, A.L.; Orrego, S.; Yang, M.; Lelkes, P.I. On the road to smart biomaterials for bone research: Definitions, concepts, advances, and outlook. *Bone Res.* **2021**, *9*, 12. [CrossRef]
36. Shekhawat, D.; Singh, A.; Bhardwaj, A.; Patnaik, A. A short review on polymer, metal and ceramic based implant materials. *IOP Conf. Ser. Mater. Sci. Eng.* **2021**, *1017*, 012038. [CrossRef]
37. Wall, E.J.; Jain, V.; Vora, V.; Mehlman, C.T.; Crawford, A.H. Complications of titanium and stainless steel elastic nail fixation of pediatric femoral fractures. *J. Bone Jt. Surg.* **2008**, *90*, 1305–1313. [CrossRef]
38. Ghunawat, S.; Relhan, V.; Garg, V.K. Hypersensitivity Reactions to Metal Implants: Clinical, Diagnostic and Treatment Overview. *Indian J. Clin. Dermatol.* **2019**, *2*, 29–34.
39. Chen, Q.; Thouas, G.A. Metallic implant biomaterials. *Mater. Sci. Eng. R Rep.* **2015**, *87*, 1–57. [CrossRef]
40. Brogini, S.; Sartori, M.; Giavaresi, G.; Cremascoli, P.; Alemani, F.; Bellini, D.; Martini, L.; Maglio, M.; Pagani, S.; Fini, M. Osseointegration of additive manufacturing Ti–6Al–4V and Co–Cr–Mo alloys, with and without surface functionalization with hydroxyapatite and type I collagen. *J. Mech. Behav. Biomed. Mater.* **2021**, *115*, 104262. [CrossRef]
41. Solanke, S.; Gaval, V.; Sanghavi, S. In vitro tribological investigation and osseointegration assessment for metallic orthopedic bioimplant materials. *Mater. Today Proc.* **2021**, *44*, 4173–4178. [CrossRef]
42. Aherwar, A.; Singh, A.K.; Patnaik, A. Cobalt Based Alloy: A Better Choice Biomaterial for Hip Implants. *Trends Biomater. Artif. Organs* **2016**, *30*, 50–55.
43. Anene, F.A.; Aiza Jaafar, C.N.; Zainol, I.; Azmah Hanim, M.A.; Suraya, M.T. Biomedical materials: A review of titanium based alloys. *Proc. Inst. Mech. Eng. Part C J. Mech. Eng. Sci.* **2021**, *235*, 3792–3805. [CrossRef]
44. Thomas, P.; Thomas, M.; Summer, B.; Dietrich, K.; Zauzig, M.; Steinhauser, E.; Krenn, V.; Arnholdt, H.; Flaig, M.J. Impaired wound-healing, local eczema, and chronic inflammation following titanium osteosynthesis in a nickel and cobalt-allergic patient: A case report and review of the literature. *J. Bone Jt. Surg.* **2011**, *93*, e61. [CrossRef]
45. Nasibi, S.; Alimohammadi, K.; Bazli, L.; Eskandarinezhad, S.; Mohammadi, A.; Sheysi, N. TZNT alloy for surgical implant applications: A systematic review. *J. Compos. Compd.* **2020**, *2*, 62–68. [CrossRef]
46. Xu, W.; Lu, X.; Tian, J.; Huang, C.; Chen, M.; Yan, Y.; Wang, L.; Qu, X.; Wen, C. Microstructure, wear resistance, and corrosion performance of Ti35Zr28Nb alloy fabricated by powder metallurgy for orthopedic applications. *J. Mater. Sci. Technol.* **2020**, *41*, 191–198. [CrossRef]
47. Findik, F. Recent developments of metallic implants for biomedical applications. *Period. Eng. Nat. Sci.* **2020**, *8*, 33–57.
48. Lin, X.; Saijilafu; Wu, X.; Wu, K.; Chen, J.; Tan, L.; Witte, F.; Yang, H.; Mantovani, D.; Zhou, H.; et al. Biodegradable Mg-based alloys: Biological implications and restorative opportunities. *Int. Mater. Rev.* **2022**, *67*, 1–39. [CrossRef]
49. Barber, C.C.; Burnham, M.; Ojameruaye, O.; McKee, M.D. A systematic review of the use of titanium versus stainless steel implants for fracture fixation. *OTA Int.* **2021**, *4*, e138. [CrossRef]
50. Mahyudin, F.; Widhiyanto, L.; Hermawan, H. Biomaterials in orthopaedics. In *Biomaterials and Medical Devices*; Springer: Cham, Switzerland, 2016; pp. 161–181.
51. Ghosh, S.; Sanghavi, S.; Sancheti, P. Metallic biomaterial for bone support and replacement. In *Fundamental Biomaterials: Metals*; Woodhead Publishing: Cambridge, UK, 2018; pp. 139–165.
52. Mehta, H.; Kaur, G.; Chaudhary, G.R.; Prabhakar, N.; Kaul, S.; Singhal, N.K. Evaluation of corrosion resistant, antimicrobial and cytocompatible behaviour of cobalt based metallosurfactants self-assembled monolayers on 316L stainless steel surface. *Surf. Coat. Technol.* **2022**, *444*, 128657. [CrossRef]
53. Orlov, A.A.; Sarychev, S.M.; Orlov, A.A. The Effect of Surface Conditions on Corrosion Resistance of a Cobalt-Chromium Alloy. In *Key Engineering Materials*; Trans Tech Publications Ltd.: Bäch, Switzerland; Volume 887, pp. 358–363.
54. Nagay, B.E.; Cordeiro, J.M.; Barão, V.A.R. Alloy materials for biomedical applications. In *Alloy Materials and Their Allied Applications*; Scrivener Publishing LLC: Austin, TX, USA, 2020; pp. 159–189.
55. Baltatu, M.S.; Burduhos-Nergis, D.D.; Burduhos-Nergis, D.P.; Vizureanu, P. *Advanced Metallic Biomaterials*; Materials Research Forum LLC: Millersville, PA, USA, 2022.

56. Kaur, M.; Singh, K. Review on titanium and titanium based alloys as biomaterials for orthopaedic applications. *Mater. Sci. Eng. C* **2019**, *102*, 844–862. [CrossRef]
57. Baltatu, M.S.; Tugui, C.A.; Perju, M.C.; Benchea, M.; Spataru, M.C.; Sandu, A.V.; Vizureanu, P. Biocompatible titanium alloys used in medical applications. *Rev. Chim.* **2019**, *70*, 1302–1306. [CrossRef]
58. Rony, L.; Lancigu, R.; Hubert, L. Intraosseous metal implants in orthopedics: A review. *Morphologie* **2018**, *102*, 231–242. [CrossRef]
59. İzmir, M.; Ercan, B. Anodization of titanium alloys for orthopedic applications. *Front. Chem. Sci. Eng.* **2019**, *13*, 28–45. [CrossRef]
60. Jin, W.; Chu, P.K. Orthopedic implants. *Encycl. Biomed. Eng.* **2019**, *1*, 3.
61. Maitz, M.F. Applications of synthetic polymers in clinical medicine. *Biosurface Biotribology* **2015**, *1*, 161–176. [CrossRef]
62. Singh, D.K.; Verma, R.K. Contemporary Development on the Performance and Functionalization of Ultra High Molecular Weight Polyethylene (UHMWPE) for Biomedical Implants. *Nano LIFE* **2021**, *11*, 2130009. [CrossRef]
63. Zivic, F.; Affatato, S.; Trajanovic, M.; Schnabelrauch, M.; Grujovic, N.; Choy, K.L. (Eds.) *Biomaterials in Clinical Practice: Advances in Clinical Research and Medical Devices*; Springer: Cham, Switzerland, 2017.
64. Jefferies, C.; Al-Malaika, S.; Sheena, H.H. New and novel stabilisation approach for radiation-crosslinked Ultrahigh Molecular Weight Polyethylene (XL-UHMWPE) targeted for use in orthopeadic implants. *Polym. Degrad. Stab.* **2021**, *183*, 109462. [CrossRef]
65. Boschetto, F.; Ngoc Doan, H.; Phong Vo, P.; Zanocco, M.; Zhu, W.; Sakai, W.; Adachi, T.; Ohgitani, E.; Tsutsumi, N.; Mazda, O.; et al. Antibacterial and Osteoconductive Effects of Chitosan/Polyethylene Oxide (PEO)/Bioactive Glass Nanofibers for Orthopedic Applications. *Appl. Sci.* **2020**, *10*, 2360. [CrossRef]
66. Bistolfi, A.; Giustra, F.; Bosco, F.; Sabatini, L.; Aprato, A.; Bracco, P.; Bellare, A. Ultra-high molecular weight polyethylene (UHMWPE) for hip and knee arthroplasty: The present and the future. *J. Orthop.* **2021**, *25*, 98–106. [CrossRef]
67. Wahed, S.B.; Dunstan, C.R.; Boughton, P.C.; Ruys, A.J.; Faisal, S.N.; Wahed, T.B.; Salahuddin, B.; Cheng, X.; Zhou, Y.; Wang, C.H.; et al. Functional Ultra-High Molecular Weight Polyethylene Composites for Ligament Reconstructions and Their Targeted Applications in the Restoration of the Anterior Cruciate Ligament. *Polymers* **2022**, *14*, 2189. [CrossRef]
68. Said, A. Ultra-High-Molecular-Weight-Polyethylene (UHMWPE) as Desired Polymer Material for Biomedical. *Khalij-Libya J. Dent. Med. Res.* **2022**, *6*, 11–16. [CrossRef]
69. Hinz, N.; Dehoust, J.; Schroeter, J.; Schulz, A.P.; Hartel, M.J.; Lutz, C.; Frosch, K.H.; Wendlandt, R. Biomechanical in vitro analysis of a novel flexible implant for pubic symphysis disruption using an ultra-high molecular weight polyethylene fiber cord. *Clin. Biomech.* **2022**, *95*, 105652. [CrossRef]
70. Kasser, M.J. Regulation of UHMWPE biomaterials in total hip arthroplasty. *J. Biomed. Mater. Res. Part B Appl. Biomater.* **2013**, *101*, 400–406. [CrossRef]
71. Kammerlander, C.; Neuerburg, C.; Verlaan, J.J.; Schmoelz, W.; Miclau, T.; Larsson, S. The use of augmentation techniques in osteoporotic fracture fixation. *Injury* **2016**, *47*, S36–S43. [CrossRef]
72. Scolaro, J.A.; Lackman, R.D. Surgical management of metastatic long bone fractures: Principles and techniques. *JAAOS-J. Am. Acad. Orthop. Surg.* **2014**, *22*, 90–100.
73. Grechenig, S.; Gänsslen, A.; Gueorguiev, B.; Berner, A.; Müller, M.; Nerlich, M.; Schmitz, P. PMMA-augmented SI screw: A biomechanical analysis of stiffness and pull-out force in a matched paired human cadaveric model. *Injury* **2015**, *46*, S125–S128. [CrossRef]
74. Yaacobi, E.; Sanchez, D.; Maniar, H.; Horwitz, D.S. Surgical treatment of osteoporotic fractures: An update on the principles of management. *Injury* **2017**, *48*, S34–S40. [CrossRef]
75. Moazen, M.; Mak, J.H.; Etchels, L.W.; Jin, Z.; Wilcox, R.K.; Jones, A.C.; Tsiridis, E. The effect of fracture stability on the performance of locking plate fixation in periprosthetic femoral fractures. *J. Arthroplast.* **2013**, *28*, 1589–1595. [CrossRef]
76. Kumar, P.A.; Irudhayam, J.S.; Naviin, D. A review on importance and recent applications of polymer composites in orthopaedics. *Int. J. Eng. Res. Dev.* **2012**, *5*, 40–43.
77. Nixon, A.J.; Auer, J.A.; Watkins, J.P. Principles of fracture fixation. In *Equine Fracture Repair*; John Wiley & Sons, Inc.: Hoboken, NJ, USA, 2019; pp. 127–155.
78. Corró, S.; García-Albó, E.; Andrés-Peiró, J.V.; Teixidor, J.; Tomás, J. Bone Defect Management and Augumentation of Distal Femoral Fractures with Polymethylmethacrylate Bone Cement. *J. Musculoskelet. Res.* **2022**, *25*, 2250013. [CrossRef]
79. Grant, K.D.; Busse, E.C.; Park, D.K.; Baker, K.C. Internal fixation of osteoporotic bone. *JAAOS-J. Am. Acad. Orthop. Surg.* **2018**, *26*, 166–174. [CrossRef]
80. Hsu, R.Y.; Ramirez, J.M.; Blankenhorn, B.D. Surgical considerations for osteoporosis in ankle fracture fixation. *Orthop. Clin.* **2019**, *50*, 245–258. [CrossRef]
81. Allizond, V.; Comini, S.; Cuffini, A.M.; Banche, G. Current Knowledge on Biomaterials for Orthopedic Applications Modified to Reduce Bacterial Adhesive Ability. *Antibiotics* **2022**, *11*, 529. [CrossRef]
82. Sa, Y.; Yang, F.; Wang, Y.; Wolke, J.G.; Jansen, J.A. Modifications of poly (methyl methacrylate) cement for application in orthopedic surgery. In *Cutting-Edge Enabling Technologies for Regenerative Medicine*; Springer: Singapore, 2018; Volume 1078, pp. 119–134.
83. Shirvan, A.R.; Nouri, A.; Wen, C. Structural polymer biomaterials. In *Structural Biomaterials*; Woodhead Publishing: Cambridge, UK, 2021; pp. 395–439.
84. Hamdy, T.M. Polymers and ceramics biomaterials in orthopedics and dentistry: A review article. *Egypt. J. Chem.* **2018**, *61*, 723–730. [CrossRef]
85. Kenny, S.M.; Buggy, M. Bone cements and fillers: A review. *J. Mater. Sci. Mater. Med.* **2003**, *14*, 923–938. [CrossRef]

86. Soles, G.L.; Ferguson, T.A. Fragility fractures of the pelvis. *Curr. Rev. Musculoskelet. Med.* **2012**, *5*, 222–228. [CrossRef]
87. Davis, F.J.; Mitchell, G.R. Polyurethane based materials with applications in medical devices. In *Bio-Materials and Prototyping Applications in Medicine*; Springer: Boston, MA, USA, 2008; pp. 27–48.
88. Mathew, E.; Domínguez-Robles, J.; Larrañeta, E.; Lamprou, D.A. Fused Deposition Modelling as a Potential Tool for Antimicrobial Dialysis Catheters Manufacturing: New Trends vs. Conventional Approaches. *Coatings* **2019**, *9*, 515. [CrossRef]
89. Pruitt, L.; Furmanski, J. Polymeric biomaterials for load-bearing medical devices. *JOM* **2009**, *61*, 14–20. [CrossRef]
90. Mayet, N.; Choonara, Y.E.; Kumar, P.; Tomar, L.K.; Tyagi, C.; Du Toit, L.C.; Pillay, V. A comprehensive review of advanced biopolymeric wound healing systems. *J. Pharm. Sci.* **2014**, *103*, 2211–2230. [CrossRef]
91. Bezuidenhout, D.; Williams, D.F.; Zilla, P. Polymeric heart valves for surgical implantation, catheter-based technologies and heart assist devices. *Biomaterials* **2015**, *36*, 6–25. [CrossRef]
92. Blaheta, R.A.; Oertl, A.; Freisleben, H.J.; Nelson, K.; Ackermann, H.; Haferkamp, A.; Engl, T. Detection of early DJ-stent encrustation by sonographic twinkling-artifacts–A pilot study. *Cent. Eur. J. Urol.* **2017**, *70*, 107.
93. Ong, K.L.; Yun, B.M.; White, J.B. New biomaterials for orthopedic implants. *Orthop. Res. Rev.* **2015**, *7*, 107–130. [CrossRef]
94. Giza, E.; Frizzell, L.; Farac, R.; Williams, J.; Kim, S. Augmented tendon Achilles repair using a tissue reinforcement scaffold: A biomechanical study. *Foot Ankle Int.* **2011**, *32*, 545–549. [CrossRef]
95. Faroni, A.; Mobasseri, S.A.; Kingham, P.J.; Reid, A.J. Peripheral nerve regeneration: Experimental strategies and future perspectives. *Adv. Drug Deliv. Rev.* **2015**, *82*, 160–167. [CrossRef] [PubMed]
96. Zafar, F.; Sharmin, E. (Eds.) *Polyurethane*; BoD–Books on Demand, IntechOpen: Rijeka, Croatia, 2012.
97. Thompson, M.S.; McCarthy, I.D.; Lidgren, L.; Ryd, L. Compressive and shear properties of commercially available polyurethane foams. *J. Biomech. Eng.* **2003**, *125*, 732–734. [CrossRef]
98. Calvert, K.L.; Trumble, K.P.; Webster, T.J.; Kirkpatrick, L.A. Characterization of commercial rigid polyurethane foams used as bone analogs for implant testing. *J. Mater. Sci. Mater. Med.* **2010**, *21*, 1453–1461. [CrossRef] [PubMed]
99. Bredbenner, T.L.; Haug, R.H. Substitutes for human cadaveric bone in maxillofacial rigid fixation research. *Oral Surg. Oral Med. Oral Pathol. Oral Radiol. Endodontol.* **2000**, *90*, 574–580. [CrossRef] [PubMed]
100. Piconi, C. Ceramics for joint replacement: Design and application of commercial bearings. In *Advances in Ceramic Biomaterials*; Woodhead Publishing: Cambridge, UK, 2017; pp. 129–179.
101. Piconi, C.; Sprio, S. Oxide Bioceramic Composites in Orthopedics and Dentistry. *J. Compos. Sci.* **2021**, *5*, 206. [CrossRef]
102. Wang, W.; Ouyang, Y.; Poh, C.K. Orthopaedic implant technology: Biomaterials from past to future. *Ann. Acad. Med.* **2011**, *40*, 237.
103. Ma, H.; Suonan, A.; Zhou, J.; Yuan, Q.; Liu, L.; Zhao, X.; Zhang, Y.G. PEEK (Polyether-ether-ketone) and its composite materials in orthopedic implantation. *Arab. J. Chem.* **2021**, *14*, 102977. [CrossRef]
104. Badulescu, O.V.; Filip, N.; Sirbu, P.D.; Bararu-Bojan, I.; Vladeanu, M.; Bojan, A.; Ciocoiu, M. Current practices in haemophilic patients undergoing orthopedic surgery-a systematic review. *Exp. Ther. Med.* **2020**, *20*, 1–4. [CrossRef]
105. Tsakiris, V.; Tardei, C.; Clicinschi, F.M. Biodegradable Mg alloys for orthopedic implants—A review. *J. Magnes. Alloy.* **2021**, *9*, 1884–1905. [CrossRef]
106. Pattanayak, S.; Sahoo, S.K. Micro engraving on 316L stainless steel orthopedic implant using fiber laser. *Opt. Fiber Technol.* **2021**, *63*, 102479. [CrossRef]
107. Badulescu, O.V.; Sirbu, P.D.; Ungureanu, C.; Pînzariu, A.; Cojocaru, E.; Filip, N.; Bararu-Bojan, I.; Vladeanu, M.; Ciocoiu, M. Orthopedic surgery in hemophilic patients with musculoskeletal disorders: A systematic review. *Exp. Ther. Med.* **2021**, *22*, 995. [CrossRef]
108. Badulescu, O.V.; Ciocoiu, M.; Filip, N.; Veringa, V. Tranexamic acid-major antifibrinolytic agent used to achieve hemostasis in hemophilic patients with anti-factor VIII antibodies who must undergo total joint replacement. *Rev. Chim.* **2019**, *70*, 638–641. [CrossRef]
109. Filip, A.; Alexa, O.; Sirbu, P.D.; Filip, C.; Andrusca, L.; Pascal, I.A.; Oprea, S.; Badulescu, O.V. Assessment of the Mechanical Properties of Orthopedic Screws Coated with Polyurethane Acrylate Containing Hydroxyapatite, Intended to Fix the Fragility Fractures. *Mater. Plast.* **2019**, *56*, 1028. [CrossRef]
110. Filip, A.; Badulescu, O.V.; Sirbu, P.D.; Veliceasa, B.; Puha, B.; Pascal, I.A.; Andrusca, L.; Oprea, D.; Filip, C.; Alexa, O. Preliminary investigation on mechanical properties of polymer coating screws for the future fragility fracture fixation. *Mater. Plast.* **2019**, *56*, 559. [CrossRef]
111. Li, H.; Xia, P.; Pan, S.; Qi, Z.; Fu, C.; Yu, Z.; Kong, W.; Chang, Y.; Wang, K.; Wu, D.; et al. The advances of ceria nanoparticles for biomedical applications in orthopaedics. *Int. J. Nanomed.* **2020**, *15*, 7199. [CrossRef]
112. Kalyanaraman, V.; Naveen, S.V.; Mohana, N.; Balaje, R.M.; Navaneethakrishnan, K.R.; Brabu, B.; Murugan, D.S.; Kumaravel, T.S. Biocompatibility studies on cerium oxide nanoparticles–Combined study for local effects, systemic toxicity and genotoxicity via implantation route. *Toxicol. Res.* **2019**, *8*, 25–37. [CrossRef]
113. Soni, M.; Mehta, P.; Soni, A.; Goswami, G.K. Green nanoparticles: Synthesis and applications. *IOSR J. Biotechnol. Biochem.* **2018**, *4*, 78–83.
114. Nikolova, M.P.; Chavali, M.S. Metal oxide nanoparticles as biomedical materials. *Biomimetics* **2020**, *5*, 27. [CrossRef]
115. Luo, J.; Zhu, S.; Tong, Y.; Zhang, Y.; Li, Y.; Cao, L.; Kong, M.; Luo, M.; Bi, Q.; Zhang, Q. Cerium Oxide Nanoparticles Promote Osteoplastic Precursor Differentiation by Activating the Wnt Pathway. *Biol. Trace Elem. Res.* **2022**, *201*, 1–9. [CrossRef]

116. Castiglioni, S.; Cazzaniga, A.; Locatelli, L.; Maier, J.A.M. Silver Nanoparticles in Orthopedic Applications: New Insights on Their Effects on Osteogenic Cells. *Nanomaterials* **2017**, *7*, 124. [CrossRef]
117. Samanta, A.; Podder, S.; Kumarasamy, M.; Ghosh, C.K.; Lahiri, D.; Roy, P.; Bhattacharjee, S.; Ghosh, J.; Mukhopadhyay, A.K. Au nanoparticle-decorated aragonite microdumbbells for enhanced antibacterial and anticancer activities. *Mater. Sci. Eng. C* **2019**, *103*, 109734. [CrossRef]
118. Holweg, P.; Labmayr, V.; Schwarze, U.; Soomer, N.; Ornig, M.; Leithner, A. Osteotomy after medial malleolus fracture fixed with magnesium screws ZX00—A case report. *Trauma Case Rep.* **2022**, *42*, 100706. [CrossRef]
119. Antoniac, I.; Miculescu, M.; Manescu, V.; Stere, A.; Quan, P.H.; Paltânea, G.; Robu, A.; Earar, K. Magnesium-Based Alloys Used in Orthopedic Surgery. *Materials* **2022**, *15*, 1148. [CrossRef]
120. Memarian, P.; Pishavar, E.; Zanotti, F.; Trentini, M.; Camponogara, F.; Soliani, E.; Gargiulo, P.; Isola, M.; Zavan, B. Active Materials for 3D Printing in Small Animals: Current Modalities and Future Directions for Orthopedic Applications. *Int. J. Mol. Sci.* **2022**, *23*, 1045. [CrossRef]
121. Feltz, K.P.; MacFadden, L.N.; Gieg, S.D.; Lough, C.P.; Bezold, W.A.; Skelley, N.W. Mechanical properties of 3D-printed orthopedic one-third tubular plates and cortical screws. *J. 3D Print. Med.* **2022**, *6*, 129–145. [CrossRef]
122. Li, B.; Zhang, M.; Lu, Q.; Zhang, B.; Miao, Z.; Li, L.; Zheng, T.; Liu, P. Application and Development of Modern 3D Printing Technology in the Field of Orthopedics. *BioMed Res. Int.* **2022**, *2022*, 8759060. [CrossRef]

MDPI AG
Grosspeteranlage 5
4052 Basel
Switzerland
Tel.: +41 61 683 77 34

Coatings Editorial Office
E-mail: coatings@mdpi.com
www.mdpi.com/journal/coatings

www.ingramcontent.com/pod-product-compliance
Lightning Source LLC
Chambersburg PA
CBHW061521210926
51850CB00008B/1991
* 9 7 8 3 7 2 5 8 5 5 3 1 5 *